Learn MongoDB 4.x

A guide to understanding MongoDB development and
administration for NoSQL developers

Doug Bierer

BIRMINGHAM - MUMBAI

Learn MongoDB 4.x

Copyright © 2020 Packt Publishing

Commissioning Editor: Pravin Dhandre
Acquisition Editor: Savia Lobo
Content Development Editor: Pratik Andrade
Senior Editor: Ayaan Hoda
Technical Editor: Sarvesh Jaywant
Copy Editor: Safis Editing
Project Coordinator: Neil Dmello
Proofreader: Safis Editing
Indexer: Rekha Nair
Production Designer: Alishon Mendonca

First published: September 2020

Production reference: 1030920

Published by Packt Publishing Ltd.
Livery Place
35 Livery Street
Birmingham
B3 2PB, UK.

ISBN 978-1-78961-938-6

www.packt.com

This year has been extraordinarily tough, not only due to the global pandemic, but also as a result of deteriorating conditions worldwide as a result of climate change and increasing economic inequality across the globe. That having been said, on a personal note, three near and dear friends have died in the period of time it took to write this book. I would like to dedicate this book to those three: Jeff Abel, Daryl Holdridge, and Brad Saunders. Dear friends, I hope you all find yourselves in a better place. May you rest in peace.

– Doug Bierer

`Packt.com`

Subscribe to our online digital library for full access to over 7,000 books and videos, as well as industry leading tools to help you plan your personal development and advance your career. For more information, please visit our website.

Why subscribe?

- Spend less time learning and more time coding with practical eBooks and Videos from over 4,000 industry professionals

- Improve your learning with Skill Plans built especially for you

- Get a free eBook or video every month

- Fully searchable for easy access to vital information

- Copy and paste, print, and bookmark content

Did you know that Packt offers eBook versions of every book published, with PDF and ePub files available? You can upgrade to the eBook version at `www.packt.com` and as a print book customer, you are entitled to a discount on the eBook copy. Get in touch with us at `customercare@packtpub.com` for more details.

At `www.packt.com`, you can also read a collection of free technical articles, sign up for a range of free newsletters, and receive exclusive discounts and offers on Packt books and eBooks.

Foreword

Learning something new is a part of a developer's everyday job. Whether you are a beginner trying to find a book that will quickly teach you all the essential concepts of the latest version of MongoDB, or a senior developer trying to set up your company's new MongoDB server this morning, you will need a dependable source of information that will teach you all that you need to know to get things up and running as quickly as possible. For these reasons and many more, you need a book that will get you to where you want to be, without the burden of doing lengthy research just to get the important stuff down pat.

This is where my good friend, Doug Bierer, comes in. Doug is not only a great developer, but he is a natural-born teacher. Doug has this extraordinary ability to get you up to speed and ready to go in no time, even with the most advanced programming concepts. He will not waste your time by giving you meaningless recipes that you are expected to apply blindly from now on, but he will rather give you what is needed, whether it be by way of computer theory or history, in order to help you understand what you are doing. His code examples are always very enlightening because they are chosen in order to get you working as quickly as possible with the tools that you wish to better understand and master.

In a few moments from now, you will grasp what NoSQL is in its most essential expression, without having to go through many bulky books. You will start by installing and configuring your MongoDB server; wrap your mind around MongoDB's more advanced concepts, such as security, replication, and sharding; and use it like a pro. Moreover, you will learn how to design your new NoSQL databases while avoiding the common pitfalls that go along with having an SQL mindset and background.

Once you are done reading this book, I know that you will be well on your way to mastering MongoDB and to becoming a better developer than you were before.

Andrew Caya,
CEO Foreach Code Factory
https://andrewscaya.net/

Contributors

About the author

Doug Bierer has been writing code since the early 1970s. His first project was a terrain simulator that ran on a Digital Equipment Corporation PDP-8 with 4 K of RAM. Since then, he's written applications for a variety of customers worldwide, in various programming languages, including BASIC, Assembler, PL/I, C, C++, FORTH, Prolog, Pascal, Java, PERL, PHP, and Python. His work with database technology spans a similar range in terms of years and includes Ingres, d:Base, Clipper, FoxBase, r:Base, ObjectVision, Btrieve, Oracle, MySQL, and, of course, MongoDB. Doug runs his own company, *unlikelysource*, and is also CTO at *Foreach Code Factory*, being involved in both training (php-cl) and a new cloud services platform (Linux for PHP Cloud Services).

I would like to thank my wife, Fon, who keeps me young!

About the reviewer

Blaine Mincey is a principal solutions architect with MongoDB and is based in Atlanta, GA. He is responsible for guiding MongoDB customers to design and build reliable, scalable systems using MongoDB technology. Prior to MongoDB, Blaine was a software engineer with several notable organizations/start-ups in the Atlanta area before joining the Red Hat/JBoss solutions architecture team.

When Blaine is not talking about technology or his family (wife, two boys, and a dog), he can be found at one of three locations: Fado's Midtown, Atlanta, watching his beloved Manchester United; Mercedes-Benz stadium, while supporting Atlanta United; or on the local football pitch coaching either of his sons' soccer squads.

Packt is searching for authors like you

If you're interested in becoming an author for Packt, please visit `authors.packtpub.com` and apply today. We have worked with thousands of developers and tech professionals, just like you, to help them share their insight with the global tech community. You can make a general application, apply for a specific hot topic that we are recruiting an author for, or submit your own idea.

Table of Contents

Section 2: Building a Database-Driven Web Application

Section 4: Replication, Sharding, and Security in a Financial Environment

Preface

Aging legacy database technologies used in current **relational database management systems** (**RDBMS**) are no longer able to handle the needs of modern web applications. MongoDB, built upon a NoSQL foundation, has taken the world by storm and is rapidly gaining market share over inadequate RDBMS-based websites. It is critical for DevOps, technical managers, and those whose careers are in transition to gain knowledge on how to build dynamic database-driven web applications based upon MongoDB. This book takes a deep dive into management and the application development of a web application that uses MongoDB 4.x as its database.

The book takes a hands-on approach, building a set of useful Python classes, starting with simple operations, and adding more complexity as the book progresses. When you have finished going through the book, you'll have mastered not only basic day-to-day MongoDB database operations, but will have gained a firm understanding of development using complex MongoDB features such as replication, sharding, and the aggregation framework. Complex queries are presented in both JavaScript form as well as the equivalent in Python. The JavaScript queries presented will allow DBAs to manage and manipulate data directly, as well as being able to generate ad hoc reports for management. Throughout the book, practical examples of each concept are covered in detail, along with potential errors and gotchas.

All examples in the book are based upon realistic real-world scenarios ranging from a small company that sells sweets online, all the way to a start-up bank that offers micro-financing to customers worldwide. With each scenario, you'll learn how to develop a usable MongoDB database structure based on realistic customer requirements. Each major part of the book, based upon a different scenario, guides you through progressively more complex tasks, ranging from generating simple reports, to directing queries through a global network of sharded MongoDB database clusters.

Who this book is for

This book is designed for anyone who roughly fits any of these descriptions:

- **DevOps**: **Information technology** (**IT**) staff involved with development and/or operations pertaining to a MongoDB 4.x database.
- **Database administrators (DBAs)**: Database administrators who need to know how to install, configure, and manage a MongoDB 4.x database.

- **Systems administrators**: Technical staff in charge of one or more servers who might be called upon to perform DBA activities and generate ad hoc reports for management.
- **Technical managers**: Managers who need to keep up with and understand the terminology and tasks of the DevOps they manage.
- **Students**: Anybody attending an institution of higher education, or anybody whose career is in transition, looking for a market niche from which they can carve out a decent career with a solid future.
- **IT professionals**: Any professional associated with the IT industry looking to transition from legacy RDBMS-based applications to modern NoSQL-based apps.

What this book covers

Chapter 1, *Introducing MongoDB 4.x*, provides a general introduction to MongoDB 4.x with a focus on new features and a brief high-level overview of the technology and the differences between MongoDB 4.x and MongoDB 3.

Chapter 2, *Setting Up MongoDB 4.x*, covers MongoDB 4.x and Python programming language drivers. In addition, this chapter shows you how to set up a demo test environment based upon Docker, in which you can practice working with MongoDB.

Chapter 3, *Essential MongoDB Administration Techniques*, shows how to perform administration critical to a properly functioning database using the mongo shell.

Chapter 4, *Fundamentals of Database Design*, describes moving from a set of customer requirements to a working MongoDB database design.

Chapter 5, *Mission-Critical MongoDB Database Tasks*, presents a series of common tasks and then shows you how to define domain classes that perform critical database access services including performing queries, adding, editing, and deleting documents.

Chapter 6, *Using AJAX and REST to Build a Database-Driven Website*, moves on to what is needed to have the application respond to AJAX queries and REST requests.

Chapter 7, *Advanced MongoDB Database Design*, covers how to define a dataset with complex requirements. You are also shown how MongoDB can be integrated with the popular Django web framework.

Chapter 8, *Using Documents with Embedded Lists and Objects*, covers working with the documents designed in the previous chapter, learning how to handle create, read, update and delete operations that involve embedded lists (arrays) and objects.

Chapter 9, *Handling Complex Queries in MongoDB*, introduces you to the aggregation framework, a feature unique to MongoDB. In this chapter, you also learn about map-reduce, geospatial queries, generating financial reports, and performing risk analysis.

Chapter 10, *Working with Complex Documents across Collections*, starts with a brief discussion on how to handle monetary data, after which the focus shifts to working with complex documents across multiple collections. Lastly, GridFS technology is introduced as a way of handling large files and storing documents directly in the database.

Chapter 11, *Administering MongoDB Security*, focuses on the administration needed to secure the database by creating database users, implement role-based access control and implement transport layer (SSL/TLS) security based upon x.509 certificates.

Chapter 12, *Developing in a Secured Environment*, looks into how to write applications that access a secured database using role-based access control and TLS security using x.509 certificates.

Chapter 13, *Deploying a Replica Set*, starts with an overview of MongoDB replication, after which you'll learn how to configure and deploy a replica set.

Chapter 14, *Replica Set Runtime Management and Development*, focuses on replica set management, monitoring, backup, and restore. In addition, it addresses how your application program code might change when accessing a replica set.

Chapter 15, *Deploying a Sharded Cluster*, starts with an overview of sharding, after which you'll learn how to configure and deploy a sharded cluster.

Chapter 16, *Sharded Cluster Management and Development*, covers how to manage a sharded cluster as well as learning about the impact of sharding on application program code.

To get the most out of this book

In order to test the sample queries and coding examples presented in this book, you need the following:

Minimum recommended hardware:

- Desktop PC or laptop
- 2 GB free disk space
- 4 GB of RAM
- 500 Kbps or faster internet connection

Software requirements:

- OS (Linux or Mac): Docker, Docker Compose, Git (optional)
- OS (Windows): Docker for Windows and Git for Windows

In order to successfully understand and work through the examples presented in this book, the following background knowledge will be helpful:

- Basic knowledge of Python and/or Javascript
- Basic knowledge of what is a database and what a database is used for
- Working knowledge of HTML and web applications in general
- Knowledge of Docker and Linux is useful but not mandatory as all necessary commands are shown

 If you're not proficient in the Python language, do not worry! A reference to the original MongoDB query is always made first, after which the actual implementation in Python is described. It is easy enough to extrapolate the MongoDB query into your programming language of choice. In addition, the actual Python code is kept simple: no complex Python constructs are used.

Python 3.x, the PyMongo driver, an Apache web server, and various Python libraries are already installed in the Docker container used for the book. Chapter 2, *Setting Up MongoDB 4.x*, gives you instructions on how to install the demo environment used in the book.

If you are using the digital version of this book, we advise you to type the code yourself or access the code via the GitHub repository (link available in the next section). Doing so will help you avoid any potential errors related to the copying and pasting of code.

Download the example code files

You can download the example code files for this book from your account at www.packt.com. If you purchased this book elsewhere, you can visit www.packtpub.com/support and register to have the files emailed directly to you.

You can download the code files by following these steps:

1. Log in or register at www.packt.com.
2. Select the **Support** tab.
3. Click on **Code Downloads**.
4. Enter the name of the book in the **Search** box and follow the onscreen instructions.

Once the file is downloaded, please make sure that you unzip or extract the folder using the latest version of:

- WinRAR/7-Zip for Windows
- Zipeg/iZip/UnRarX for Mac
- 7-Zip/PeaZip for Linux

The code bundle for the book is also hosted on GitHub at `https://github.com/PacktPublishing/Learn-MongoDB-4.x`. In case there's an update to the code, it will be updated on the existing GitHub repository. We also have other code bundles from our rich catalog of books and videos available at `https://github.com/PacktPublishing/`. Check them out!

Download the color images

We also provide a PDF file that has color images of the screenshots/diagrams used in this book. You can download it here: `http://www.packtpub.com/sites/default/files/downloads/9781789619386_ColorImages.pdf`.

Conventions used

There are a number of text conventions used throughout this book.

`CodeInText`: Indicates code words in text, database collection and field names, folder names, filenames, file extensions, pathnames, dummy URLs, user input, and Twitter handles. Here is an example: "Finally, the `$sort` stage reorders the final results by the match field _id (that is, the country code)."

A block of code is set as follows:

```
db.bookings.aggregate([
    { $match : { "bookingInfo.paymentStatus" : "confirmed" } },
    { $group: { "_id"    : "$customer.customerAddr.country",
                "total" : { $sum : "$totalPrice" } } },
    { $sort : { "_id" : 1 } }
]);
```

If a line of code or a command needs to be all on a single line, but the book's page width prevents this, the line will be broken up into two lines. A backslash (\) is placed at the break point. The remainder of the line appears indented on the next line as follows:

```
this.command('has', 'many', 'arguments', 'and', \
  'would', 'occupy', 'a', 'single', 'line')
```

Any command-line input or output is written as follows:

```
cd /path/to/repo
docker-compose build
docker-compose up -d
docker exec -it learn-mongo-server-1 /bin/bash
```

Bold: Indicates a new term, an important word, or words that you see onscreen. For example, words in menus or dialog boxes appear in the text like this. Here is an example: "You can also select **Fill in connection fields individually**, in which case these are the two tab screens you can use."

 Warnings or important notes appear like this.

 Tips and tricks appear like this.

Get in touch

Feedback from our readers is always welcome.

General feedback: If you have questions about any aspect of this book, mention the book title in the subject of your message and email us at customercare@packtpub.com.

Errata: Although we have taken every care to ensure the accuracy of our content, mistakes do happen. If you have found a mistake in this book, we would be grateful if you would report this to us. Please visit www.packtpub.com/support/errata, selecting your book, clicking on the Errata Submission Form link, and entering the details.

Piracy: If you come across any illegal copies of our works in any form on the Internet, we would be grateful if you would provide us with the location address or website name. Please contact us at copyright@packt.com with a link to the material.

If you are interested in becoming an author: If there is a topic that you have expertise in and you are interested in either writing or contributing to a book, please visit `authors.packtpub.com`.

Reviews

Please leave a review. Once you have read and used this book, why not leave a review on the site that you purchased it from? Potential readers can then see and use your unbiased opinion to make purchase decisions, we at Packt can understand what you think about our products, and our authors can see your feedback on their book. Thank you!

For more information about Packt, please visit `packt.com`.

Section 1: Essentials

This section is designed to give you a high-level overview of MongoDB 4.x. In this section, you are introduced to the first of three realistic scenarios, a fictitious company called *Sweets Complete, Inc.* These scenarios are representative of real-life companies and are designed to show how MongoDB can be incorporated into various types of businesses.

In this section, you'll learn about the key differences between MongoDB 3 and 4. You will learn not only how to install MongoDB 4.x for a future customer, but also how to set up an environment based upon Docker technology that you can use throughout the remainder of the book in order to run the examples. You will also learn basic MongoDB administration using the *mongo* shell.

This section contains the following chapters:

- Chapter 1, *Introducing MongoDB 4.x*
- Chapter 2, *Setting Up MongoDB 4.x*
- Chapter 3, *Essential MongoDB Administration Techniques*

Introducing MongoDB 4.x

In this book, we cover how to work with a MongoDB 4.x database, starting with the simplest concepts and moving on to more complex ones. The book is divided into parts or sections, each of which looks at a different scenario.

In this chapter, you are given a general introduction to MongoDB 4.x with a focus on new features and a brief high-level overview of the technology. We also discuss security enhancements, along with backward-incompatible changes that might cause an application written for MongoDB 3 to break after an upgrade to MongoDB 4.x.

In the next chapter, a simple scenario is introduced: a fictitious company called *Sweets Complete Inc.* that sells confections online to a small base of international customers. In the next two parts that follow, you are introduced to *BookSomething.com*, another fictitious company with a large database of hotel listings worldwide. Finally, in the last part, you are introduced to *BigLittle Micro Finance Ltd.*, a fictitious company that connects lenders with borrowers and deals with a massive volume of geographically dispersed data.

In this chapter, the following topics are covered:

- A high-level technology overview of MongoDB 4.x
- Significant new features introduced in MongoDB 4.x
- Important security enhancements
- Spotting and avoiding potential problems when migrating from MongoDB 3 to 4

High-level technology overview of MongoDB 4.x

When it was first introduced in 2009, MongoDB took the database world by storm, and since that time it has rapidly gained in popularity. According to the 2019 StackOverflow developer survey (`https://insights.stackoverflow.com/survey/2019#technology-_-databases`), MongoDB is ranked fifth, with 26% of professional developers and 25.5% of all respondents saying they use MongoDB. DB-Engines (`https://db-engines.com/en/ranking`) also ranks MongoDB as the fifth most widely used database, using an algorithm that takes into account the frequency of search, DBA Stack Exchange and StackOverflow references, and the frequency with which MongoDB appears in job postings. What is of even more interest is that the trend graph generated by DB-Engines shows that the score (and therefore ranking) of MongoDB has grown by 200% since 2013. You can refer to `https://db-engines.com/en/ranking_trend` for more details. In 2013, MongoDB was not even in the top 10!

There are many key features of MongoDB that account for its rise in popularity. Subsequent chapters in this book cover the most important of these features in detail. In this section, we present you with a brief, *big-picture* overview of three key aspects of MongoDB.

MongoDB is based upon documents

One of the most important distinctions between MongoDB and the traditional **relational database management systems** (**RDBMS**) is that instead of tables, rows, and columns, the basis for storage in MongoDB is a *document*. In a certain sense, you can think of the traditional RDBMS system as *two dimensional*, whereas MongoDB is *three dimensional*. Documents are typically modeled using JSON formatting and then inserted into the database where they are converted to a binary format for storage (more on that in later chapters!).

Related to the document basis for storage is the fact that MongoDB documents have *no fixed schema*. The main benefit of this is *vastly reduced overhead*. Database restructuring is a piece of cake, and doesn't cause the massive problems, website crashes, and security breaches seen in applications reliant upon a traditional RDBMS database restructuring.

The really great news for developers is that most modern programming applications are based on classes representing information that needs to be stored. This has spawned the creation of a large number of *object-relational mapping* (**ORM**) libraries for the various programming languages. In MongoDB, on the other hand, the need for a complex ORM infrastructure is completely eliminated as programmatic objects can be directly stored in the database as-is:

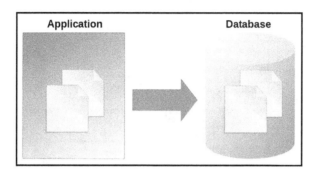

So instead of columns, MongoDB documents have *fields*. Instead of tables, there are *collections* of documents. Let's now have a look at replication in MongoDB.

High availability

Another feature that causes MongoDB to stand out from other database technologies is its ability to ensure *high availability* through a process known as *replication*. A server running MongoDB can have copies of its databases duplicated across two more servers. These copies are known as *replica sets*. Replica sets are organized through an election process whereby the members of the replica *vote* on which server becomes the *primary*. Other servers are then assigned the role of *secondary*.

This arrangement not only ensures that the database is continuously available, but that it can also be used by application code by way of *read preferences*. A read preference tells the replica set which servers in the replica set are preferred. If the read preferences are set less restrictively, then the first server in the set to respond might be able to satisfy the request, thereby implementing a form of parallel process that has the potential to greatly enhance performance. This setup is illustrated in the following diagram:

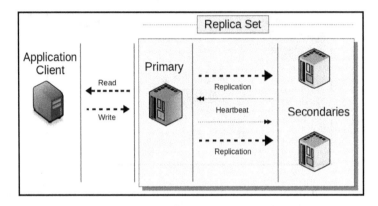

This topic is covered in extensive detail in `Chapter 13`, *Deploying a Replica Set*. Lastly, we have a look at *sharding*.

Horizontal scaling

One more feature, among many, is MongoDB's ability to handle a massive amount of data. This is accomplished by splitting up a sizeable collection across multiple servers, creating a *sharded cluster*. In the process of splitting the collection, a *shard key* is chosen from among the fields present with the collection's documents. The shard key is then used by the *sharded cluster balancer* to determine the appropriate distribution of documents. Application program code is then able, by its knowledge of the value of the shard key, to direct queries to specific members of the sharded cluster, achieving potentially enormous performance gains. This setup is illustrated in the following diagram:

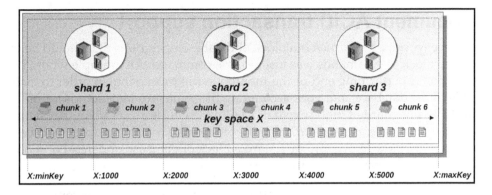

This topic is covered in extensive detail in `Chapter 15`, *Deploying a Sharded Cluster*. In the next section of this chapter, we have a look at the major differences between MongoDB 3 and MongoDB 4.x.

Discovering what's new and different in MongoDB 4.x

What's new and different in the MongoDB 4.x release can be broken down into two main categories: new features and internal enhancements. Let's look at the most significant new features first.

Significant new features

The most significant new features introduced in MongoDB 4.x include the following:

- Multidocument ACID transaction support
- Nonblocking secondary replica reads
- In-progress index build interruption

Transactions, secondary replica reads, and the aggregation pipeline are covered in detail in later chapters of this book. For an excellent brief overview of the major changes from MongoDB 3 to 4, go to `https://www.mongodb.com/blog/post/mongodb-40-release-candidate-0-has-landed`.

Multidocument ACID transaction support

In the database world, a *transaction* is a block of database operations that should be treated as if the entire block of commands was just a single command. An example would be where your application is performing end-of-the-month payroll processing. In order to maintain the integrity of the database, and your accounting files, you would need this set of operations to be safeguarded in the event of a failure. ACID is an acronym that stands for *atomicity, consistency, isolation*, and *durability*. It represents a set of principles that the database needs to follow in order to safeguard a block of database updates. For more information on ACID, you can refer to `https://en.wikipedia.org/wiki/ACID`.

In MongoDB 3, a write operation on a single document, even a document containing other embedded documents, was considered *atomic*. In MongoDB 4.x, multiple documents can be included in a single atomic transaction. Although invoking this support negatively impacts performance, the gain in database integrity might prove attractive. It's also worth noting that the lack of such support prior to MongoDB 4.x was a major criticism leveled against MongoDB, and slowed its adoption at the corporate level.

Invoking transaction support impacts *read preferences* and *write concerns*. For more information, you can refer to `https://docs.mongodb.com/manual/core/transactions/ #transaction-options-read-concern-write-concern-read-preference`. Although these topics are covered in detail later in the book, we can briefly summarize them by stating that read preferences allow you to direct operations to specific members of a replica set. For example, you might want to indicate a preference for the *primary* member server in a replica set rather than allowing any member, including *secondaries*, to be used in a read operation. Write concerns allow you to adjust the level of acknowledgement when writing to the database, thereby ensuring that data integrity is maintained. In MongoDB 4.x, you are able to set read preferences and write concerns at the transaction level, that in turn influences individual document operations.

In MongoDB version 4.2 and above, the 16 MB limit on transaction size is removed. Also, as of MongoDB 4.2, full support for multidocument transactions is added for sharded clusters. In addition, full transaction support is extended to replica sets whose secondary members are using the *in-memory* storage engine (`https://docs.mongodb.com/manual/core/ inmemory/`).

 Replica sets are discussed in `Chapter 13`, *Deploying a Replica Set*. Sharded clusters are covered in `Chapter 15`, *Deploying a Sharded Cluster*.

Nonblocking secondary reads

MongoDB developers have often included read concerns (as mentioned previously) in their operations in order to shift the burden of response from the primary server in a replica set to its secondaries instead. This frees up the primary to process write operations. Traditionally, MongoDB, prior to version 4, blocked such secondary reads while an update from the primary was in progress. The block ensured that any data read from the secondary would appear exactly the same as data read from the primary.

The downside to this approach, however, was that while the block was in place, the secondary read operation had to wait, which in turn negatively impacted read performance. Likewise, if a read operation was requested prior to a write, it would hold up update operations between the primary and secondary, negatively impacting write performance.

Because of internal changes introduced in MongoDB 4.x to support multidocument transactions, storage engine timestamps and snapshots are now used, which has the side effect of eliminating the need to block secondary reads. The net effect is an overall improvement in consistency and lower latency in terms of reads and writes. Another way to view this enhancement is that it allows an application to read from a secondary at the same time writes are being applied without delay.

In-progress index build interruption

A big problem with versions of MongoDB prior to 4.4 is that the following commands error out if an *index build* (`https://docs.mongodb.com/master/core/index-creation/#index-builds-on-populated-collections`) operation is in progress:

- `db.dropDatabase()`
- `db.collection.drop()`
- `db.collection.dropIndexes()`

In MongoDB 4.4, when this happens, an attempt to force the in-progress index build operation is made. If successful, the index build is halted, and the `drop*()` operation continues without error. In the case of a `drop*()` performed on a replica set, the abort attempt is made on the primary. Once the primary commits to the abort, it then synchronizes to the secondaries.

Other noteworthy new features

There are a number of other new features that do not represent a massive paradigm shift, but are extremely useful nonetheless. These include improvements to the aggregation pipeline, field-level encryption, `password()` prompt, and wildcard indexes. Let's first have a look at aggregation pipeline improvements.

Aggregation pipeline type conversions

Another major new feature we discuss here involves the introduction of `$convert`, a new aggregation pipeline operator. This new operator allows the developer to change the data type of a document field while being processed in the pipeline. Target data types include *double, string, Boolean, date, integer, long,* and *decimal*. In addition, you can convert a field in the pipeline to the data type `objectId`, which is critically useful when you need direct access to the autogenerated unique identification `field _id`. For more information on the aggregation pipeline operator and `$convert`, go to `https://docs.mongodb.com/ master/core/aggregation-pipeline/#aggregation-pipeline`.

Client-side field-level encryption

The official programming language drivers for MongoDB 4.2 now support *client-side field-level encryption*. The implications for security improvements are enormous. This enhancement means that your applications can now provide end-to-end encryption for transmitted data down to the field level. So you could have a transmission of data from your application to MongoDB that includes, for example, an encrypted national identification number mixed in with otherwise plain-text data.

> You can refer to `https://docs.mongodb.com/master/core/security-client-side-encryption/#driver-compatibility-table` to access the drivers for MongoDB 4.2. For client-side field-level encryption, you can refer to `https://docs.mongodb.com/master/core/security-client-side-encryption/#client-side-field-level-encryption`.

Password prompt

In many cases, it is highly undesirable to include a hard-coded password in a *Mongo* script. Starting with MongoDB 4.2, in place of a hard-coded password value, you can substitute a built-in JavaScript function `passwordPrompt()`. You can refer to `https://docs.mongodb. com/master/reference/method/passwordPrompt/#passwordPrompt` for more details on the function. This causes the Mongo shell to pause and wait for manual user input before proceeding. The password that is entered is then used as the password value.

Wildcard indexes

Starting with MongoDB 4.2, support has been added for *wildcard indexes* (https://docs. mongodb.com/master/core/index-wildcard/#wildcard-indexes). This feature is useful for situations where the index is either not yet available or is unknown. For situations where the field is known and well established, it's best to create a normal index. There are cases, however, where you have a subset of documents that contain a particular field otherwise lacking in other documents in the collection. You might also be in a situation where a field is added later, and where the DBA has not yet had a chance to create an index on the new field. In these cases, adding a wildcard index allows MongoDB to perform a query more efficiently.

Extended JSON v2 support

Starting with MongoDB 4.2, support for the Extended JSON v2 (https://docs.mongodb. com/master/reference/mongodb-extended-json/#mongodb-extended-json-v2) specification has been enabled for the following utilities:

- bsondump
- mongodump
- mongoexport
- mongoimport

Improved logging and diagnostics

Starting with MongoDB 4.2, there are now five verbosity log levels, each revealing increasing amounts of information. In MongoDB 4.0.6, you can now set a threshold on the maximum time it should take for data to replicate between members of a replica set. It's now possible to get the information from the MongoDB log file if that time is exceeded.

A further enhancement to diagnostics capabilities includes additional fields that are added to the output of the db.serverStatus() command that can be issued from a Mongo shell.

Hedged reads

MongoDB 4.4 adds the ability to perform a *hedged read* (https://docs.mongodb.com/ master/core/sharded-cluster-query-router/#hedged-reads) on a sharded cluster. By setting a *hedged read preference* option, applications are able to direct read requests to servers in replica sets other than the primary. The advantage of this approach is that the application simply takes the first result response, improving performance. The potential cost, of course, is that if a secondary responds, the data might be slightly out of date.

TCP fast open support

MongoDB version 4.4 introduces support for **TCP Fast Open** (**TCO**) connections. For this to work, it must be supported by the operating system hosting MongoDB. The following configuration file (and command line) parameters have been added to enable and control support under the `setParameter` configuration option: `tcpFastOpenServer`, `tcpFastOpenClient`, and `tcpFastQueueSize`. In addition, four new TCO-related information counters have been added to the output of the `serverStatus()` database command.

 You can refer to `https://tools.ietf.org/html/rfc7413` for more details on TCO. For more information on the parameter, you can refer to `https://docs.mongodb.com/master/reference/parameters/#param.tcpFastOpenServer`. Refer to `https://docs.mongodb.com/master/reference/command/serverStatus/#serverstatus` for more information on `serverStatus`.

Natural sort

MongoDB version 4.4 introduces a new operator, `$natural`, which is used in a `cursor.hint()` operation. This operator causes the results of a sort operation to return a list in natural (also called human-readable) order. As an example, take these values:

```
['file19.txt','file10.txt','file20.txt','file8.txt']
```

An ordinary sort would return the list in this order:

```
['file10.txt','file19.txt','file20.txt','file8.txt']
```

Whereas a *natural* sort would return the following:

```
['file8.txt','file10.txt','file19.txt','file20.txt']
```

Internal enhancements

The first enhancement we examine is related to nonblocking secondary reads (as mentioned earlier). After that, we cover shard migration, authentication, and stream enhancements.

Timestamps in the storage engine

One of the major new features introduced in MongoDB version 3 was the integration of the *WiredTiger* storage engine. Prior to December 2014, *WiredTiger Inc.* was a company that specialized in database storage engine technology. Its impressive list of customers included *Amazon Inc.* In December 2014, *WiredTiger* was acquired by MongoDB after partnering with them on multiple projects.

In MongoDB version 3, multidocument transactions were not supported. Furthermore, in the replication process (`https://docs.mongodb.com/manual/replication/#replication`), changes accepted by the primary server in a replica set were pushed out to the secondary servers in the set, which were largely controlled through *oplogs* (`https://docs.mongodb.com/manual/core/replica-set-oplog/#replica-set-oplog`) and programming logic outside of the storage engine. Simply stated, the oplog represents changes made to the database. When a secondary synchronizes with a primary, it creates a copy of the oplog and then applies the changes to its own local copy of the database. The logic in place that controlled this process in MongoDB 3 was quite complicated and consumed resources that could otherwise have been used to satisfy user requests.

In MongoDB 4, the internal update mechanism of *WiredTiger*, the storage engine, was rewritten to include a timestamp in each update document. The main reason for this change was to provide support for multidocument transactions. It was soon discovered, however, that this seemingly simple change could potentially revolutionize the entire replication process.

In MongoDB 4, much of the logic required to ensure data integrity during replication synchronization has now been shifted to the storage engine itself, which in turn frees up resources to service user requests. The net effect is threefold: improved data integrity, read operations producing a more up-to-date set of documents, and improved performance.

For an excellent in-depth explanation of how timestamps work in the WiredTiger storage engine, have a look at the video at `https://www.mongodb.com/presentations/wiredtiger-timestamps-enforcing-correctness-in-operation-ordering-across-the-distributed-storage-layer`, which features Dr. Michael Cahill, formerly of *WiredTiger, Inc.*, now Director of Engineering at MongoDB.

Shard migration

Typically, DevOps engineers distribute the database into shards to support a massive amount of data. There comes a time, however, when the data needs to be moved. For example, let's say a host server needs to be upgraded or replaced. In MongoDB version 3.2 and earlier, this process could be quite daunting. In one documented case, a 500 GB shard took *13 days* to migrate. In MongoDB 3.4, parallelism support was provided that sped up the migration process. Part of the reason for the improvement was that the *chunk balancer* logic was moved to the *config server* (`https://docs.mongodb.com/manual/core/sharded-cluster-config-servers/#config-servers`), which must be configured as part of a replica set.

Another improvement, available with MongoDB 4.0.3, allows the *sharded cluster balancer* (`https://docs.mongodb.com/manual/core/sharding-balancer-administration/#sharded-cluster-balancer`) to preallocate *chunks* if *zones and ranges* have been defined, which facilitates rapid capacity expansion. DevOps engineers are able to add and remove nodes from a sharded cluster in real time. The sharded cluster balancer handles the work of rebalancing data between the nodes, thereby alleviating the need for manual intervention. This gives DevOps engineers the ability to scale database capacity up or down on demand. This feature is especially needed in environments that experience seasonal shifts in demand. An example would be a retail outlet that needs to scale up its database capacity to support increased consumer spending during holidays.

For more information on chunks, refer to `https://docs.mongodb.com/manual/core/sharding-data-partitioning/#data-partitioning-with-chunks`, and for more information on zones and ranges, refer to `https://docs.mongodb.com/manual/core/zone-sharding/`.

Change streams

As the database is updated, changes are recorded in the *oplog* maintained by the primary server in the replica set, which is then used to replicate changes to the secondaries. Trying to read a list of changes via the oplog is a tedious and resource-intensive process, so many developers choose to use *change streams* (`https://docs.mongodb.com/manual/changeStreams/?jmp=blog_ga=2.5574835.1698487790.1546401611-137143613.1528093145#change-streams`) to subscribe to all changes on a collection. For those of you who are familiar with software design patterns, this is a form of the publish/subscribe pattern.

Aside from their obvious use in troubleshooting and diagnostics, changing streams can also be used to give an indicator of whether or not data changes are durable.

What is new and different in MongoDB 4.x is the introduction of a startAtOperationTime parameter that allows you to specify the timestamp at which you wish to tap into the change stream. This timestamp can also be in the past, but cannot extend beyond what is recorded in the current oplog.

If you enter 4.0 as a value of another parameter, featureCompatibilityVersion, then the streams return token data, used to restart the stream, in the form of a hex-encoded string, which gives you greater flexibility when comparing blocks of token data. An interesting side effect of this is that a replica set based on MongoDB 4.x could theoretically make use of a change stream token opened on a replica set based on MongoDB 3.6. Another new feature in MongoDB 4 is that change streams that are opened on multidocument transactions include the transaction number.

Important new security enhancements

There were many security improvements introduced in MongoDB 4, but here, we highlight the two most significant changes: support for SHA-256 and **transport layer security (TLS)** handling.

SHA-256 support

SHA stands for *secure hash algorithm*. SHA-256 is a hash function (https://csrc.nist.gov/Projects/Hash-Functions) derivative of the SHA-2 family. The significance of offering SHA-256 support is based on the difference between the SHA-1, which MongoDB supports, and SHA-2 families of hash algorithms. SHA-1, introduced in 1995, used algorithms similar to an older family of hash functions: MD2, MD4, and MD5. SHA-1, however, produces a hash value of 160 bits compared with 128 for the MDx series. SHA-256, introduced in 2012, increases the hash value size to 256, which makes it exponentially more difficult to crack. Attack vectors that could compromise communications based upon SHA-1 and SHA-2 include the preimage attack, the collision attack, and the length-extension attack.

The first attack relies upon *brute-force* attack methods to reverse the hash. In the past, this required computational power beyond the reach of anyone other than a well-funded organization (for example, a government agency or a large corporation). Today, a normal desktop computer could have multiple cores, plenty of memory, and a graphics processing unit (GPU) that are easily capable of such attacks. To launch the attack, the attacker would need to be in a place where access to the database itself is possible, which means that other layers of security (such as the firewall) would have first been breached.

A collision attack uses two different messages that produce the same hash. Once the match has been found, it is mathematically possible to interfere with TLS communications. The attacker could, for example, start forging signatures, which would wreak havoc on systems dependent on digitally signed documents. The danger of this form of attack is that it can theoretically be successfully launched in half the number of iterations compared with a preimage attack.

At the time of writing, the SHA-256 hash function is immune to both preimage and collision attacks; however, both the SHA-1 and SHA-2 family of hash functions, including SHA-256, are vulnerable to length-extension attacks. This attack involves adding to the message, thereby extending its length and then recalculating the hash. The modified message is then seen as valid, allowing the attacker a way into the communication stream. Unfortunately, even though SHA-256 is *resistant* to this form of attack, it is still vulnerable.

It might be of interest to note that Bitcoin uses SHA-256 for the verification of transactions.

TLS handling

Transport layer security (**TLS**) was introduced in 1999 to address serious vulnerabilities inherent in all versions of the Secure Sockets Layer (SSL). It is highly recommended that you secure your MongoDB installations with TLS 1.1 or above (covered later in this book). Once you have configured your *mongod* instances to use TLS, all communications are affected. These include communications between clients, drivers, and the server, as well as internal communications between members of a replica set and between nodes in a sharded cluster.

TLS security depends on which block cipher algorithm and mode are selected. For example, the *3DES* (Data Encryption Standard 3) algorithm with the **Cipher Block Chaining** (**CBC**) mode are considered vulnerable to attack even in TLS version 1.2! The **Advanced Encryption Standard** (**AES**) algorithm and **Galois Counter Mode** (**GCM**) are considered a secure combination, but are only supported in TLS versions 1.2 and 1.3 (ratified in 2018). It should be noted, however, that the AES-256 and GCM combination is not supported when running the MongoDB Enterprise edition on a Windows server.

Using any form of SSL with MongoDB is now deprecated. TLS 1.0 support is also disabled in MongoDB 4.x and above. Ultimately, the version of TLS you end up using in your MongoDB installation completely depends on what cryptographic libraries are available for the server's operating system. This means that as you upgrade your OS and refresh your MongoDB 4+ installation, TLS support is also automatically upgraded. Currently, MongoDB 4+ uses *OpenSSL* on Linux hosts, *Secure Channel* on Windows, and *Secure Transport* on the Mac.

As of MongoDB 4.4, the *Mongo* shell now issues a warning if the x.509 certificate is due to expire within the next 30 days. Likewise, you now see log file messages if there is a pending certificate expiration between *mongod* instances in a sharded cluster or replica set.

 See `https://www.mongodb.com/blog/post/exciting-new-security-features-in-mongodb-40` for a good discussion of security features that were introduced with MongoDB 4.0.

Avoiding problems when upgrading from MongoDB 3.x to 4.x

If you are not familiar with the concept of *backward incompatibilities*, then you have probably not yet survived a major upgrade! To give you an idea of how important it is to be aware of this when reviewing the *change log* for MongoDB, this concept is also referred to as *code breaks*...as in *things that can break your code*.

 The article at `https://docs.mongodb.com/manual/release-notes/4.0-compatibility/` covers the full list of compatibility changes in MongoDB 4.0. For information on MongoDB 4.2 compatibility changes, go to `https://docs.mongodb.com/master/release-notes/4.2-compatibility/#compatibility-changes-in-mongodb-4-2`. For information on MongoDB 4.4 compatibility changes, go to `https://docs.mongodb.com/master/release-notes/4.4/#changes-affecting-compatibility`.

MMAPv1 storage engine

MMAPv1 (`https://docs.mongodb.com/manual/storage/#mmapv1-storage-engine`), the original MongoDB storage engine, has been deprecated in MongoDB 4.0 and removed as of MongoDB 4.2. It has been replaced by *WiredTiger*, which has been available since MongoDB version 3.0. WiredTiger has been the default since MongoDB version 3.2, so there is a good chance that this backward-incompatible change does not affect your applications nor installation.

If your installation was using the MMAPv1 storage engine before the upgrade, then you immediately notice more efficient memory and disk-space allocation. Simply put, MMAPv1 grabbed as much free memory as it could, and would allocate additional disk space with an insatiable appetite. This made a MongoDB 3 installation using MMAPv1 a *bad neighbor* on a server that is also doing other things!

Another difference that DevOps engineers appreciate is that embedded documents no longer continue to grow in size after being created. This was an unwanted side effect produced by the MMAPv1 storage engine, which could have potentially affected write performance and lead to data fragmentation.

If, as a result of an upgrade from MongoDB 3 to 4, you are in the unfortunate position of having to update a database that is stored using the MMAPv1 storage engine, then you *must* manually back up the data on each server prior to the MongoDB 4.x upgrade. After the upgrade has completed, you then need to perform a manual restore.

A detailed discussion on backup and restore is given in `Chapter 3`, *Essential MongoDB Administration Techniques*.

For a good nonbiased comparison of MMAPv1 and WiredTiger, go to `https://stackoverflow.com/questions/37985134/how-to-choose-from-mmapv1-wiredtiger-or-in-memory-storageengine-for-mongodb`.

Replica set protocol version

Communications between members of a replica set are governed by an internal protocol simply referred to as pv0 or pv1. pv0 was the original protocol. Prior to MongoDB 3.2, the only version available was pv0. MongoDB 4.x dropped support for pv0 and only supports pv1. Accordingly, before you perform an upgrade to MongoDB 4, you must reconfigure all replica sets to pv1 (https://docs.mongodb.com/manual/reference/replica-set-protocol-versions/#modify-replica-set-protocol-version).

Fortunately, this process is quite easy, and can be accomplished by this simple procedure. These steps must then be repeated for each replica set: verify oplog entry replication and upgrade to pv1. Let's go into more detail regarding these two steps.

Verifying that at least one oplog entry has replicated

These steps must be performed on each *secondary* in the replica set:

1. Use the mongo shell to connect to each *secondary* in the replica set:

   ```
   mongo --host <address of secondary>
   ```

2. Once connected to the *secondary*, run the rs.status(); command and check the values of the optimes::appliedOpTime::t key.
3. Repeat this for each *secondary* and confirm that the t value is greater than -1. This tells us that at least one oplog entry has replicated from the *primary* to all secondaries.

Upgrading the primary to protocol version 1

You can now upgrade the replica set protocol version to pv1:

1. Use the mongo (https://docs.mongodb.com/manual/mongo/#the-mongo-shell) shell to connect to the primary in the replica set:

   ```
   mongo --host <address of primary>
   ```

2. Once connected to the *primary*, run these commands to update the protocol version for the replica set:

   ```
   cfg = rs.conf();
    cfg.protocolVersion=1;
    rs.reconfig(cfg);
   ```

Feature compatibility

A number of the new features available in MongoDB 4.x only work if you update the `setFeatureCompatibilityVersion` parameter (https://docs.mongodb.com/master/ reference/command/setFeatureCompatibilityVersion/ #setfeaturecompatibilityversion) in the `admin` database. The new features affected include the following:

- SCRAM-SHA-256
- New type conversion operators and enhancements
- Multidocument transactions
- `$dateToString` option changes
- New change stream methods
- Change stream resume token data type changes

To view the current `featureCompatibility` setting, go through the following steps:

1. Use the `mongo` shell to connect to your database as a user who has the rights to modify the admin database:

```
mongo --username <name of user> --password
```

2. Use this command to view the current setting:

```
db.adminCommand({getParameter:1, featureCompatibilityVersion:1})
```

To perform the update, go through the following steps:

1. Use the `mongo` shell to connect to your database as a user who has the rights to modify the admin database:

```
mongo --username <name of user> --password
```

2. You can then update this parameter as follows, substituting 4.0, 4.2, 4.4, and so on in place of <VERSION>:

```
db.adminCommand( { setFeatureCompatibilityVersion: "<VERSION>" } )
```

It is important to note that the `mongos` binary crashes where the binary version and/or feature compatibility version of `mongos` is lower than the connected mongod instances.

User authentication

When establishing security for database users, you have a choice of several different approaches. One of the most popular approaches is *challenge-response*. Simply put: the database *challenges* the user to prove their identity. The response (in most cases), is a username and password combination. In MongoDB 3, this popular approach was implemented by default using MONGODB-CR (MongoDB Challenge Response). As of MongoDB 4, this mechanism is no longer available. This means that when you upgrade from MongoDB 3 to MongoDB 4, you *must* implement at least its replacement, **Salted Challenge Response Authentication Method** (**SCRAM**).

An alternative would be to use x.509 certificates, covered in `Chapter 11`, *Administering MongoDB Security*.

If your user credentials are in MONGODB-CR format, then you must use the following command to upgrade to *SCRAM* format:

```
db.adminCommand({authSchemaUpgrade: 1});
```

It is critical that you perform this upgrade *while still running MongoDB 3*. The reason for this is that the `authSchemaUpgrade` parameter has been removed in MongoDB 4! Another side effect of upgrading to SCRAM is that your application driver might also be affected. Have a look at the SCRAM Driver Support table in the documentation at `https://docs.mongodb.com/manual/core/security-scram/#driver-support` to be sure. The minimum programming language driver for Python that supports SCRAM authentication, for example, is version 2.8.

It is *extremely important* to note that when upgrading to SCRAM from MONGODB-CR, the old credentials are *discarded*, which means that this process is *irreversible*. A good overview of the upgrade from MONGODB-CR to SCRAM can be found at `https://docs.mongodb.com/manual/release-notes/3.0-scram/index.html#upgrade-to-scram`. Please note that this upgrade process *only works* on MongoDB 3. You need to perform this upgrade *before* you upgrade the version of MongoDB.

For even more security when implementing SCRAM, you now have the option of using SHA-256 instead of SHA-1 (see the previous information).

Removed and deprecated items

Any command or executable binary that is *removed* can potentially cause problems if you are relying on these as part of your application or automated maintenance procedures. Any application that relies upon an item that is been *deprecated* should be examined and scheduled to be rewritten in a timely manner.

Removed items

The following table shows the removed items:

Item	Type	Notes
mongoperf	Binary	Used to measure disk I/O performance without having to enter a mongo shell or otherwise access MongoDB.
$isolated	Operator	This operator was used in previous versions of MongoDB during update operations to prevent multidocument updates from being read until all changes took place. MongoDB 4.x uses transaction support instead. Any commands that include this operator need to be rewritten.

Significant removed items

The following table shows the removed items:

Item	Notes	Type
copyDb clone	Command	Copies an entire database. Although this command is still available, you cannot use it to copy a database managed by a mongod version 4 instance to one managed by a mongod version 3.4 or earlier instance. Use the external binaries mongodump and mongorestore instead after upgrading to MongoDB 4.
db.copyDatabase()	Mongo shell command	This is a wrapper for the copyDb command. The same notes for copyDb clone apply.
db.cloneDatabase()	Mongo shell command	This is a wrapper for the clone command. The same notes for copyDb clone apply.
geoNear	Command	Reads geospatial information (that is, latitude and longitude) and returns documents in order, based on their proximity to the source point. Instead of this command, in MongoDB 4.x, you would use the $geoNear aggregation stage operator or the $near or $nearSphere query operators, depending on the nature of the query you wish to construct.

These commands were *deprecated* in MongoDB 4.0 and *removed* as of MongoDB 4.2.

Deprecated SSL configuration options

Another compatibility issue comes from the TLS/SSL configuration options. In MongoDB 3 and MongoDB 4.0, in the configuration files for both *mongod* (MongoDB database daemon) and *mongos* (which is used to control a sharded cluster), you could add a series of options under the `net.ssl` (`https://docs.mongodb.com/v4.0/reference/configuration-options/#net-ssl-options`) key. As of MongoDB 4.2, these options are deprecated in favor of the `net.tls` options. The `net.tls` options have enhanced functionality compared with the `net.ssl` options. These are covered in detail in `Chapter 11`, *Administering MongoDB Security*.

Summary

In this chapter, you learned about the most important features that were added to MongoDB version 4. These were broken down into three categories: new features, security enhancements, and things to avoid during an upgrade from MongoDB 3 to 4.

One important new feature was adding timestamps to the WiredTiger storage engine, which opened the doors for multidocument transaction support and nonblocking secondary read enhancements. Other internal enhancements include improvements in the shard-migration process, which significantly cuts down the time required for this operation to complete.

In the realm of security, you learned about how SHA-256 support gives you greater security when communicating with the MongoDB database, and also with communications between servers within a replica set or sharded cluster. You also learned that TLS 1.0 support has been removed, and that the new default is TLS 1.1. MongoDB 4.x even provides support for the latest version of TLS, version 1.3, but only if the underlying operating system libraries provide support.

Finally, as the original MongoDB storage engine, MMAPv1, has been removed in favor of WiredTiger, if your original MongoDB 3 installation had data that used MMAPv1, you need to back up while still running MongoDB 3 and then restore after the MongoDB 4.x upgrade has occurred. You were also presented with a list of the most significant items, including binary executables, parameters, and commands, which have been removed, and those which have been deprecated.

In the next chapter, you learn how to install MongoDB 4.x and its Python programming language driver.

Setting Up MongoDB 4.x

2

This chapter helps you install everything you need to use MongoDB 4.x with the Python programming language. In this chapter, you first learn about installing MongoDB on a customer site, including requirements for **random-access memory** (**RAM**), the **central processing unit** (**CPU**), and storage. Next, you learn the differences between a MongoDB 3 and a MongoDB 4.x installation. After that, you learn to install and configure MongoDB 4.x along with the PyMongo driver. At the end of the chapter, you learn how to download the source code and sample data for the book, as well as set up a demo test environment you can use to follow the book examples.

In this chapter, we cover the following topics:

- Installing MongoDB on a customer site
- Installing MongoDB 4.x
- Configuring MongoDB 4.x
- Installing the PyMongo driver package
- Loading the sample data
- Creating the demo environment using Docker and Docker Compose

Technical requirements

The minimum recommended hardware are:

- Desktop PC or laptop
- 1 **gigabyte** (**GB**) free disk space
- 4 GB of RAM
- 500 **kilobits per second** (**Kbps**) or faster internet connection

The software requirements are:

- OS (Linux or macOS): Docker, Docker Compose, `git` (optional)
- OS (Windows): Docker for Windows and Git for Windows

Installation of the required software and how to restore the code repository for the book are explained in this chapter. The code used in this chapter can be found in the book's GitHub repository at `https://github.com/PacktPublishing/Learn-MongoDB-4.x/tree/master/chapters/02`.

Installing MongoDB on a customer site

In this section, you are given an overview of what is needed to install MongoDB on a server running Linux, Windows, or macOS, located at a customer site. Later, in another section, you are shown how to create a virtual test environment in which you can practice the techniques discussed in this book. Before you can start on the installation, however, it is important to review which version of MongoDB is desired, as well as host computer hardware requirements.

 A good overview of MongoDB installation can be found here: `https://docs.mongodb.com/manual/administration/install-community/#install-mongodb-community-edition`.

Available versions of MongoDB 4.x

There are a number of versions of MongoDB 4.x available, including the following:

- MongoDB Atlas and Stitch for the cloud environment
- MongoDB Enterprise Edition
- MongoDB Community Edition

The last product listed is free and tends to have a faster release cycle. Interestingly, MongoDB asserts that the *Community Edition*, unlike the model used by other open source companies, uses the same software pool for its beta versions as does the *Enterprise Edition*. The Enterprise Edition, when purchased, also includes support as well as advanced security features, such as Kerberos and **Lightweight Directory Access Protocol (LDAP)** authentication.

Let's now look at the RAM, CPU, and storage requirements.

Understanding RAM and CPU requirements

The minimum CPU requirements for MongoDB are quite light. The MongoDB team recommends that each `mongod` instance be given access to two cores in a cloud environment, or a multi-core CPU when running on a physical server. The WiredTiger engine is multithreaded and is able to take advantage of multiple cores. Generally speaking, the number of CPU cores available improves throughput and performance. When the number of concurrent users increases, however, if operations occupy all available cores, a drastic downturn in throughput and performance is observed.

 Have a look at the *WiredTiger* section in the MongoDB *Administration - Production Notes* (`https://docs.mongodb.com/manual/administration/ production-notes/#hardware-considerations`) for more information.

As of MongoDB 3.4, the WiredTiger internal cache will default either to 50% of RAM (1 GB), or 256 **megabytes (MB)**, whichever is larger.

Accordingly, you need to make sure that you have enough RAM available on your server to handle other OS demands. So, following this formula, if your server has 64 GB of RAM, MongoDB will grab 31.5 GB! You can adjust this parameter, of course, by setting the `cacheSizeGB` parameter in the MongoDB configuration file. Values for `<number>`, as illustrated in the following code snippet, represent the size in GB, and can range from 0.25 to 10,000:

```
storage.wiredTiger.engineConfig.cacheSizeGB : <number>
```

Another consideration is what type of RAM is being used. In many cases, especially in corporate environments where server purchasing and server maintenance are done by different groups, memory might have been swapped around to the point where your server suffers from **non-uniform memory access (NUMA)** problems (`https://queue.acm.org/ detail.cfm?id=2513149`).

It's worth noting that the size of the index and internal MongoDB cache used for a given operation needs to fit comfortably into RAM. If the index cannot fit, it spills over onto disk, degrading the performance. Have a look at the *Performance monitoring* section in `Chapter 3`, *Essential MongoDB Administration Techniques*.

Examining storage requirements

The amount of disk space used by MongoDB depends on how much data you plan to house, so it's difficult to give you an exact figure. The type of drive, however, will have a direct impact on read and write performance. For example, if your data is able to fit into a relatively smaller amount of drive space, you might consider using a **Solid State Drive** (**SSD**) rather than a standard hard disk drive due to radical differences in speed and performance. On the other hand, for massive amounts of data, an SSD might not prove economical.

Here are some other storage recommendations for MongoDB:

- **Swap space**: Use the tools available for your OS to create a swap partition or swap file. Many DevOps professionals no longer follow the traditional formula that states the swap space should be equal to twice the size of RAM. Although increasing the size of swap on your server will not have a direct impact on performance, follow the traditional formula to ensure all other OS services are performing well, which, in turn, will have a positive effect on MongoDB. Thus, if your server has 64 GB of RAM, the swap space would be 128 GB.
- **RAID**: The official MongoDB recommendation for **Redundant Array of Inexpensive Drives** or **Redundant Array of Independent Disks** (**RAID**) is RAID 10 and is also called RAID 1+0 at a minimum. This is a *nested* RAID level, which involves a RAID 0 array of RAID 1 arrays of disks. RAID levels 0 to 6 by themselves are considered inadequate. RAID levels 50, 60, or 100 would also be considered adequate.
- **Remote filesystems**: Although you can host your MongoDB database files on a remote filesystem, it is not a recommended arrangement. Remote filesystems can take the form of **Network File System** (**NFS**) and mount points in `*nix` (that is, Unix, Linux, or macOS) and Windows shares (Windows networks).

A critical aspect of storage requirements has to do with how many **Input/Output Operations Per Second (IOPS)** the disk supports. If your use case requires 1,000 operations per second and your disk supports 500 IOPS, you will have to shard to achieve your ideal performance requirements. Additionally, you should consider the compression feature of the WiredTiger storage engine. As an example, if you have 1 **terabyte (TB)** of data, MongoDB will typically achieve a 30% compression factor, reducing the size of your physical disk requirements.

Choosing the filesystem type

Each OS offers a choice of filesystem types. Some filesystem types are older and less performant. However, they are not good candidates to run MongoDB. There are two main reasons for this: the need to support legacy applications and legacy files. The other reason is that OS vendors want to give their customers more choice.

The recommended filesystem types are summarized in this table:

OS	Filesystem	Notes
Linux kernel 2.6.25+	XFS	This is the preferred filesystem.
Linux kernel 2.6.28+	EXT4	The next best choice for MongoDB on Linux.
Windows Server 2019, 2016, 2012 R2	ReFS	**Resilient File System (ReFS)**.

For information on currently supported OS versions, please have a look at this documentation reference: `https://docs.mongodb.com/manual/installation/#supported-platforms`.

Let's now look at other server considerations.

Other server considerations

In addition to the hardware considerations listed previously, here are some additional points to consider:

- **Verifying needed OS libraries**: MongoDB relies upon various OS libraries. One very common example is the need for an up-to-date version of the `OpenSSL` library if you plan to configure MongoDB to use **Secure Sockets Layer/Transport Layer Security (SSL/TLS)** in its communications.
- **Clock synchronization**: As any DevOp is aware, internal computer clocks are notorious for their *drift*. Accordingly, many schemes have been used over the years to keep computer clocks in sync. The most prevalent technology used today is the **Network Time Protocol (NTP)**. Servers run an internal daemon (or service) that makes occasional checks to one or more primary NTP servers. An internal *drift* algorithm is used to make micro-adjustments to the server clocks such that the need to check with an NTP server declines over time, as the NTP daemon running locally *learns* to adjust more and more accurately.

The next section in this chapter expands upon this discussion, taking into consideration the differences between installing MongoDB 3 and installing MongoDB 4.x.

In earlier versions of NTP, the size of the timestamp field was only 64 bits, divided between seconds and picoseconds. Given that the starting point was January 1, 1980, under the older version, NTP servers would *run out of time* on February 8, 2038! It might be of some comfort to know that NTP version 4—the current version—now supports a 128-bit timestamp field, with an estimated span of *584 billion years*. See **Request for Comments (RFC) 5905** (`https://www.ietf.org/rfc/rfc5905.txt`) for more information.

Installing MongoDB 4.x

Now that you have an idea what the requirements are in terms of the OS, filesystem, and server hardware, and understand the most important differences between MongoDB 3 and MongoDB 4.x installation, it's time to have a look at the actual installation process. In this section, we cover a general overview of MongoDB installation that can be applied either at a customer site or inside a demo environment running in a Docker container on your own computer.

 IMPORTANT: The version of MongoDB 4.x featured in this book is the *MongoDB Community Edition* running on Docker.

For those of you who have experience in MongoDB 3, in the next section, we briefly discuss major differences in installation between version 3 and 4.x.

MongoDB 3 and MongoDB 4.x installation differences

As mentioned in the first chapter, one of the main differences between MongoDB 3.x and MongoDB 4.x is that the WiredTiger storage engine is now the default. Accordingly, you will need to ensure that you take advantage of WiredTiger's multithreaded capabilities by adjusting the RAM and CPU usage. There is an excellent guideline given here: `https://docs.mongodb.com/manual/administration/production-notes/#allocate-sufficient-ram-and-cpu`.

One of the main differences between installing MongoDB 3.x compared to MongoDB 4.x is the list of operating systems supported. When considering upgrading a server currently running MongoDB 3.x, consult the supported platforms guide (`https://docs.mongodb.com/manual/installation/#supported-platforms`) in the MongoDB documentation. Generally speaking, each new version of MongoDB removes support for older OS versions and adds support for new versions. Also, as a general consideration, where versions of MongoDB below v3.4 supported 32-bit x86 platforms, in MongoDB 3.4 and later—including MongoDB 4.x—there is *no support* for 32-bit x86 platforms.

Another difference you will observe is that as of MongoDB 4.4, *time zone* data is now incorporated into the installation program, as seen in this screenshot:

```
File  Edit  View  Search  Terminal  Help
Configuring tzdata
------------------

Please select the geographic area in which you live. Subsequent configuration questions will narrow this down by presenting a list of cities, represent
ing the time
zones in which they are located.

  1. Africa  2. America  3. Antarctica  4. Australia  5. Arctic  6. Asia  7. Atlantic  8. Europe  9. Indian  10. Pacific  11. SystemV  12. US  13. Etc
Geographic area: 6

Please select the city or region corresponding to your time zone.

  1. Aden      11. Baku       21. Damascus   31. Hong_Kong  41. Kashgar      51. Makassar      61. Pyongyang  71. Singapore       81. Ujung_Pandang
  2. Almaty    12. Bangkok    22. Dhaka      32. Hovd       42. Kathmandu    52. Manila        62. Qatar      72. Srednekolymsk   82. Ulaanbaatar
  3. Amman     13. Barnaul    23. Dili       33. Irkutsk    43. Khandyga     53. Muscat        63. Qostanay   73. Taipei          83. Urumqi
  4. Anadyr    14. Beirut     24. Dubai      34. Istanbul   44. Kolkata      54. Nicosia       64. Qyzylorda  74. Tashkent        84. Ust-Nera
  5. Aqtau     15. Bishkek    25. Dushanbe   35. Jakarta    45. Krasnoyarsk  55. Novokuznetsk  65. Rangoon    75. Tbilisi         85. Vientiane
  6. Aqtobe    16. Brunei     26. Famagusta  36. Jayapura   46. Kuala_Lumpur 56. Novosibirsk   66. Riyadh     76. Tehran          86. Vladivostok
  7. Ashgabat  17. Chita      27. Gaza       37. Jerusalem  47. Kuching      57. Omsk          67. Sakhalin   77. Tel_Aviv        87. Yakutsk
  8. Atyrau    18. Choibalsan 28. Harbin     38. Kabul      48. Kuwait       58. Oral          68. Samarkand  78. Thimphu         88. Yangon
  9. Baghdad   19. Chongqing  29. Hebron     39. Kamchatka  49. Macau        59. Phnom_Penh    69. Seoul      79. Tokyo           89. Yekaterinburg
 10. Bahrain   20. Colombo    30. Ho_Chi_Minh 40. Karachi   50. Magadan      60. Pontianak     70. Shanghai   80. Tomsk           90. Yerevan
Time zone: 12
```

The next section shows you how to install MongoDB 4.x on a Linux-based platform.

Installing MongoDB 4.x on Linux

It's a fact that there are literally hundreds of Linux *distributions* available. Most are free, but there are others for which you must pay. Many of the free Linux vendors also offer enterprise Linux versions that are stable, tested, and offer support. Obviously, for the purposes of this book, it would be impossible to detail the MongoDB installation process on every single distribution. Instead, we will concentrate on two main *families* of Linux: the **RPM-** (formerly **RedHat Package Manager**: https://rpm.org/) based and the Debian-based distributions.

Of the RPM-based distributions, we find *RedHat*, *Fedora*, and *CentOS*, among others, which are extremely popular. On the Debian side, we have *Kali Linux* and *Ubuntu* (and its offshoots!). There are many other distributions—and even other major families—that are not covered in this book. Among the more well-known distributions not covered are *SUSE*, *pacman*, *Gentoo*, and *Slackware*.

For more information on different Linux distributions, have a look at the following websites: Debian (https://www.debian.org/), SUSE (https://www.suse.com/), pacman (https://www.archlinux.org/pacman/), Gentoo (https://www.gentoo.org/), and Slackware (http://www.slackware.com/).

MongoDB Linux installations all tend to follow the same general procedure, detailed as follows:

1. Import the public key of the MongoDB package management server.
2. Add a MongoDB-specific **Uniform Resource Locator** (**URL**) to the package management sources list (websites from which package files are downloaded).
3. Update the package manager from the newly added sources.
4. Use the package manager to install either all MongoDB packages (`mongodb-org`) or specific components (for example, `mongodb-org-server`).

For specifics on a given Linux distribution, see this page: `https://docs.mongodb.com/manual/installation/#mongodb-community-edition-installation-tutorials`.

Do not use the version of MongoDB that is available through precompiled binaries ordinarily available through your package management system. These versions are not the latest versions and can cause problems down the road. It's best to grab the package directly from `https://www.mongodb.com/`.

Here is a screenshot of the start of the main installation on an Ubuntu 20.04 server:

```
File Edit View Search Terminal Help
root@installation1:/# apt-get install -y mongodb-org=4.4.0~rc2
Reading package lists... Done
Building dependency tree
Reading state information... Done
The following additional packages will be installed:
  krb5-locales libbrotli1 libcurl4 libgssapi-krb5-2 libk5crypto3 libkeyutils1 libkrb5
-3 libkrb5support0 libnghttp2-14 librtmp1 libssh-4 mongodb-database-tools
  mongodb-org-database-tools-extra mongodb-org-mongos mongodb-org-server mongodb-org-
shell mongodb-org-tools tzdata
Suggested packages:
  krb5-doc krb5-user
The following NEW packages will be installed:
  krb5-locales libbrotli1 libcurl4 libgssapi-krb5-2 libk5crypto3 libkeyutils1 libkrb5
-3 libkrb5support0 libnghttp2-14 librtmp1 libssh-4 mongodb-database-tools
  mongodb-org mongodb-org-database-tools-extra mongodb-org-mongos mongodb-org-server
mongodb-org-shell mongodb-org-tools tzdata
0 upgraded, 19 newly installed, 0 to remove and 0 not upgraded.
Need to get 103 MB of archives.
```

Once the main installation has completed, you will note that the installation script creates a user and group named `mongodb`. Here is a screenshot of the end of the main installation process:

```
File Edit View Search Terminal Help
Current default time zone: 'Asia/Bangkok'
Local time is now:      Wed Apr 29 10:44:13 +07 2020.
Universal Time is now:  Wed Apr 29 03:44:13 UTC 2020.
Run 'dpkg-reconfigure tzdata' if you wish to change it.

Setting up librtmp1:amd64 (2.4+20151223.gitfa8646d.1-2build1) ...
Setting up libk5crypto3:amd64 (1.17-6ubuntu4) ...
Setting up mongodb-org-database-tools-extra (4.4.0~rc3) ...
Setting up libkrb5-3:amd64 (1.17-6ubuntu4) ...
Setting up libgssapi-krb5-2:amd64 (1.17-6ubuntu4) ...
Setting up mongodb-database-tools (100.0.1) ...
Setting up libssh-4:amd64 (0.9.3-2ubuntu2) ...
Setting up libcurl4:amd64 (7.68.0-1ubuntu2) ...
Setting up mongodb-org-server (4.4.0~rc3) ...
Adding system user `mongodb' (UID 101) ...
Adding new user `mongodb' (UID 101) with group `nogroup' ...
Not creating home directory `/home/mongodb'.
Adding group `mongodb' (GID 101) ...
```

The script also creates a `/etc/mongod.conf` default configuration file, covered later in this chapter. To start the `mongod` instance using this configuration file, use this command:

```
mongod -f /etc/mongod.conf &
```

To access the database, issue the `mongo` command, which starts a shell, giving direct access to the database, as shown here:

```
File Edit View Search Terminal Help
root@installation1:/# mongo
{"t":{"$date":"2020-04-29T04:46:14.589Z"},"s":"I", "c":"NETWORK", "id":4648601,"ctx":"main","msg":"Implicit TCP FastOpen unavailable
. If TCP FastOpen is required, set tcpFastOpenServer, tcpFastOpenClient, and tcpFastOpenQueueSize."}
MongoDB shell version v4.4.0-rc3
connecting to: mongodb://127.0.0.1:27017/?compressors=disabled&gssapiServiceName=mongodb
Implicit session: session { "id" : UUID("9f0dbe07-e18f-46ff-ae4d-193bc6170139") }
MongoDB server version: 4.4.0-rc3
Welcome to the MongoDB shell.
For interactive help, type "help".
For more comprehensive documentation, see
        http://docs.mongodb.org/
Questions? Try the support group
        http://groups.google.com/group/mongodb-user
Server has startup warnings:
{"t":{"$date":"2020-04-29T04:46:11.935+00:00"},"s":"W", "c":"CONTROL", "id":22120, "ctx":"initandlisten","msg":"Access control is n
ot enabled for the database. Read and write access to data and configuration is unrestricted","tags":["startupWarnings"]}
{"t":{"$date":"2020-04-29T04:46:11.935+00:00"},"s":"W", "c":"CONTROL", "id":22138, "ctx":"initandlisten","msg":"You are running thi
s process as the root user, which is not recommended","tags":["startupWarnings"]}
---
Enable MongoDB's free cloud-based monitoring service, which will then receive and display
metrics about your deployment (disk utilization, CPU, operation statistics, etc).

The monitoring data will be available on a MongoDB website with a unique URL accessible to you
and anyone you share the URL with. MongoDB may use this information to make product
improvements and to suggest MongoDB products and deployment options to you.

To enable free monitoring, run the following command: db.enableFreeMonitoring()
To permanently disable this reminder, run the following command: db.disableFreeMonitoring()
---
>
```

Let's now look at the installation of MongoDB on Windows and macOS.

Installing on Windows and macOS

Installing MongoDB directly on a Windows server is actually easier than with Linux, as all you need to do is to go to the downloads website for MongoDB and choose the Windows installer script matching your Windows version. To start MongoDB, you would go to the Windows *Task Manager*, locate MongoDB, and choose **start**. You can also configure MongoDB to start automatically as a service. One final consideration is that the MongoDB Windows installation also automatically installs *MongoDB Compass* (discussed in more detail in `Chapter 9`, *Handling Complex Queries in MongoDB*).

 To install MongoDB on a macOS, it is recommended you use the *Homebrew* package manager. Otherwise, the actual installation procedure closely follows that when installing on Linux. For more information on installing MongoDB on Windows, see this documentation reference: `https://docs.mongodb.com/master/tutorial/install-mongodb-on-windows/#install-mongodb-community-edition-on-windows`. For information on macOS installation, have a look here: `https://docs.mongodb.com/manual/tutorial/install-mongodb-on-os-x/#install-mongodb-community-edition-on-macos`.

Next, we look at the MongoDB configuration.

Configuring MongoDB 4.x

In the previous section, you learned how to install and then start MongoDB. When starting MongoDB in this manner, you were actually using the default configuration options. It is extremely important that you understand some of the different startup options located in a file we will refer to as `mongod.conf`, or the *MongoDB configuration file*, throughout the rest of the book. In this section, we cover MongoDB configuration concepts that can be applied either at a customer site or inside a demo environment running in a Docker container on your own computer.

The name and location of the MongoDB configuration file

The name and location will vary slightly between operating systems. The table shown here describes the defaults for the four major operating systems covered in the previous section on installation:

OS	MongoDB configuration file
DEB-based	`/etc/mongod.conf`
RPM-based	`/etc/mongod.conf`
Windows	`C:\Program Files\MongoDB\Server\4.x\bin\mongod.cfg`
Mac OS X	`/usr/local/etc/mongod.conf`

Before getting into the details, let's have a look at the syntax.

YAML configuration file syntax

The MongoDB configuration file is an **American Standard Code for Information Interchange (ASCII)** text file using the **YAML Ain't Markup Language (YAML)** style of formatting, with the following features:

- Key-value pairs are specified as follows:

  ```
  key : value
  ```

- Section headings are followed by a colon and a line feed, as follows:

  ```
  heading:
  ```

- Key-value pairs that are subsets of a section are indented by two spaces, as follows:

  ```
  heading:
    key1: value1
    key2: value2
  ```

- Comments can be added by starting the line with a # symbol, as follows:

  ```
  # This is a comment
  ```

IMPORTANT: YAML does not support tabs. Be sure to use spaces when indenting.

We will now examine some of the more important configuration options.

Configuration options

It is important to provide configuration to direct how the mongod server daemon starts. Although you could theoretically start mongod using command-line options, most DevOps prefer to place a permanent configuration into a configuration file. Configuration options fall into these major categories:

- Core: Storage, logging and network
- Process management
- Security
- Profiling
- Replication and sharding

In addition, the *Enterprise* (that is, paid) version of MongoDB features **auditing** and **Simple Network Management Protocol** (**SNMP**) options. Of these options in this chapter, we will only focus on the Core category. Other options will be covered in later chapters as those topics arise.

NOTE: The name and location of the configuration file vary depending on the operating system. See the earlier section on installing MongoDB to get an idea of the default name and location of this file.

Storage configuration options

Under the `storage` key, the following table summarizes the more widely used options:

Key	Sub key	Notes
dbPath		Directory where the actual database files reside.
journal	enabled	Possible values: `true` | `false`. If set `true`, journaling is enabled.
journal	commitIntervalMs	This represents the interval between journal commits. Values can be from 1 to 500 milliseconds. The default is 100.
engine		The default is `wiredTiger`. You can also specify `inMemory`, which activates the in-memory storage engine.

There are a number of other parameters as well; however, these are covered in other chapters in the book as appropriate.

Logging configuration options

It is important for troubleshooting and diagnostics purposes to activate MongoDB logging. Accordingly, there are a number of parameters that can be activated under the `systemLog` key, as summarized in this table:

Key	Notes
path	Directory where log files will be created and stored.
logAppend	When set to `true`, after a MongoDB server restart, new entries are appended to the existing log entries. If set to `false` (default), upon restart, the existing log file is backed up and a new one is created.
verbosity	Determines how much information goes into the log. Verbosity levels range from 0 to 5, with 0 as the default. Level 0 includes informational messages. Setting `verbosity` to 1 through 5 increases the amount of information in the log message as well as debug information.

component	This option opens up the possibility of creating more refinements in logging. As an example, to set logging verbosity to 1 for queries, but 2 for storage, add these settings:
	`systemLog:`
	`    component:`
	`        query:`
	`        verbosity:  1`
	`        storage:`
	`            verbosity:  2`

Let's now look at different network configuration options.

Network configuration options

The `net` configuration file key allows you to adjust network-related settings. Here are some of the core options:

Key	Notes
port	The port on which the MongoDB server daemon listens. The default is 2017.
bindIp	You can specify one or more **Internet Protocol** (**IP**) addresses as values. Instead of 127.0.0.1, you can also indicate localhost. If you wish to bind to all IP addresses, specify 0.0.0.0. Alternatively, starting with MongoDB 4.2, you can specify "*" (double quotes required). IP addresses can be IPv4 or IPv6.
bindIpAll	As an alternative to bindIp: 0.0.0.0, you can set a value of true for this key instead.
ipV6	Set this parameter to true if you wish to use IPv6 addressing.
maxIncomingConnections	Set an integer value to limit the number of network connections allowed. The default is 65536.

Another important networking option is `net.tls`; however, this is not covered in this chapter, but rather in the next chapter, `Chapter 3`, *Essential MongoDB Administration Techniques*.

Example MongoDB configuration file

Here is an example using the three directives discussed in this section:

```
storage:
  dbPath: /var/lib/mongodb
  journal:
    enabled: true

systemLog:
  destination: file
  logAppend: true
  path: /var/log/mongodb/mongod.log

net:
  port: 27017
  bindIp: 0.0.0.0
```

In this example, journaling is enabled, and the database files are stored in the `/var/lib/mongodb` directory. The system log is recorded in the `/var/log/mongodb/mongod.log` file, and new entries are added to the existing file. The MongoDB server daemon listens on port `27017` to requests from all IP addresses.

 For more information on the `mongod.conf` file configuration options, have a look at the documentation here: `https://docs.mongodb.com/manual/reference/configuration-options/index.html#configuration-file-options`. For information on journaling, see `https://docs.mongodb.com/manual/core/journaling/index.html#journaling`. For information on log message components, see `https://docs.mongodb.com/manual/reference/log-messages/#components`.

Command-line options

It's important to note that there is a one-to-one correspondence between configuration file settings and command-line switches. Thus, for quick testing, you can simply start the MongoDB server daemon with a command-line switch. If successful, this can then be added to the configuration file.

For example, let's say you are testing IPv6 routing. To confirm that the MongoDB server will listen to requests on a given IPv6 address, from the command line you could start the server daemon as follows:

```
# mongod --ipv6 --bind_ip fe80::42:a6ff:fe0c:a377
```

The `--ipv6` option enables IPv6 support, and the `bind_ip` option specifies the address. If successful, these commands could then be added to the `mongod.conf` file.

 For a complete list of MongoDB configuration file settings mapped to command line options, refer to the *Configuration File Settings and Command-Line Options Mappings* web page (`https://docs.mongodb.com/manual/reference/configuration-file-settings-command-line-options-mapping/#configuration-file-settings-and-command-line-options-mapping`).

Expansion directives

First introduced in MongoDB 4.2, there are two *magic* keys added that are referred to as *externally sourced* or *expansion* directives. The two directives are __rest, allowing the use of a **representational state transfer** (**REST**) endpoint to obtain further configuration, and __exec, which loads configuration from the output of a shell or OS command.

IMPORTANT: In order to use the __rest expansion directive, you *must* start the mongod or mongos instance with this command-line option: --configExpand none|exec|rest|rest,exec. For more information, see the documentation on expansion directives here: https://docs.mongodb.com/manual/reference/expansion-directives/#expansion-directives-reference.

Using the __rest expansion directive

The __rest expansion directive allows mongod or mongos to start with additional directives, or an entire configuration block, from an external REST request. Using parameters supplied with the directive, the server will issue an HTTP GET request to the REST **application programming interface** (**API**) endpoint specified and will receive its configuration in turn. The advantage of this approach is that it enables the centralization of startup directives across a large distributed MongoDB system.

An obvious disadvantage when making requests to remote resources is that it opens a potentially huge security vulnerability. Thus, at a minimum, requests to any remote endpoint must use HTTPS. In addition, operating system permissions must restrict *read* access of the configuration file containing the __rest expansion directive to only the mongod or mongos user.

Here is a summary of parameters needed in conjunction with the __rest expansion directive:

Parameter	Notes
__rest	A string that represents the REST endpoint **Uniform Resource Identifier** (**URI**).
type	Indicates the formatting of the return value. Currently accepted values are either string or yaml.
trim	Defaults to none. You can otherwise set this value to whitespace, which causes leading and/or trailing whitespace to be removed from the returned value.
digest	The SHA-256 digest of the result. digest_key must also be supplied (next).
digest_key	Hex string representing the key used to produce the SHA-256 digest.

Here is a brief example:

```
net:
  tls:
    certificateKeyFilePassword:
      __rest: "https://continuous-learning.com/api/cert/certPwd"
      type: "string"
```

In the preceding code example, the certificate key file password is provided by a REST request to continuous-learning.com.

 For more information on REST, have a look here: https://www.ics.uci.edu/~fielding/pubs/dissertation/rest_arch_style.htm#sec_5_2_1_1.

Using the __exec expansion directive

The __exec expansion directive operates in much the same manner as the __rest directive. The main difference is that instead of consulting a REST endpoint, an operating system command is referenced, which in turn produces the result. Another difference is that the __exec directive could be used to load an entire configuration file. This gives DevOps the ability to automate mongod or mongos startup options.

As with the __rest directive, a number of parameters are associated with the __exec directive, as summarized here:

Parameter	Notes
__exec	The OS command to be executed. Note that this command could well be a Python script.
type	Indicates the formatting of the return value. Currently accepted values are either string or yaml.
trim	Defaults to none. You can otherwise set this value to whitespace, which causes leading and/or trailing whitespace to be removed from the returned value.
digest	The SHA-256 digest of the result. digest_key must also be supplied (next).
digest_key	Hex string representing the key used to produce the SHA-256 digest.

Here is a brief example:

```
net:
  tls:
    certificateKeyFilePassword:
      __exec: "python /usr/local/scripts/cert_pwd.py"
      type: "string"
```

In the preceding code example, the password for the certificate key file is provided by a `cert_pwd.py` Python script.

Installing the PyMongo driver package

We do not cover the actual installation of Python itself, as that is beyond the scope of this book. However, we will cover the installation of the PyMongo driver, which sits between Python and MongoDB. Before we dive into the details of installing the PyMongo driver package, it's important to discuss potential driver compatibility issues.

For more information about how to install Python, please consult the documentation at `https://docs.python.org/3/using/index.html`. For information on the PyMongo driver, see `https://github.com/mongodb/mongo-python-driver`.

In many cases, you will find *multiple versions* of Python installed on the server you plan to use to host MongoDB. If that is the case, you must ensure that the version of PyMongo chosen not only supports the version of MongoDB you are using but the version of Python you plan to use as well!

Python driver compatibility

Before you proceed to install the PyMongo driver, you might need to deal with driver compatibility issues. These issues are twofold, and are detailed as follows:

- Driver version support for the version of MongoDB you are using
- Driver version support for the version of Python you are using

There is an excellent documentation reference that cross-references the version of PyMongo against the version of MongoDB. The table can be found here: `https://docs.mongodb.com/ecosystem/drivers/pymongo/#mongodb-compatibility`. In the table, for example, you will notice that MongoDB version 2.6 is supported by PyMongo driver versions 2.7 all the way up to 3.9. However, do you *really* want to be using MongoDB version 2? Probably not!

More relevant to this book is the support for MongoDB 4.x. As of the time of writing, the PyMongo driver version 3.9 supports both MongoDB 4.0 and 4.2. On the other hand, if you find that you need a slightly older version of the PyMongo driver, perhaps due to issues with the version of Python you are running (discussed next), you will find that MongoDB 4.0 is supported by PyMongo driver versions 3.7, 3.8, and 3.9.

In the same section of the MongoDB documentation, there is a follow-up table that lists which version of Python supports the various versions of the PyMongo driver. That table can be found here: `https://docs.mongodb.com/ecosystem/drivers/pymongo/#language-compatibility`.

Many operating systems automatically provide Python; however, more often than not, it's version 2. Interestingly, you will note that the PyMongo driver version 3.9 does in fact support Python version 2.7. If you are using Python version 3, again, you will find that the PyMongo driver version 3.9 supports Python 3.4 through 3.7.

So, in summary, the version of the PyMongo driver that you choose to install is dictated by both the version of MongoDB you plan to use and the version of Python. Next, we'll have a look at a tool called `pip`.

pip3

The Python package manager is called `pip`. You will need `pip` to install the PyMongo driver, but, just as with Python, there is `pip` for Python release 2 and also `pip3` for Python 3. The `pip` command does not work with Python 3. If you do not have `pip3` available, use the appropriate installation procedure for the operating system that will house the Python code you plan to create, which will communicate with MongoDB.

 For further information, please consult the official `pip` documentation here: `https://pip.pypa.io/en/stable/`.

Installing PyMongo using pip

There are a number of ways in which you can install the PyMongo driver package, detailed as follows:

- Compiling the source code
- Installing a pre-built binary
- Using `pip3`

This section describes how to use `pip3` to install the PyMongo driver package as well as where PyMongo files are stored in the filesystem. For more information on the latest PyMongo driver, you can consult the **Python Package Index** (**PyPI**) (see `https://pypi.org/`). Enter `pymongo` in the search dialog box, and a list of pertinent results is displayed. Clicking on the `https://pypi.org/project/pymongo/` link, you will see the current version of PyMongo.

To use `pip3` to install the PyMongo driver, open a Terminal window, and run the following command. Please note that you will most likely need either root or administrator privileges to perform this command:

```
pip3 install pymongo
```

Here is the output from an Ubuntu Linux 20.04 server:

```
File Edit View Search Terminal Help
root@installation1:~# pip3 install pymongo
Collecting pymongo
  Downloading pymongo-3.10.1-cp38-cp38-manylinux2014_x86_64.whl (480 kB)
    |████████████████████████████████| 480 kB 854 kB/s
Installing collected packages: pymongo
Successfully installed pymongo-3.10.1
root@installation1:~# pip3 show pymongo
Name: pymongo
Version: 3.10.1
Summary: Python driver for MongoDB <http://www.mongodb.org>
Home-page: http://github.com/mongodb/mongo-python-driver
Author: Mike Dirolf
Author-email: mongodb-user@googlegroups.com
License: Apache License, Version 2.0
Location: /usr/local/lib/python3.8/dist-packages
Requires:
Required-by:
root@installation1:~#
```

We have now successfully installed PyMongo using `pip`. The next step is to test the connection.

Testing the connection between Python and MongoDB

For this simple test, all we need is the `MongoClient` class from the `pymongo` module. The example shown in this sub-section can be found at `/path/to/repo/chapters/02/test_connection.py`. Let's have a look at the code, as follows:

1. We start by defining sample data and importing the needed classes, like this:

```
data = { 'first_name' : 'Fred', 'last_name'  : 'Flintstone'}
import pprint
from pymongo import MongoClient
```

2. Next, we create a `MongoClient` instance and a reference to the new database test, as follows:

```
client = MongoClient('localhost')
test_db = client.test
```

In order to maintain the purity of the test, as we are likely to run it many times, we add a statement that removes all documents from the new collection, `test_collection`.

3. We then insert and retrieve a single document using the `insert_one()` and `find_one()` pymongo collection methods. A typical return value after an `insert` operation would be the last autogenerated ID, as shown here:

```
test_db.test_collection.delete_many({})
insert_result =
test_db.test_collection.insert_one(data).inserted_id
find_result = test_db.test_collection.find_one()
```

4. Finally, we use `pprint` to display the results, as follows:

```
pprint.pprint(insert_result)
pprint.pprint(find_result)
```

Here is a screenshot of how the output might appear:

```
File  Edit  View  Search  Terminal  Help

Insert Result:
ObjectId('5eb7a6210a13427dbf753beb')

Find Result:
{'_id': ObjectId('5eb7a6210a13427dbf753beb'),
 'first_name': 'Fred',
 'last_name': 'Flintstone'}

------------------
(program exited with code: 0)
Press return to continue
```

The three pymongo collection methods shown—delete_many(), insert_one(), and find_one()—will be described in greater detail in later chapters of this book. For more information, see this web page: https://api.mongodb.com/python/current/api/pymongo/collection.html.

In the next section, you will learn to retrieve and load the source code and sample data provided in the *GitHub* repository that accompanies this book.

Loading the sample data

Although the focus of this chapter is on installation, the remainder of the book teaches you how to implement and manage the major features of MongoDB (for example, replica sets, sharded clusters, and more). In addition, the book focuses on mastering MongoDB queries and developing applications based on Python. Accordingly, we have provided you with a Docker image that includes the following:

- The latest version of MongoDB preinstalled
- An Apache web server installation
- Python 3
- The PyMongo driver

In this section, we will show you how to restore the source code accompanying this book from a GitHub repository and then bring online the demo environment using Docker and Docker Compose. First, we will walk you through the process of restoring the source code and sample data from the GitHub repository associated with this book.

Installing the source code and sample data using Git

In order to load the sample data and Python code needed to run the examples shown in this book, you will need to either *clone* the GitHub repository for the book or download and unzip it. The code repository for the book is located here:

```
https://github.com/PacktPublishing/Learn-MongoDB-4.x
```

How you can go about restoring the source code and sample data depends on whether or not you have `git` installed on your computer. Accordingly, we'll look at two techniques, with and without using `git`.

Here are a series of steps to restore the sample data using `git`:

1. From a browser, enter the following URL:

   ```
   https://github.com/PacktPublishing/Learn-MongoDB-4.x
   ```

2. Locate the green button that says **Code**.
3. Click on the down arrow, and select the clipboard icon to copy the repository URL:

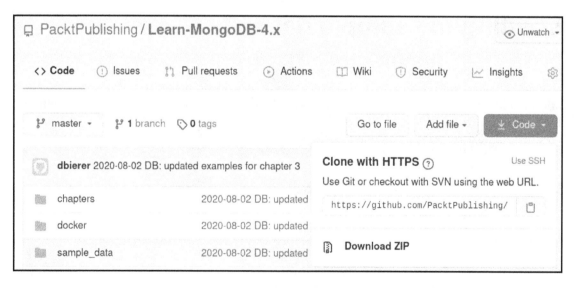

4. Open a Terminal window (or, if using *Git for Windows*, open the provided `Bash` shell) and make and/or change to a directory under which you are prepared to *clone* the sample data repository for this book.

5. Enter the following command to *clone* the repository:

```
git clone https://github.com/PacktPublishing/Learn-MongoDB-4.x.git
/path/to/repo
```

6. The result should appear as shown in this screenshot. Note that we are using `/home/ned/Repos/learn-mongodb` in place of `/path/to/repo`:

```
ned@ned: ~/Repos
ned@ned:~/Repos$ git clone https://github.com/PacktPublishing/Learn-MongoDB-4.x.git
Cloning into 'Learn-MongoDB-4.x'...
remote: Enumerating objects: 95, done.
remote: Counting objects: 100% (95/95), done.
remote: Compressing objects: 100% (68/68), done.
remote: Total 1265 (delta 16), reused 73 (delta 14), pack-reused 1170
Receiving objects: 100% (1265/1265), 9.77 MiB | 2.06 MiB/s, done.
Resolving deltas: 100% (560/560), done.
ned@ned:~/Repos$
```

7. The path to this repository will be referred to throughout this book as `/path/to/repo`. Thus, in the example shown in the screenshot, `/path/to/repo` corresponds to `/home/ned/Repos/learn-mongodb`.

> If you are planning on installing MongoDB and installing the sample data on a Windows computer, it is advised that you install *Git for Windows* (https://gitforwindows.org/). This gives you a `Bash` shell in which you can enter Linux commands, including `git`.

Installing source and sample data by downloading and unzipping

If you prefer not to use `git`, another technique is to simply download the entire repository in the form of a ZIP file. You can then extract its contents wherever is appropriate on your computer. To use this approach, proceed as follows:

1. From a browser, open this URL:
   https://github.com/PacktPublishing/Learn-MongoDB-4.x.
2. Locate the green button that says **Code**.
3. Click on the down arrow, and select the **Download ZIP** option.
4. Use the tools associated with your operating system to open the downloaded ZIP file, and extract it into the subdirectory of your creation.

Now, let's have a look at the directory structure of the sample data.

Understanding the source code directory structure

The source code and sample data for the examples shown in the book are located in these directories under `/path/to/repo`:

- `chapters`: Under this subdirectory, you will find source code for each chapter, by number.
- `www`: This directory structure contains source code based upon Django.
- `sample_data`: In this directory, you will find JavaScript files that can be executed to populate the database with sample data. There is no need to insert data just yet: as the need arises, you will be instructed in the chapter to insert the sample data appropriate for that section of the book.

Next, we will have a look at creating a demo environment you can use to test the examples shown in the book using Docker and Docker Compose.

Creating the demo environment using Docker and Docker Compose

Docker is a virtualization technology that has significant advantages over others (for example, *VirtualBox* or *VMware*) in that it directly uses resources from the host computer and is able to share those resources between **virtual machines** (**VMs**) (referred to as *containers* in the Docker reference guide). In this section, you learn how to configure a Docker container in which you can practice installing MongoDB following the discussion in this chapter. Following that, you learn how to configure a Docker container that can be used to practice using the source code and MongoDB query examples found in `Chapter 3`, *Essential MongoDB Administration Techniques*, to `Chapter 12`, *Developing in a Secured Environment*, of this book.

 As with many software environments available these days, Docker is free to install and use for your own personal purposes (for example, learning MongoDB!). Please see the *Docker Terms of Service* (`https://www.docker.com/legal/docker-terms-service`) if you plan to include Docker in a project for hire. The other virtual environments mentioned are VirtualBox (`https://www.virtualbox.org/`) and VMware (`https://www.vmware.com/`).

Installing Docker and Docker Compose

Before you can recreate the demo environment used to follow examples from this book, you will need to install *Docker*. In this section, we will not cover the actual installation of Docker as it's already well documented. Please note that in all cases, you will need to make sure that *hardware virtualization* support has been enabled for your computer's **basic input/output system** (**BIOS**). Also, for Windows, you will install *Docker Desktop on Windows* rather than the straight binary executable, as is the case for Linux- or Mac-based operating systems. In addition, for Windows, you will need to *enable* the Windows *hypervisor*. The Windows installation steps will give you more information on this.

In addition, to recreate the demo environment, you will need to install *Docker Compose*. Docker Compose is an open source tool provided by Docker that allows you to install a Docker container defined by a configuration file. As with Docker itself, installation instructions vary slightly between Linux, Windows, and macOS computer hosts.

Docker installation instructions can be found here: `https://docs.docker.com/engine/install/`. The primary installation guide for Docker Compose can be found here: `https://docs.docker.com/compose/install/`. For information about hardware virtualization, consult this article from Intel: `https://www.intel.com/content/www/us/en/virtualization/virtualization-technology/intel-virtualization-technology.html`. For more information on the Windows hypervisor, have a look at this article: `https://docs.microsoft.com/en-us/virtualization/hyper-v-on-windows/quick-start/enable-hyper-v`.

Configuring Docker for Chapter 2

A special Docker Compose configuration has been created that allows you to test your skills installing MongoDB. This container is drawn from the latest **long-term support** (**LTS**) version of the Ubuntu Linux image on Docker Hub. It also has the necessary prerequisites for installing MongoDB. The files to create and run this container are located in the source code repository for the book, at `/path/to/repo/chapters/02`.

Let's first have a look at the `docker-compose.yml` file first. It is in YAML file format (as is the `mongod.conf` file), so you must pay careful attention to indents. The first line indicates the version of Docker Compose. After that, we define `services`. The only service defined creates a container named `learn-mongo-installation-1` and an image named `learn-mongodb/installation-1`. Two volumes are mounted to allow data to be retained even if the container is restarted.

Also, a file in the `common_data` directory is mounted as `/etc/hosts`, as illustrated in the following code block:

```
version: "3"
services:
 learn-mongodb-installation:
 container_name: learn-mongo-installation-1
 hostname: installation1
 image: learn-mongodb/installation-1
 volumes:
 - db_data_installation:/data/db
 - ./common_data:/data/common
 - ./common_data/hosts:/etc/hosts
```

Continuing with the `services` definition, we also need to map the MongoDB port `27017` to something else so that there will be no port conflicts on your host computer. We indicate the directory containing the Docker build information (for example, the location of the Dockerfile), and tell the container to stay open by setting `stdin_open` and `tty` both to `true`. Finally, still in the `services` block, we assign an IP address of `171.16.2.11` to the container, as illustrated in the following code block:

```
    ports:
     - 27111:27017
    build: ./installation_1
    stdin_open: true
    tty: true
    networks:
      app_net:
        ipv4_address: 172.16.2.11
```

In the last two definition blocks, we define a Docker virtual network, allowing the container to route through your local host computer. We also instruct Docker to create a volume called `db_data_installation` that resides outside the container. In this manner, data is retained on your host computer between container restarts. The code for this can be seen in the following snippet:

```
networks:
  app_net:
    ipam:
      driver: default
      config:
        - subnet: "172.16.2.0/24"

volumes:
  db_data_installation: {}
```

Next, we look at the main build file,
`/path/to/repo/chapters/02/installation_1/Dockerfile`. As you can see from the
following code block, this file draws from the latest stable version of Ubuntu and installs
tools needed for the MongoDB installation:

```
FROM ubuntu:latest
RUN \
    apt-get update && \
    apt-get -y upgrade && \
    apt-get -y install vim && \
    apt-get -y install inetutils-ping && \
    apt-get -y install net-tools && \
    apt-get -y install wget && \
    apt-get -y install gnupg
```

To build the container, use this command:

```
docker-compose build
```

To run the container, use this command (the -d option causes the container to run in the
background):

```
docker-compose up -d
```

To access the container, use this command:

```
docker exec -it learn-mongo-installation-1 /bin/bash
```

Once the command shell into the installation container is established, you can then practice
installing MongoDB following the instructions given in this chapter. When you are done
practicing MongoDB installation, be sure to stop the container, using this command:

```
docker-compose down
```

Let's now have a look at creating the demo environment used in Chapter 3, *Essential
MongoDB Administration Techniques*, to Chapter 12, *Developing in a Secured Environment*, of
this book.

You will need to stop the container described here before you start the container described in the next sub-section; otherwise, you might encounter an error.

For more information about the directives in the `docker-compose.yml` file, please review the *Overview of Docker Compose* documentation article (`https://docs.docker.com/compose/`). Here is a useful list of Docker commands: `https://docs.docker.com/engine/reference/commandline/container/`.

Configuring Docker for Chapters 3 to 12

In the root directory of the repository source code, you will see a `docker-compose.yml` file. This file contains instructions to create the infrastructure for the demo server you can use to test files discussed in `Chapter 3`, *Essential MongoDB Administration Techniques*, to `Chapter 12`, *Developing in a Secured Environment*, of the book. The virtual environment created through this `docker-compose` file contains not only MongoDB fully installed, but also, *Python*, `pip`, and the *PyMongo* driver have all been installed as well. The heart of this Docker container is based upon the *official MongoDB image* hosted on Docker Hub.

`Chapter 13`, *Deploying a Replica Set*, to `Chapter 16`, *Sharded Cluster Management and Development*, have their own demo configuration, as you will discover when reading those chapters.

After the version identification, the next several lines contain `services` definitions, as follows:

```
version: 3
services:
  learn-mongodb-server-1:
    container_name: learn-mongo-server-1
    hostname: server1
    image: learn-mongodb/member-server-1
    volumes:
      - db_data_server_1:/data/db
      - .:/repo
      - ./docker/hosts:/etc/hosts
      - ./docker/mongod.conf:/etc/mongod.conf
    ports:
```

```
    - 8888:80
     - 27111:27017
    build: ./docker
    restart: always
    command: -f /etc/mongod.conf
    networks:
      app_net:
        ipv4_address: 172.16.0.11
```

In this example, there is only one definition, `learn-mongodb-server-1`, which is used for most of the examples in this book. In this definition is the name of the Docker container, its hostname, the name of the image, volumes to be mounted, and additional configuration information. You will also notice that an IP address of `172.16.0.11` has been assigned to the container. Once it is up and running, you can access it using this IP address.

The remaining two directives define the overall Docker virtual network and an external disk allocation where the Docker container will store MongoDB data files, as illustrated in the following code snippet:

```
networks:
  app_net:
    ipam:
      driver: default
      config:
        - subnet: "172.16.0.0/24"

volumes:
  db_data_server_1: {}
```

Docker Compose uses a `/path/to/repo/docker/Dockerfile` file for instructions on how to build the new image. Here are the contents of that file:

```
FROM unlikelysource/learning_mongodb
COPY ./init.sh /tmp/init.sh
RUN  chmod +x /tmp/init.sh
RUN  /tmp/init.sh
COPY ./run_services.sh /tmp/run_services.sh
RUN  chmod +x /tmp/run_services.sh
CMD /tmp/run_services.sh
```

The first directive tells Docker to pull the base image from the image created expressly for this book. After that, an initialization script is copied and executed. Here are the contents of the initialization script:

```
#!/bin/bash
echo '****************************************************'
echo ' '
echo 'To shell into the Docker container, do this:'
echo '    docker exec -it learn-mongo-server-1 /bin/bash'
echo ' '
echo '(1) To restore sample data run this script:'
echo '    /path/to/repo/restore_data_inside.sh'
echo '(2) To restart Apache run this script:'
echo '    /path/to/repo/restart_apache_inside.sh'
echo ' '
echo '****************************************************'
```

The second script referenced starts MongoDB and Apache and then loops. This ensures both services remain running. Here is the other script, found at `/path/to/repo/docker/run_services.sh`. Since this script is long, let's take it block by block. The first block of scripting starts a `mongod` (that is, a MongoDB daemon) running in the background. If it fails to start, the script exits, stopping the Docker container from running. The code for this can be seen in the following snippet:

```
#!/bin/bash
/usr/bin/mongod -f /etc/mongod.conf --bind_ip_all &
status=$?
if [ $status -ne 0 ]; then
  echo "Failed to start mongod: $status"
  exit $status
fi
```

In a similar vein, the second block of code starts Apache running, as follows:

```
/usr/sbin/apachectl -k start &
status=$?
if [ $status -ne 0 ]; then
  echo "Failed to start apache: $status"
  exit $status
fi
```

The remainder of the script simply loops, and checks the process status for both `mongod` and Apache 2. If either of the process checks fail to return results, the script ends, causing the container to stop. The code can be seen in the following snippet:

```
while sleep 60; do
  ps aux |grep httpd |grep -v grep
  PROCESS_1_STATUS=$?
  ps aux |grep apache2 |grep -v grep
  PROCESS_2_STATUS=$?
  # If the greps above find anything, they exit with 0 status
  # If they are not both 0, then something is wrong
  if [ -f $PROCESS_1_STATUS -o -f $PROCESS_2_STATUS ]; then
    echo "One of the processes has already exited."
    exit 1
  fi
done
```

For more information on the official MongoDB Docker image, have a look here: `https://hub.docker.com/_/ubuntu`.

Now, let's have a look at getting the demo environment up and running.

Running the demo environment

To bring the demo environment online, open a Terminal window, and change to the directory where you restored the source code for the book. To build the new Docker image to be used to test examples from `Chapter 3`, *Essential MongoDB Administration Techniques*, to `Chapter 12`, *Developing in a Secured Environment*, run this command:

```
docker-compose build
```

To run the image, simply issue this command (the `-d` option causes the container to run in the background):

```
docker-compose up -d
```

After that, you can issue this command to open up a command shell inside the running container:

```
docker exec -it learn-mongo-server-1 /bin/bash
```

You should see a /repo directory that contains all the source code from the repository. If you then change to the /repo/chapters/02 directory, you can run the Python test program that confirms whether or not a connection to MongoDB is successful. Here is a screenshot of what you might see:

```
File  Edit  View  Search  Terminal  Help
jed@jed:~$ docker exec -it learn-mongo-server-1 /bin/bash
root@server1:/# cd /repo/chapters/02
root@server1:/repo/chapters/02# python3 test_connection.py

Insert Result:
<pymongo.results.InsertOneResult object at 0x7f73da6923c8>

Find Result:
{'_id': ObjectId('5e9d2e8462d17942c4a1cbfe'),
 'first_name': 'Fred',
 'last_name': 'Flintstone'}
root@server1:/repo/chapters/02#
```

If you need to restart either MongoDB or Apache, simply stop the running container and restart it. The startup script will then restart both MongoDB and Apache.

Finally, we will have a look at establishing communications between your local computer that is hosting the Docker container, and the container's Apache web server.

Connecting to the container from your host computer

In a production environment (that is, on a server that is directly accessible via the internet), you would normally add a **Domain Name Service (DNS)** entry using the services of your **Internet Service Provider** (ISP). For development purposes, however, you can simulate a DNS address by adding an entry to the hosts file on your own local computer. For Linux and macOS computers, the file is located at /etc/hosts. For Windows computers, the file is located at C:\Windows\System32\drivers\etc\hosts. For our fictitious company *Book Someplace*, we add the following entry to the local hosts file:

```
172.16.0.11     learning.mongodb.local
```

At this point, from your host computer, in a Terminal window or Command Prompt from outside the Docker container, test the connection using `ping learning.mongodb.local`, as shown here:

```
File  Edit  View  Search  Terminal  Help
jed@jed:~/Repos/learn-mongodb$ ping -c3 learning.mongodb.local
PING learning.mongodb.local (172.16.0.11) 56(84) bytes of data.
64 bytes from learning.mongodb.local (172.16.0.11): icmp_seq=1 ttl=64 time=0.159 ms
64 bytes from learning.mongodb.local (172.16.0.11): icmp_seq=2 ttl=64 time=0.173 ms
64 bytes from learning.mongodb.local (172.16.0.11): icmp_seq=3 ttl=64 time=0.144 ms

--- learning.mongodb.local ping statistics ---
3 packets transmitted, 3 received, 0% packet loss, time 2046ms
rtt min/avg/max/mdev = 0.144/0.158/0.173/0.018 ms
jed@jed:~/Repos/learn-mongodb$
```

And that concludes the setup instructions for the book.

Summary

In this chapter, you learned to recognize some of the hardware prerequisites for a MongoDB 4.x installation. Among the considerations discussed were how much memory was required, the type and nature of storage, and the file and operating systems. You then learned some of the differences between a MongoDB 3 and a MongoDB 4.x installation. The most important consideration is that support for some of the older operating systems (for example, Ubuntu 14.04 and Windows 8.1) has been—or will shortly be—removed.

You were given a general overview of how to install MongoDB on Linux, Windows, and macOS. You then learned about the more important configuration options and the location and nature of the configuration file. After that, you learned how to install the recommended Python driver, PyMongo. You learned that although there are many ways in which this driver can be installed, the preferred method is to use the `pip3` Python package manager. The last section showed you how to download the source code and sample data for the book. You also learned how to set up a demo environment using Docker technology, to be used as a reference throughout the remainder of the book.

In the next chapter, you learn how to use the `mongo` shell to access the MongoDB database.

3
Essential MongoDB Administration Techniques

This chapter is aimed primarily at system administrators and **database administrators** (**DBAs**), but the information contained within it will also prove useful to developers. In this chapter, you learn how to perform administration critical to a properly functioning database. In this chapter, we will learn to use the `mongo` shell and generate ad hoc queries for management. We will then cover the validation and cleanup of data, and, by the end of the chapter, we will learn to perform backup and restore operations, as well as performance monitoring.

The following topics are addressed in this chapter:

- Connecting with the mongo shell
- Using the mongo shell for common database operations
- Performing backup and restore operations
- Performance monitoring

Technical requirements

Minimum recommended hardware:

- Desktop PC or laptop
- 2 **gigabytes** (**GB**) free disk space
- 4 GB of **random-access memory** (**RAM**)
- 500 kilobits per second (**Kbps**) or faster internet connection

Software requirements:

- OS (Linux or Mac): Docker, Docker Compose, Git (optional)
- OS (Windows): Docker for Windows and Git for Windows

Installation of the required software and how to restore the code repository for the book are explained in Chapter 2, *Setting up MongoDB 4.x*. To run the code examples in this chapter, open a Terminal window (Command Prompt) and enter these commands:

```
cd /path/to/repo
docker-compose build
docker-compose up -d
docker exec -it learn-mongo-server-1 /bin/bash
```

When you are finished practicing the commands covered in this chapter, be sure to stop Docker as follows:

```
docker-compose down
```

The code used in this chapter can be found in the book's GitHub repository at https://github.com/PacktPublishing/Learn-MongoDB-4.x/tree/master/chapters/03.

Connecting with the mongo shell

After installing MongoDB on a server, a separate executable named `mongo` is also installed. Here is an example of opening the shell without any parameters from the Docker container mentioned at the start of this chapter:

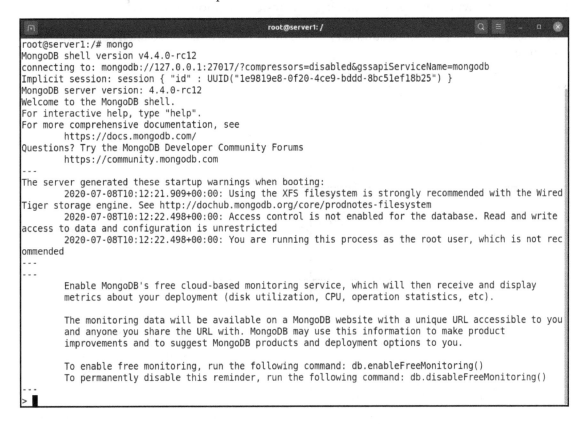

From this prompt, you are now able to view database information and query collections and perform other vital database operations.

IMPORTANT: As the MongoDB Docker team updates the official image, the version of MongoDB and initial `mongo` shell output will differ from what you see in the preceding screenshot.

The mongo shell command-line switches

There are a number of command-line switches associated with the mongo shell (`https://docs.mongodb.com/manual/mongo/#the-mongo-shell`), the most important of which are summarized here. You should note that the mongo shell is used to connect to a mongod server instance, which is typically running as a system daemon or service.

The following table presents a summary of the more important mongo shell command-line switches:

Option	Example	Notes
`--port`	`mongo --port 28888`	Lets you connect to a mongod instance that uses a non-default port. Note that the port can also be specified when using the `--host` flag (described next).
`--host`	`mongo --host 192.168.2.57:27017` `mongo --host mongo.unlikelysource.net`	This option lets you connect to a mongod instance running on another server. `:port` is optional. If left off, the port value defaults to `27017`.
`--username` `--password`	`mongo --username doug --password 12345`	Connects to a mongod instance as an authenticated user. If the `--password` switch is left off, you will be prompted for the password. Use the `--authenticationDatabase` switch if you are not authenticating via the default admin database.
`--tls`	`mongo --tls`	Makes a secure connection using **Transport Layer Security (TLS)** (version 4.2 and above).
`--help`	`mongo --help`	Displays a help screen for running the mongo command.
`--nodb`	`mongo --nodb`	Lets you enter the shell without connecting to a database.
`--shell`	`mongo --shell <javascript.js>`	Enters the mongo shell after running an external script.

All of the command-line switches summarized in the preceding table can be mixed together. Here is an example that connects to a remote `mongod` instance on IP address `192.168.2.57`, which listens on port `28028`. The user `doug` is authenticated using the security database. The connection is secured by TLS. Have a look at the following code snippet:

```
mongo --host 192.168.2.57:28028 --username doug \
    --authenticationDatabase security --tls
```

> All command-line switches can alternatively be assembled into a single connection string instead of using individual command-line switches. In this case, you would use **Uniform Resource Locator** (**URL**)-style syntax. Here is the connect string for the last example shown previously:
> mongo
> mongodb://doug@192.168.2.57:28028/?authSource=security&tls=true.

Customizing the mongo shell

There are a certain number of customization options you can perform on the `mongo` shell. Simply by creating a `.mongorc.js` file (`https://docs.mongodb.com/manual/reference/program/mongo/#mongo-mongorc-file`) in your user home directory, you can customize the prompt to display dynamic information such as the uptime, hostname, database name, number of commands executed, and so forth. This is accomplished by defining a JavaScript function to the `prompt` reserved variable name, whose return value is echoed to form the prompt.

> You can also make the `mongo` shell prompt global by creating the file as `/etc/mongorc.js`.

Here is a sample `.mongorc.js` script that displays the MongoDB version, hostname, database name, uptime, and command count:

```
host = db.serverStatus().host;
cmd  = 1;
prompt = function () {
    upt  = db.serverStatus().uptime;
    dbn  = db.stats().db;
    return "\nMongoDB " + host + "@" + dbn
        + "[up:" + upt + " secs]\n" + cmd++ + ">";
}
```

By placing this file in the user home directory, subsequent use of the `mongo` command will display the modified prompt, as shown:

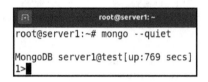

Let's now look at the shell command helpers.

Shell command helpers

When using the `mongo` shell interactively, there is a handy set of *shell command helpers* that facilitate discovery and navigation while in the shell. The command helpers are summarized here:

Command	Options	Notes
use	<dbname>	Switches default database to <dbname>
help		Displays a list of commands with brief descriptions
help	admin	List of administrative commands
help	connect	Help pertaining to the database connection
help	keys	Keyboard shortcuts
help	misc	Other uncategorized help
help	mr	Displays help on Map-Reduce
show	dbs	Shows database names
show	collections	When using a database, shows a list of collections
show	users	List of users for the current database
show	logs	List of accessible logger names
show	log <name>	Displays last entry of log <name>; global is the default
db.help()		Help on database methods
db.<coll>.help()		Help on collection methods, where <coll> is a collection in the currently used database
sh.help()		Help with sharding
rs.help()		Help with replica set commands

Unfortunately, shell helper commands cannot be used in shell scripts (discussed in the next section in this chapter). For more information, refer to https://docs.mongodb.com/manual/reference/mongo-shell/#command-helpers.

In the following example, we use shell helper commands to get a list of databases, use the admin database, and list collections in this database:

```
root@server1: /repo
MongoDB server1@test[up:1078 secs]
1>show dbs;
admin    0.000GB
config   0.000GB
local    0.000GB

MongoDB server1@test[up:1080 secs]
2>use admin;
switched to db admin

MongoDB server1@admin[up:1090 secs]
3>show collections;
system.version

MongoDB server1@admin[up:1093 secs]
4>
```

Let's now learn to use the mongo shell methods.

Using mongo shell methods

The mongo shell makes available a rich set of shell methods that include the following categories:

Category	Includes methods that affect ...	Examples
Database	... an entire database	db.serverStatus()
Collections	... a collection within a database	db.<collection_name>.find()
Cursor	... a cursor (result of a query)	db.<collection_name>.find().sort()
User management	... database users	db.createUser()
Role management	... the ability of a user role to perform operations	db.grantPrivilegesToRole()
Replication	... servers that are members of a *replica set*	rs.initiate()

Sharding	... servers that hold fragments (*shards*) of a collection	`sh.addShard()`
Free monitoring	... monitoring of your overall MongoDB deployment	`db.enableFreeMonitoring()`
Constructors	... the production of commonly needed information	`ObjectId.toString()`
Connection	... the connection to the MongoDB instance	`Mongo.getDB()`
Native	... commands that give limited access to the local server	`hostname()`

A *cursor* is an iteration produced by collection query methods. The most notable of these are `db.collection.find()` and `db.collection.aggregate()`. For information on cursors, refer to `https://docs.mongodb.com/manual/tutorial/iterate-a-cursor/#iterate-a-cursor-in-the-mongo-shell`. For information on replica sets, see `Chapter 13`, *Deploying a Replica Set*. For information on sharding, see `Chapter 15`, *Deploying a Sharded Cluster*. More information on shell methods can be found here: `https://docs.mongodb.com/manual/reference/method/#mongo-shell-methods`.

When developing a MongoDB application in a programming language, make sure your database commands work properly by testing the proposed commands in the `mongo` shell. You can then convert the JavaScript code used in the shell command to match the command set offered by your database adapter.

In this chapter, we will focus on basic database, collection, cursor, and monitoring methods. In other chapters in this book, we will cover user and role management, replication, and sharding methods. Examples of shell methods that allow you to perform common database operations are covered in the next section of this chapter.

Running scripts using the shell

You can use the mongo shell to run scripts that use JavaScript syntax. This is a good way to permanently save complex shell commands or perform bulk operations. There are two ways to run JavaScript-based scripts using the mongo shell: directly from the command line, or from within the mongo shell.

Running a script from the command line

Here is an example of running a shell script that inserts documents into the sweetscomplete database from the command line:

```
mongo /path/to/repo/sample_data/sweetscomplete_customers_insert.js
```

NOTE: If you're still in the shell, type exit to return to the command line!

Here is the resulting output:

```
root@server1:~# mongo /repo/sample_data/sweetscomplete_customers_insert.js
MongoDB shell version v4.4.0-rc12
connecting to: mongodb://127.0.0.1:27017/?compressors=disabled&gssapiServiceName=mongodb
Implicit session: session { "id" : UUID("293c769f-4672-4ffc-af7a-a97972283109") }
MongoDB server version: 4.4.0-rc12
root@server1:~#
```

Let's now run a script from inside the shell.

Running a script from inside the shell

You can also run a shell script from inside the mongo shell using the load(<filename.js>) command helper. After having first returned to the mongo shell, issue the use sweetscomplete; command. You can run the sweetscomplete_products_insert.js script that is located in the /path/to/repo/sample_data directory, as illustrated in the following code snippet:

```
mongo --quiet
use sweetscomplete;
show collections;
load("/path/to/repo/sample_data/sweetscomplete_products_insert.js");
show collections;
```

When you examine the output from this command next, note that you first need to use the database. We then look at the list of collections before and after running the script. Notice in the following screenshot that a new `products` collection has been added:

```
root@server1:~# mongo --quiet

MongoDB server1@test[up:700 secs]
1>use sweetscomplete;
switched to db sweetscomplete

MongoDB server1@sweetscomplete[up:704 secs]
2>show collections;
customers

MongoDB server1@sweetscomplete[up:707 secs]
3>load("/repo/sample_data/sweetscomplete_products_insert.js");
true

MongoDB server1@sweetscomplete[up:723 secs]
4>show collections;
customers
products

MongoDB server1@sweetscomplete[up:726 secs]
5>
```

Let's now look at the shell script syntax.

Shell script syntax

It's very important to understand that you *cannot use shell command helpers* from an external JavaScript script, even if you run the script from inside the shell! Many of the shell methods you may wish to use reference the db object. When you are running shell methods from inside the shell, this is not a problem: when you issue the use command, this anchors the db object to the database currently in use. Since the use command is a shell command helper, however, it cannot be used inside an external script. Accordingly, the first set of commands you need to add to an external shell script is needed to set the db object to the desired database.

This can be accomplished as follows, where `<DATABASE>` is the name of the target database you wish to operate upon:

```
conn = new Mongo();
db = conn.getDB("<DATABASE>");
```

From this point on, you can use the db object as you would with any other shell method. Here is a sample shell script that assigns the db object to the sweetscomplete database, removes the common collection, and then inserts a document:

```
conn = new Mongo();
db = conn.getDB("sweetscomplete");
db.common.drop();
db.common.insertOne(
{
    "key" : "gender",
    "data" : { "M" : "male", "F" : "female" }
});
```

We now turn our attention to common database operations: inserting, updating, deleting, and querying.

Using the Mongo shell for common database operations

Among the most important shell methods you will need to master, whether a database administrator or a developer, is to be able to create a database, add documents to collections, and perform queries. We will start by creating a new learn database.

Creating a new database and collection

Oddly, the shell method used to create a new database is not listed among the database commands! Instead, a new database is created by simply inserting a document into a collection. Using two simple commands, you will create the database, create the collection within the new database, and add a document!

Inserting documents

To insert a document, you can use either the `db.collection.insertOne()` or the `db.collection.insertMany()` command. The syntax is quite simple: you specify the name of the collection into which you plan to insert a document, followed by the document to be inserted, using **JavaScript Object Notation (JSON)** syntax, as illustrated in the following diagram:

In this example, you will create a new `learn` database and a collection called `chapters`, as follows:

```
use learn;
db.chapters.insertOne({
    "chapterNumber" : 3,
    "chapterName"    : "Essential MongoDB Administration Techniques"
});
```

Here is the output:

If the `featureCompatibilityVersion` parameter is set to `4.2` or less (for example, you're running MongoDB `4.2` or below), the maximum size of the combined database name plus the collection name cannot exceed `120` bytes. If, however, it is set to `4.4` or greater (for example, you're running MongoDB `4.4` or above), the maximum size of the combined database name plus the collection name can go up to `255` bytes.

Querying a collection

The main command used to query a collection from the `mongo` shell is `db.<collection_name>.find()`. If you are only interested in a single result, you can alternatively use `db.<collection_name>.findOne()`. The `find()` and `findOne()` commands have two basic parameters: the `query` (also referred to as filter) and the `projection`, as illustrated in the following diagram:

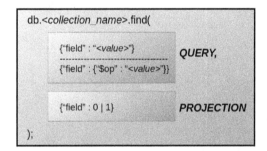

The result of a query is a *cursor*. The cursor is an *iteration*, which means you need to type the `it` helper command in the `mongo` shell to see the remaining results.

As mentioned here, the `db.collection.find()` shell method produces a *cursor* over which you can iterate to retrieve all results. There are a number of *cursor methods* that can be applied, which has the effect of modifying the final results. A number of examples in this section will demonstrate a few of the most important such methods. For more information, refer to `https://docs.mongodb.com/manual/reference/method/js-cursor/#cursor-methods`.

If you are using a programming language driver, you can use a loop to iterate through a cursor result.

Simple queries

As you can see from the preceding diagram, the query (also referred to as *query criteria* or *filter*) can be as simple as using JSON syntax to identify a document field, and the value it equals. As an example, we will use the `customers` collection in the `sweetscomplete` database. When you specify a query filter, you can provide a hardcoded value, as shown here:

```
db.customers.findOne({"phoneNumber":"+44-118-652-0519"});
```

However, you can also simply wrap the target value into a *regular expression* by surrounding the target string with delimiters (`"/"`). Here is a simple example, whereby we use the `db.collection.findOne()` command to return a single customer record where the target phone number is located using a `/44-118-652-0519/` regular expression:

```
File Edit View Search Terminal Help
MongoDB server1@sweetscomplete[up:336 secs]
3>db.customers.findOne({"phoneNumber":/44-118-652-0519/});
{
        "_id" : ObjectId("5eaa96cfcbab62398c8a7266"),
        "customerKey" : "REANMCCA0519",
        "firstName" : "Reanna",
        "lastName" : "Mccarthy",
        "phoneNumber" : "+44-118-652-0519",
        "email" : "rmccarth118@Telecom.com",
        "streetAddressOfBuilding" : "304 Blue Mountain Avenue",
        "buildingName" : null,
        "floor" : null,
        "roomApartmentCondoNumber" : null,
        "city" : "Newby",
        "stateProvince" : "England",
        "locality" : "Cumbria",
        "country" : "GB",
        "postalCode" : "CA10",
        "latitude" : "54.5843",
        "longitude" : "-2.6369",
        "username" : "rmccarth",
        "password" : "$2y$10$0jV4OQd6LGwG5Jf5YqUFd.IodwGyG/q7pWAJgpRfsx5qdfIqfGxRS",
        "secondaryPhoneNumbers" : [
                "+44-329-910-7898"
        ],
        "secondaryEmailAddresses" : [
                "rmccarth@CUBENet.net"
        ],
        "socialMedia" : [ ],
        "gender" : "F",
        "dateOfBirth" : "1955-03-09"
}

MongoDB server1@sweetscomplete[up:366 secs]
4>
```

Let's say that the management has asked you to produce a count of customers. In this case, you can use a *cursor method* to operate on the results. To form this query, simply execute `db.collection.find().count()` without any parameters, as shown here:

```
File  Edit  View  Search  Terminal  Help
MongoDB server1@sweetscomplete[up:460 secs]
5>db.customers.find().count();
286

MongoDB server1@sweetscomplete[up:466 secs]
6>
```

IMPORTANT: The `db.collection.find()` command returns a *cursor*, whereas `db.collection.findOne()` does not. For more information, consult the *Iterate a Cursor in the mongo Shell* documentation article (`https://docs.mongodb.com/manual/tutorial/iterate-a-cursor/#iterate-a-cursor-in-the-mongo-shell`).

Building complex queries using query operators

Below the dotted line in the preceding diagram, you will note a more complex form of stating the query. In the diagram, `$op` would represent one of the *query operators* (also referred to as *query selectors*) that are available in MongoDB.

Query operators fall into the following categories:

- Comparison (`$eq`, `$gt`, `$gte`, `$in`, `$lt`,`$lte`, `$ne`, `$nin`)
- Logical (`$and`, `$not`, `$or`, `$nor`)
- Element (`$exists`, `$type`)
- Evaluation (`$expr`, `$jsonSchema`, `$mod`, `$regex`, `$text`, `$where`)
- Geospatial (`$geoIntersects`, `$geoWithin`, `$near`, `$nearSphere`)
- Array (`$all`, `$elemMatch`, `$size`)
- Bitwise (`$bitsAllClear`, `$bitsAllSet`, `$bitsAnyClear`, `$bitsAnySet`)
- Comments (`$comment`)

As an example, let's say that the management wants a count of customers from non-English-speaking majority countries over the age of 50. The first thing we can do is create a JavaScript array variable of majority-English-speaking country codes. As a reference, we draw a list of countries from a document produced by the British government (https://www.gov.uk/english-language/exemptions). The code can be seen in the following snippet:

```
maj_english = ["AG","AU","BS","BB","BZ","CA","DM","GB","GD",
               "GY","IE","JM","NZ","KN","LC","VC","TT","US"];
```

Next, we create two sub-phrases. The first addresses customers *not in* ($nin) the preceding list, as follows:

```
{"country" : { "$nin": maj_english }}
```

The second clause addresses the date of birth. Note the use of the *less-than* ($lt) operator in the following code snippet:

```
{"dateOfBirth" : {"$lt":"1968-01-01"}}
```

We then join the two clauses, using the $and query operator, for the final complete query, as follows:

```
db.customers.find(
{
    "$and" : [
        {"country" : { "$nin": maj_english }},
        {"dateOfBirth" : {"$lt":"1968-01-01"}}
    ]
}).count();
```

Here is the result:

```
File  Edit  View  Search  Terminal  Help
MongoDB server1@sweetscomplete[up:656 secs]
7>db.customers.find(
... {
...      "$and" : [
...          {"country" : { "$nin": maj_english }},
...          {"dateOfBirth" : {"$lt":"1968-01-01"}}
...      ]
... }).count();
76

MongoDB server1@sweetscomplete[up:675 secs]
8>
```

 Subsequent chapters in this book will cover other query operators of increasing degrees of complexity. For more information on query operators, see this page: https://docs.mongodb.com/manual/reference/operator/query/#query-and-projection-operators.

Applying projections to your query

The second argument to the `db.collection.find()` and `db.collection.findOne()` commands, also using JSON syntax, is the `projection` argument. The `projection` argument allows you to control which fields appear or do not appear in the final output.

Here is a summary of the projection options:

Projection expression	What appears in the output
`{ }` (or blank)	All fields
`{ "field1" : 1 }`	Only `field1` and the `_id` fields
`{ "field1" : 0 }`	All fields *except for* `field1`
`{ "field1" : 1, "_id" : 0 }`	Only `field1`
`{ "field1" : 1, "field2" : 0 }`	Not allowed! Cannot have a mixture of `include` and `exclude` (except for the `_id` field)

For this example, we will assume that the management has asked for the name and email address of all customers from Quebec. Accordingly, you construct a query that uses the `projection` argument to only include fields pertinent to the name and address. You also suppress the `_id` field as it would only confuse those in management.

As an example, we first define a JavaScript variable that represents the *query* portion. Next, we define a variable that represents the *projection*. Finally, we issue the `db.collection.find()` command, adding the `pretty()` cursor modifier (discussed next) to improve the output appearance. Here is how that query might appear:

```
query = { "stateProvince" : "QC" }
projection = { "_id":0, "firstName":1, "lastName":1, "email":1,
"stateProvince":1, "country":1 }
db.customers.find(query,projection).pretty();
```

Here is the output:

```
File  Edit  View  Search  Terminal  Help
MongoDB server1@sweetscomplete[up:1004 secs]
16>db.customers.find(query,projection).pretty();
{
        "firstName" : "Dominick",
        "lastName" : "Lester",
        "email" : "dlester124@KDDI.com",
        "stateProvince" : "QC",
        "country" : "CA"
}
{
        "firstName" : "Darryl",
        "lastName" : "Wilkinson",
        "email" : "dwilkins130@Unicom.com",
        "stateProvince" : "QC",
        "country" : "CA"
}
```

In addition to values of 1 or 0, you can also use *projection operators* to perform more granular operations on the projection. These are especially useful when working with embedded arrays, covered in later chapters of this book.

Later in this book, we will address more sophisticated ways of using the cursor when we discuss the MongoDB *aggregation pipeline* (see `Chapter 9, Handling Complex Queries in MongoDB`). Here is the documentation on *projection* (`https://docs.mongodb.com/manual/reference/method/db.collection.find/#projection`) and *projection operators* (`https://docs.mongodb.com/manual/reference/operator/projection/#projection-operators`).

MongoDB 4.4 introduces a new `allowDiskUse()` method that allows MongoDB to write temporary files to disk if a query sort operation exceeds the default allowed memory size of 100 **megabytes** (**MB**). `allowDiskUse()` is appended after the `sort()` method, which itself would follow a query (for example, `db.collection.find().sort().allowDiskUse()`).

Updating a document

Updating a document involves two mandatory arguments: the *filter* and the *update document* arguments. The third (optional) argument is `options`. The primary shell methods used are either `db.collection.updateOne()`, which—as the name implies—only updates a single document, or `db.collection.updateMany()`, which lets you perform updates on all documents matching the filter. The former argument is illustrated in the following diagram:

```
db.<collection_name>.updateOne(

    {"field" : "<value>"}              FILTER,
    ------------------------------------
    {"field" : {"$op" : "<value>"}}

    {"field1" : "xxx", "field2" : "yyy" }  UPDATE,

    { upsert : true |false, etc. }      OPTIONS

);
```

The filter has exactly the same syntax as the query (that is, the first argument to the `db.collection.find()` command) discussed previously, so we will not repeat that discussion here. The update usually takes the form of a partial document, containing only the fields that need to be replaced with respect to the original. Finally, *options* are summarized here:

Option	Values	Notes
upsert	true | false	If set to `true`, a new document is created if none matches the filter. The default is `false`.
writeConcern	write concern document	Influences the level of acknowledgment that occurs after a successful write. This is covered in more detail in Chapter 14, *Replica Set Runtime Management and Development*.
collation	collation document	Used to manage data from an international website capturing locale-aware data.
arrayFilters	array	Lets you create an independent filter for embedded arrays.

There is a `db.collection.replaceOne()` shell method that can be used as an alternative to `db.collection.updateOne()`. It is used in situations where you need to replace the entire document, not just update a subset of fields within an existing document.

If the update operation includes a field that does not already exist, due to the flexible nature of MongoDB document structuring, MongoDB will add the field to the document. This capability allows you to perform ad hoc database schema adjustments without loss of data. No more messy legacy **Relational Database Management System (RDBMS)** data conversions!

Documentation on the write concern document can be found at `https://docs.mongodb.com/manual/reference/write-concern/#write-concern-specification`. Documentation on the collation document can be found at `https://docs.mongodb.com/manual/reference/collation/#collation-document`. These are both described in greater detail in Chapter 13, *Deploying a Replica Set*.

Update operators

In order to perform an update, the *update document* needs to use *update operators*. At a minimum, you need to use the $set update operator to assign the new value. The update operators are broken out into categories, as follows:

- Fields ($currentDate, $inc, $min, $max, $mul, $rename, $set, $setOnInsert, $unset)
- Array ($, $[], $[<identifier>], $addToSet, $pop, $pull, $push, $pullAll)
- Bitwise ($bit)

In addition, there are *modifiers* ($each, $position, $slice, $sort) that are used in conjunction with $push so that updates can affect all or a subset of an embedded array. Further, the $each modifier can be used with $addToSet to add multiple elements to an embedded array. A more detailed discussion of the *array* and *modifier* operators is presented in Chapter 10, *Working with Complex Documents across Collections*. In this section, we will only deal with the *fields* update operators.

Bitwise update operators let you perform AND, OR, and XOR operations, at the binary level, on integer values. For more information, see the MongoDB documentation on the $bit update operator. You can find more about *update operators* at https://docs.mongodb.com/manual/reference/operator/update/#update-operators.

Updating customer contact information

The simplest—and probably most common—type of update would be to update contact information. First, it's extremely important to *test* your update filter by using it in a db.collection.find() command. The reason why we use db.collection.find() instead of db.collection.findOne() is because you need to ensure that your update only affects the one customer! Thus, if we wish to change the email address and phone number for a fictitious customer named *Ola Mann*, we start with this command:

```
db.customers.findOne({ "email" : "omann137@Chunghwa.com" });
```

Now that we have verified the existing customer email is correct and that our filter only affects one document, we are prepared to issue the update, as follows:

```
db.customers.updateOne(
    {"email" : "omann137@Chunghwa.com"},
    { $set: {
        "email" : "ola.mann22@somenet.com",
```

```
            "phoneNumber" : "+94-111-222-3333" }
    }
);
```

Here is the result:

```
File  Edit  View  Search  Terminal  Help
MongoDB server1@sweetscomplete[up:1275 secs]
21>db.customers.updateOne(
...     {"email" : "omann137@Chunghwa.com"},
...     { $set: {
...         "email" : "ola.mann22@somenet.com",
...         "phoneNumber" : "+94-111-222-3333" }
...     }
... );
{ "acknowledged" : true, "matchedCount" : 1, "modifiedCount" : 1 }

MongoDB server1@sweetscomplete[up:1277 secs]
22>
```

Let's now cover data cleanup.

Data cleanup

Updates are also often performed to *clean up* data. As an example, you notice that for some customers, the `buildingName`, `floor`, and `roomApartmentCondoNumber` fields are set to `null`. This can cause problems when customer reports are generated, and the management wants you to replace the `null` values with an empty string instead.

Because you will be using the `updateMany()` command, the potential for a total database meltdown disaster is great! Accordingly, to be on the safe side, here are the actions you can perform:

- Back up the database (discussed in the next section).
- Run a `find()` command that tests the query filter, ensuring the correct documents are affected.
- Make a note of the first customer key on the list produced by the `find()` command.

- Build an `updateOne()` command using the tested filter.
- Use `findOne()` to confirm that the first customer document was changed correctly.
- Rerun the first `find()` command, appending `count()`, to get a count of the documents remaining to be updated.
- Run an `updateMany()` command using the tested syntax, checking to see if the count matches.

As mentioned previously, you should first craft a query to ensure that only those documents where the type is `null` pass the filter. Here is the query for the `buildingName` field:

```
db.customers.find({"buildingName":{"$type":"null"}},{"customerKey":1,"build
ingName":1});
```

Here is the output:

```
File  Edit  View  Search  Terminal  Help
MongoDB server1@sweetscomplete[up:1277 secs]
22>db.customers.find({"buildingName":{"$type":"null"}},{"customerKey":1,"buildingName":1});
{ "_id" : ObjectId("5eaa96cfcbab62398c8a7254"), "customerKey" : "PATRYODE9823", "buildingName" : null }
{ "_id" : ObjectId("5eaa96cfcbab62398c8a7255"), "customerKey" : "ALTASANT6271", "buildingName" : null }
{ "_id" : ObjectId("5eaa96cfcbab62398c8a7256"), "customerKey" : "MARLDORS7479", "buildingName" : null }
{ "_id" : ObjectId("5eaa96cfcbab62398c8a7257"), "customerKey" : "WMMORE0896", "buildingName" : null }
{ "_id" : ObjectId("5eaa96cfcbab62398c8a7258"), "customerKey" : "CORICAST5104", "buildingName" : null }
{ "_id" : ObjectId("5eaa96cfcbab62398c8a7259"), "customerKey" : "JULIOROZ8091", "buildingName" : null }
{ "_id" : ObjectId("5eaa96cfcbab62398c8a725a"), "customerKey" : "GENEMYER4370", "buildingName" : null }
{ "_id" : ObjectId("5eaa96cfcbab62398c8a725b"), "customerKey" : "LIBEBRAN9462", "buildingName" : null }
{ "_id" : ObjectId("5eaa96cfcbab62398c8a725c"), "customerKey" : "DAVIHOWE0711", "buildingName" : null }
{ "_id" : ObjectId("5eaa96cfcbab62398c8a725d"), "customerKey" : "REGIIBAR8447", "buildingName" : null }
{ "_id" : ObjectId("5eaa96cfcbab62398c8a725e"), "customerKey" : "KYLESELL3321", "buildingName" : null }
{ "_id" : ObjectId("5eaa96cfcbab62398c8a7260"), "customerKey" : "IKEADAM7878", "buildingName" : null }
{ "_id" : ObjectId("5eaa96cfcbab62398c8a7261"), "customerKey" : "AUREBRAN7182", "buildingName" : null }
{ "_id" : ObjectId("5eaa96cfcbab62398c8a7262"), "customerKey" : "BIANLIND8466", "buildingName" : null }
{ "_id" : ObjectId("5eaa96cfcbab62398c8a7263"), "customerKey" : "FERMZAVA5969", "buildingName" : null }
{ "_id" : ObjectId("5eaa96cfcbab62398c8a7264"), "customerKey" : "GLENMCLA8639", "buildingName" : null }
{ "_id" : ObjectId("5eaa96cfcbab62398c8a7265"), "customerKey" : "PIERCARE0110", "buildingName" : null }
{ "_id" : ObjectId("5eaa96cfcbab62398c8a7266"), "customerKey" : "REANMCCA0519", "buildingName" : null }
{ "_id" : ObjectId("5eaa96cfcbab62398c8a7267"), "customerKey" : "TIJUHOGA5010", "buildingName" : null }
{ "_id" : ObjectId("5eaa96cfcbab62398c8a7268"), "customerKey" : "ELVASALA6803", "buildingName" : null }
Type "it" for more

MongoDB server1@sweetscomplete[up:1365 secs]
23>
```

Make a note of the `customerKey` field for the first entry on the list produced from the query. You can then run an `updateOne()` command to reset the `null` value to an empty string. You then confirm the change was successful using the `customerKey` field. Here is a screenshot of this sequence of operations:

```
File  Edit  View  Search  Terminal  Help

MongoDB server1@sweetscomplete[up:60331 secs]
30>key = "PATRYODE9823"
PATRYODE9823

MongoDB server1@sweetscomplete[up:60335 secs]
31>db.customers.updateOne(
...      {"buildingName":{"$type":"null"},"customerKey":key},
...      {"$set":{"buildingName":""}}
... );
{ "acknowledged" : true, "matchedCount" : 0, "modifiedCount" : 0 }

MongoDB server1@sweetscomplete[up:60339 secs]
32>db.customers.findOne(
...      {"buildingName":{"$type":"null"},"customerKey":key},
...      {"customerKey":1,"buildingName":1}
... );
null

MongoDB server1@sweetscomplete[up:60345 secs]
33>
```

Before the final update, perform a query to get a count of how many documents remain to be modified, as follows:

```
db.customers.find({"buildingName":{"$type":"null"}}).count();
```

You are now ready to perform the final update, using this command:

```
db.customers.updateMany({"buildingName":{"$type":"null"}}, \
    {"$set":{"buildingName":""}});
```

And here is the result:

```
File Edit View Search Terminal Help
MongoDB server1@sweetscomplete[up:60659 secs]
34>db.customers.find({"buildingName":{"$type":"null"}}).count();
261

MongoDB server1@sweetscomplete[up:60660 secs]
35>db.customers.updateMany({"buildingName":{"$type":"null"}},{"$set":{"buildingName":""}});
{ "acknowledged" : true, "matchedCount" : 261, "modifiedCount" : 261 }

MongoDB server1@sweetscomplete[up:60668 secs]
36>db.customers.find({"buildingName":{"$type":"null"}}).count();
0

MongoDB server1@sweetscomplete[up:60673 secs]
37>
```

> You could perform conditional updates on all three fields simultaneously using the MongoDB *Aggregation Framework* and the `$ifNull` aggregation pipeline operator. This topic is covered in detail in `Chapter 9`, *Handling Complex Queries in MongoDB*.

Deleting a document

The command used to delete a single document is `db.collection.deleteOne()`. If you want to delete more than one document, use `db.collection.deleteMany()` instead. If you wish to delete an entire collection, the most efficient command is `db.collection.drop()`.

Document deletion has features in common with both update and query operations. There are two arguments: the *filter*, which is the same as described in the preceding section on updates. The second argument consists of `options`, which are a limited subset of the options available during an update: only the `writeConcern` and collation options are supported by a `delete` operation.

The following diagram maps out the generic syntax for the `db.collection.deleteOne()` shell method:

```
db.<collection_name>.deleteOne(

    {"field" : "<value>"}                    FILTER,
    -----------------------------------
    {"field" : {"$op" : "<value>"}}

    { writeConcern : { xxx }, etc. }         OPTIONS

);
```

For the purposes of illustration, let's turn our attention back to the `learn` database featured in the first section in this chapter on inserting documents. First, we use the `load()` shell helper method to restore the sample data for the `learn` database, as illustrated in the following screenshot:

```
File Edit View Search Terminal Help
MongoDB server1@learn[up:62140 secs]
10>use learn;
switched to db learn

MongoDB server1@learn[up:62142 secs]
11>load("/repo/sample_data/learn_insert.js");
true

MongoDB server1@learn[up:62144 secs]
12>db.chapters.find();
{ "_id" : ObjectId("5eab8854afd2a9edf3e26414"), "chapterNumber" : 3, "chapterName" : "Essential MongoDB Administration Techniques" }
{ "_id" : ObjectId("5eab8854afd2a9edf3e26415"), "chapterNumber" : 1, "chapterName" : "Introducing MongoDB 4.x" }
{ "_id" : ObjectId("5eab8854afd2a9edf3e26416"), "chapterNumber" : 2, "chapterName" : "Setting up MongoDB 4.x Draft complete" }
{ "_id" : ObjectId("5eab8854afd2a9edf3e26417"), "chapterNumber" : 3, "chapterName" : "Essential MongoDB Administration Techniques Draft" }
{ "_id" : ObjectId("5eab8854afd2a9edf3e26418"), "chapterNumber" : 4, "chapterName" : "Fundamentals of Database Design Draft" }
{ "_id" : ObjectId("5eab8854afd2a9edf3e26419"), "chapterNumber" : 5, "chapterName" : "Mission Critical MongoDB Database Tasks Draft" }
{ "_id" : ObjectId("5eab8854afd2a9edf3e2641a"), "chapterNumber" : 6, "chapterName" : "Using AJAX and REST to Build a Database Driven Website Draft" }
{ "_id" : ObjectId("5eab8854afd2a9edf3e2641b"), "chapterNumber" : 7, "chapterName" : "Advanced MongoDB Database Design Draft" }
{ "_id" : ObjectId("5eab8854afd2a9edf3e2641c"), "chapterNumber" : 8, "chapterName" : "Using Documents with Embedded Lists and Objects Draft" }
{ "_id" : ObjectId("5eab8854afd2a9edf3e2641d"), "chapterNumber" : 9, "chapterName" : "Handling Complex Queries in MongoDB Draft" }
{ "_id" : ObjectId("5eab8854afd2a9edf3e2641e"), "chapterNumber" : 10, "chapterName" : "Working with Complex Documents Across Collections Draft" }
{ "_id" : ObjectId("5eab8854afd2a9edf3e2641f"), "chapterNumber" : 11, "chapterName" : "Administering MongoDB Security Draft" }
{ "_id" : ObjectId("5eab8854afd2a9edf3e26420"), "chapterNumber" : 12, "chapterName" : "Developing in a Secured Environment Draft" }
{ "_id" : ObjectId("5eab8854afd2a9edf3e26421"), "chapterNumber" : 13, "chapterName" : "Deploying a Replica Set Draft complete" }
{ "_id" : ObjectId("5eab8854afd2a9edf3e26422"), "chapterNumber" : 14, "chapterName" : "Deploying a Sharded Cluster" }

MongoDB server1@learn[up:62151 secs]
13>
```

You will notice that there is a duplicate document for Chapter 3. If we were to specify either the chapter number or chapter name as the query filter, we would end up deleting *both* documents. The only truly unique field in this example is the `_id` field. For this illustration, we frame a `deleteOne()` command using the `_id` field as the deletion criteria. However, as you can see from the following screenshot, you cannot simply state the value of the `_id` field: the deletion fails, the `deleteCount` value is 0, and the duplicate entry still remains:

```
File Edit View Search Terminal Help
14>db.chapters.find({"chapterNumber":3});
{ "_id" : ObjectId("5eab8854afd2a9edf3e26414"), "chapterNumber" : 3, "chapterName" :
 "Essential MongoDB Administration Techniques" }
{ "_id" : ObjectId("5eab8854afd2a9edf3e26417"), "chapterNumber" : 3, "chapterName" :
 "Essential MongoDB Administration Techniques Draft" }

MongoDB server1@learn[up:62403 secs]
15>db.chapters.deleteOne({"_id":"5eab8854afd2a9edf3e26414"});
{ "acknowledged" : true, "deletedCount" : 0 }

MongoDB server1@learn[up:62455 secs]
16>db.chapters.find({"chapterNumber":3});
{ "_id" : ObjectId("5eab8854afd2a9edf3e26414"), "chapterNumber" : 3, "chapterName" :
 "Essential MongoDB Administration Techniques" }
{ "_id" : ObjectId("5eab8854afd2a9edf3e26417"), "chapterNumber" : 3, "chapterName" :
 "Essential MongoDB Administration Techniques Draft" }

MongoDB server1@learn[up:62463 secs]
17>
```

Granted—we could frame a `delete` `query` filter using the chapter number and rely upon `deleteOne()` to only delete the first match; however, that's a risky strategy. It's much better to exactly locate the document to be deleted.

The reason why the delete operation fails is due to the fact that the `_id` field is actually an `ObjectId` instance, not a string! Accordingly, if we reframe the `delete` `query` filter as an `ObjectId`, the operation is successful. Here is the final command and result:

```
File Edit View Search Terminal Help
MongoDB server1@learn[up:62841 secs]
18>db.chapters.deleteOne({"_id":ObjectId("5eab8854afd2a9edf3e26414")});
{ "acknowledged" : true, "deletedCount" : 1 }

MongoDB server1@learn[up:62855 secs]
19>db.chapters.find({"chapterNumber":3});
{ "_id" : ObjectId("5eab8854afd2a9edf3e26417"), "chapterNumber" : 3, "chapterName" :
 "Essential MongoDB Administration Techniques Draft" }

MongoDB server1@learn[up:62857 secs]
20>
```

Let's now have a look at the essential operations of *backup* and *restore*.

Performing backup and restore operations

Close your eyes, and picture for a moment that you are riding a roller-coaster. Now, open your eyes and have a look at your MongoDB installation! Even something as simple as backup and restore becomes *highly problematic* when applied to a MongoDB installation that contains *replica sets* and where collections are split up into *shards*!

The primary tools for performing backup and restore operations are `mongodump` and `mongorestore`.

 You can also take snapshots of your data rather than attempt to dump the entire local database. This technique is considered a best practice; however, it is operating system-specific and requires setting up operating system tools that are outside of the purview of this book. For more information, see the *Backup and Restore Using Filesystem Snapshots* topic in the MongoDB documentation. For documentation on `mongodump`, see `https://docs.mongodb.com/manual/reference/program/mongodump/#mongodump`. For documentation on `mongorestore`, see `https://docs.mongodb.com/manual/reference/program/mongorestore/#mongorestore`.

Using mongodump to back up a local database

The primary backup tool is `mongodump`, which is run from the command line (not from the `mongo` shell!). This command is able to back up the local MongoDB database of any accessible host. The output from this command is in the form of **Binary Serialized Object Notation (BSON)** documents, which are compressed binary equivalents of JSON documents. When backing up, there are a number of options available in the form of command-line switches, which affect what is backed up, and how.

 If you want to back up everything in the local MongoDB database, just type `mongodump`. A directory called `dump` will be created; each database will have its own subdirectory, and each collection its own backup file.

 For documentation on BSON, see `http://bsonspec.org/`. For documentation on JSON, see `http://json.org/`.

mongodump options summary

To invoke a given option, simply add it to the command line following `mongodump`. Options can be combined, although some will be mutually exclusive. Here is a list of the `mongodump` command-line options:

Category	Option	Category	Option
General	`--help` `--version`	*Namespace*	`--db=<database-name>` `--collection=<collection-name>`
Connection	`--host <hostname>` `--port <port_number>`	*URI (connection string)*	`--uri=mongodb-uri`

Secure Sockets Layer (SSL) (MongoDB version 4.0 and below)	`--ssl`
	`--sslCAFile=<filename>`
	`--sslPEMKeyFile=<filename>`
	`--sslPEMKeyPassword=<password>`
	`--sslCRLFile=<filename>`
	`--sslAllowInvalidCertificates`
	`--sslAllowInvalidHostnames`
	`--sslFIPSMode`
		Query	`--query="<JSON string forming a filter>"`
			`--queryFile=<file containing a query filter>`
			`--readPreference=<string>|<json>`
			`--forceTableScan`

TLS (MongoDB version 4.2 and above)	`--tls`
	`--tlsCertificateKeyFile=<filename>`
	`--tlsCertificateKeyFilePassword=<value>`
	`--tlsCAFile=<filename>`
	`--tlsCRLFile=<filename>`
	`--tlsAllowInvalidHostnames`
	`--tlsAllowInvalidCertificates`
	`--tlsFIPSMode`
	`--tlsCertificateSelector <parameter>=<value>`
	`--tlsDisabledProtocols <string>`
		Verbosity	`--verbose=<1-4>  or  --quiet`

Authentication	`--username=<username>`
	`--password=<password>`
	`--authenticationDatabase=`
	`    <database-name>`
	`--authenticationMechanism=`
	`    <mechanism>`
		Output	`--out=<directory-path; default: "./dump">`
			`--gzip`
			`--repair`
			`--oplog`
			`--archive=<file-path>`
			`--dumpDbUsersAndRoles`
			`--excludeCollection=<collection-name>`
			`--excludeCollectionsWithPrefix=`
			`    <collection-prefix>`
			`--numParallelCollections=<N; default: 4)`
			`--viewsAsCollections`

Some options are available as single letters. For example, instead of using the `--verbose` option, you could alternatively add `-v`.

Type the following command to get more details on these options along with more examples of single-letter alternatives: `mongodump --help`.

Security options (authentication, SSL, and **Transport Layer Security (TLS)**) are covered in more detail in `Chapter 11`, *Administering MongoDB Security*. For a complete list of `mongodump` options, see `https://docs.mongodb.com/manual/reference/program/mongodump/#options`.

Things to consider before you back up

Here are two important considerations you should address before performing a backup:

- **Performance impact**: When you perform a backup, the amount of data being processed by the local server increases geometrically. Inevitably, this will have a negative impact on performance. You might want to consider scheduling the backup for a time when the least amount of database activity is expected.
- **Replica sets**: If the server you are backing up is part of a replica set, backing up a secondary can cause problems as the data being backed up might not be current. You can use the `--oplog` option to cause the backup to include the operations log, which can be used to identify the point in time when the backup of the primary server in a replica set occurred.
  It is highly recommended when you back up a replica set member to only back up the *primary*. This is easily accomplished by adding the following option:

```
--host=<replica_set_name>/<primary_host_address>:<port>
```

`mongodump` is useful for small MongoDB deployments. When you have a large amount of data being handled through replica sets and sharded clusters, it might be better to use alternative backup solutions such as using filesystem snapshots (described in the next sub-section) or external hardware-based solutions.

Backing up a specific database

Here is an example of the `sweetscomplete` database being backed up to a `~/backup` directory at verbosity level 2:

```
root@server1:~# mongodump --db=sweetscomplete --verbose=2 --out=/root/backup
2020-07-08T10:49:55.739+0000    will listen for SIGTERM, SIGINT, and SIGKILL
2020-07-08T10:49:55.746+0000    enqueued collection 'sweetscomplete.customers'
2020-07-08T10:49:55.746+0000    enqueued collection 'sweetscomplete.common'
2020-07-08T10:49:55.746+0000    enqueued collection 'sweetscomplete.products'
2020-07-08T10:49:55.747+0000    enqueued collection 'sweetscomplete.purchases'
2020-07-08T10:49:55.747+0000    dumping up to 4 collections in parallel
2020-07-08T10:49:55.748+0000    writing sweetscomplete.common to /root/backup/sweetscomplete/common.bson
2020-07-08T10:49:55.748+0000    writing sweetscomplete.customers to /root/backup/sweetscomplete/customers.bson
2020-07-08T10:49:55.749+0000    counted 4 documents in sweetscomplete.common
2020-07-08T10:49:55.749+0000    counted 286 documents in sweetscomplete.customers
2020-07-08T10:49:55.749+0000    writing sweetscomplete.products to /root/backup/sweetscomplete/products.bson
2020-07-08T10:49:55.749+0000    done dumping sweetscomplete.common (4 documents)
2020-07-08T10:49:55.749+0000    counted 26 documents in sweetscomplete.products
2020-07-08T10:49:55.750+0000    writing sweetscomplete.purchases to /root/backup/sweetscomplete/purchases.bson
2020-07-08T10:49:55.750+0000    counted 496 documents in sweetscomplete.purchases
2020-07-08T10:49:55.752+0000    done dumping sweetscomplete.customers (286 documents)
2020-07-08T10:49:55.753+0000    done dumping sweetscomplete.products (26 documents)
2020-07-08T10:49:55.754+0000    done dumping sweetscomplete.purchases (496 documents)
2020-07-08T10:49:55.754+0000    dump phase III: the oplog
2020-07-08T10:49:55.754+0000    finishing dump
root@server1:~#
```

> **TIP**
> In order to fully test backup and restore, first insert all the sample data from inside the Docker container, as follows:
> `/path/to/repo/restore_data_inside.sh`

Restoring using mongorestore

The primary MongoDB command to restore data is `mongorestore`. The command-line options are identical to those listed previously for `mongodump`, with the following additions:

Input options	Restore options
	`--drop`
	`--dryRun`
`--objcheck`	`--writeConcern=<write-concern>`
`--oplogReplay`	`--noIndexRestore`
`--oplogLimit=<seconds>[:ordinal]`	`--noOptionsRestore`
`--oplogFile=<filename>`	`--keepIndexVersion`
`--archive=<filename>`	`--maintainInsertionOrder`
`--restoreDbUsersAndRoles`	`--numParallelCollections=<N; default: 4>`
`--dir=<directory-name>`	`--numInsertionWorkersPerCollection=<N; default: 1>`
`--gzip`	`--stopOnError`
	`--bypassDocumentValidation`
	`--preserveUUID`

> For documentation on `mongorestore`, see `https://docs.mongodb.com/manual/reference/program/mongorestore/#mongorestore`.

Things to consider before restoring data

Before restoring data, here are a few important considerations:

- **Restoring a replica set**: Restoring a replica set is not as simple as running mongorestore on the primary. There is a very specific procedure that must be followed, which is described in detail in Chapter 13, *Deploying a Replica Set*. The procedure basically involves bringing the replica set down, restoring data on one server, and then redeploying the replica set.
- **Using the oplog**: If you performed a backup using the mongodump --oplog option, when restoring you can add the --oplogReplay option. This will ensure that the restore is performed for a precise point in time.
- **Maintaining document integrity**: The --drop option causes the collection to be dropped entirely prior to the restore operation. Bear in mind that the trade-off for added database integrity is that it takes longer to restore. Also consider adding the --objcheck option, which checks the integrity of objects (that is, documents) before database insertion.

Restoring the purchases collection

In this example, we restore the purchases collection to the sweetscomplete database, adding a verbosity level of 2, while maintaining database integrity, as follows:

```
mongorestore --db=sweetscomplete --collection=purchases --drop --objcheck \
    --verbose=2 --dir=/root/backup/sweetscomplete/purchases.bson
```

Here is the output from this command:

```
root@server1:~# mongorestore --db=sweetscomplete --collection=purchases --drop --objcheck --verbose=2 --dir=/root/backup/sweetscomplete/purchases.bson
2020-07-08T10:52:48.008+0000    using --dir flag instead of arguments
2020-07-08T10:52:48.008+0000    using write concern: &{majority false 0}
2020-07-08T10:52:48.010+0000    will listen for SIGTERM, SIGINT, and SIGKILL
2020-07-08T10:52:48.010+0000    connected to node type: standalone
2020-07-08T10:52:48.010+0000    mongorestore target is a file, not a directory
2020-07-08T10:52:48.010+0000    checking for collection data in /root/backup/sweetscomplete/purchases.bson
2020-07-08T10:52:48.010+0000    reading collection purchases for database sweetscomplete from /root/backup/sweetscomplete/purchases.bson
2020-07-08T10:52:48.010+0000    scanning directory &{0xc00025c750 /root/backup <nil>} for metadata
2020-07-08T10:52:48.011+0000    found metadata for collection at /root/backup/sweetscomplete/purchases.metadata.json
2020-07-08T10:52:48.011+0000    enqueued collection 'sweetscomplete.purchases'
2020-07-08T10:52:48.011+0000    restoring up to 4 collections in parallel
2020-07-08T10:52:48.011+0000    dropping collection sweetscomplete.purchases before restoring
2020-07-08T10:52:48.017+0000    reading metadata for sweetscomplete.purchases from /root/backup/sweetscomplete/purchases.metadata.json
2020-07-08T10:52:48.018+0000    creating collection sweetscomplete.purchases using options from metadata
2020-07-08T10:52:48.025+0000    restoring sweetscomplete.purchases from /root/backup/sweetscomplete/purchases.bson
2020-07-08T10:52:48.025+0000    using 4 insertion workers
2020-07-08T10:52:48.086+0000    no indexes to restore
2020-07-08T10:52:48.086+0000    finished restoring sweetscomplete.purchases (496 documents, 0 failures)
2020-07-08T10:52:48.086+0000    496 document(s) restored successfully. 0 document(s) failed to restore.
root@server1:~#
```

Note that if you use the `--db` and `--collection` options, you must also use `--dir` to indicate the associated BSON file.

Using the --nsInclude switch to restore

An easier way to restore is to use the `--nsInclude` switch. This allows you to specify a *namespace* pattern, which could also include an asterisk (*) as a wildcard. Thus, if you wanted to restore all collections of a single database, you could issue this command:

```
mongorestore --nsInclude sweetscomplete.*
```

Here is an example of the `products` collections being restored from the `sweetscomplete` database:

```
root@server1:~# mongorestore --nsInclude sweetscomplete.products --dir=/root/backup/
2020-07-08T10:58:44.055+0000    preparing collections to restore from
2020-07-08T10:58:44.055+0000    reading metadata for sweetscomplete.products from /root/backup/sweetscomplete/products.metadata.json
2020-07-08T10:58:44.063+0000    restoring sweetscomplete.products from /root/backup/sweetscomplete/products.bson
2020-07-08T10:58:44.069+0000    no indexes to restore
2020-07-08T10:58:44.070+0000    finished restoring sweetscomplete.products (26 documents, 0 failures)
2020-07-08T10:58:44.070+0000    26 document(s) restored successfully. 0 document(s) failed to restore.
root@server1:~#
```

For documentation on *namespace* patterns, see `https://docs.mongodb.com/manual/reference/limits/#namespaces`.

In the last section of this chapter, we have a look at performance monitoring.

Performance monitoring

Determining what is considered normal for your database requires gathering statistics over a period of time. The period of time varies widely, depending upon the frequency and volume of database usage. In this section, we examine two primary means of gathering statistics: the `db*stats()` and `db.serverStatus()` methods. First, we have a look at how to monitor MongoDB performance using built-in `mongo` shell methods that produce statistics.

Mongo shell stats() methods

The most readily available monitoring command is the `stats()` shell method, available at both the database and collection levels. At the database level, this shell method gives important information such as the number of collections and indexes, as well as information on the average document size, average file size, and information on the amount of filesystem storage used.

Here is an example of output from `db.stats()` using the `sweetscomplete` database:

```
File Edit View Search Terminal Help
22>use sweetscomplete;
switched to db sweetscomplete

MongoDB server1@sweetscomplete[up:64070 secs]
23>db.stats();
{
        "db" : "sweetscomplete",
        "collections" : 4,
        "views" : 0,
        "objects" : 812,
        "avgObjSize" : 2379.729064039409,
        "dataSize" : 1932340,
        "storageSize" : 1662976,
        "numExtents" : 0,
        "indexes" : 4,
        "indexSize" : 81920,
        "scaleFactor" : 1,
        "fsUsedSize" : 127877709824,
        "fsTotalSize" : 244529655808,
        "ok" : 1
}

MongoDB server1@sweetscomplete[up:64073 secs]
24>
```

Information given by `db.<COLLECTION>.stats()` (substitute the name of the collection in place of `<COLLECTION>`), *a wrapper for the* `collStats` *database command,* easily produces 10 times the amount of information as `db.stats()`. The output from `db.<COLLECTION>.stats()` gives the following general information:

Output key	Data type	Notes
ns	string	Shows the *namespace* of the collection. The namespace is a string that includes the database and collection, separated by a period.
size	integer	The size of the collection.

count	integer	The number of documents in the collection.
avgObjSize	integer	The average size of a document in this collection.
storageSize	integer	The storage size allocated to this collection.
capped	Boolean	Indicates whether or not this is a *capped collection*. A capped collection, as the name implies, is only allowed to grow to a specified size.
nindexes	integer	The number of indexes associated with this collection.
indexBuilds	list	An array of indexes currently being built. Once the index build process completes, that index name is removed from this list.
totalIndexSize	integer	Total size in bytes that the indexes occupy on this server.
indexSizes	document	JSON document giving a breakdown of the size for each individual index associated with this collection.
scaleFactor	integer	Indicates the value of *scale*, provided as an argument to stats() in this form: db.collection.stats({"scale":N}). If N = 1, the values returned are in bytes. If N = 1024, the values returned are in **kilobytes (KB)**, and so on.
ok	integer	Status code for this collection. A value of 1 indicates everything is functioning properly.

In addition to the general information just described, another output key is WiredTiger, which gives detailed information on database caching, operational aspects of the B-tree algorithm used by WiredTiger, compression performance, block management, and other details specific to the internals of the WiredTiger storage engine. A detailed description of the hundreds of statistics available on WiredTiger performance is well beyond the scope of this book!

If all you need to do is to find information on the amount of storage space used by MongoDB collection components, a number of shortcut methods have been created that leverage db.collection.stats(). The following fall into this category:

db.collection **method**	Outputs
db.collection.dataSize()	db.collection.stats().size
db.collection.storageSize()	db.collection.stats().storageSize
db.collection.totalIndexSize()	db.collection.stats().totalIndexSize

For more information on db.stats(), see https://docs.mongodb.com/ manual/reference/method/db.stats/#db-stats.

For more information on db.collection.stats(), see https://docs. mongodb.com/manual/reference/method/db.collection.stats/index. html#db-collection-stats.

For more information on capped collections, see https://docs.mongodb. com/manual/core/capped-collections/index.html#capped-collections.

For more information on WiredTiger, see https://docs.mongodb.com/ manual/core/wiredtiger/index.html.

Let's now turn our attention to server status.

Monitoring server status

Another extremely useful shell method used to gather information about the state of the database server is db.serverStatus(). This shell method is a wrapper for the serverStatus. database command. As with the stats() method described in the previous sub-section, this utility provides literally hundreds of statistics. An explanation of all of them is simply beyond what this book is able to cover.

For a complete list of all output information, please see the documentation under serverStatus/*Output* (https://docs.mongodb.com/manual/ reference/command/serverStatus/#output).

Here is a screenshot of the first few statistics:

```
File  Edit  View  Search  Terminal  Help
MongoDB server1@sweetscomplete[up:68203 secs]
4>db.serverStatus();
{
        "host" : "server1",
        "version" : "4.2.6",
        "process" : "mongod",
        "pid" : NumberLong(1),
        "uptime" : 68207,
        "uptimeMillis" : NumberLong(68206849),
        "uptimeEstimate" : NumberLong(68206),
        "localTime" : ISODate("2020-05-01T04:05:23.492Z"),
        "asserts" : {
                "regular" : 0,
                "warning" : 0,
                "msg" : 0,
                "user" : 15,
                "rollovers" : 0
        },
        "connections" : {
                "current" : 1,
                "available" : 838859,
                "totalCreated" : 19,
                "active" : 1
        },
        "electionMetrics" : {
```

Here is a brief summary of the more important statistics:

db.serverStatus() **statistic**	Notes
version	The version of MongoDB running on this server.
uptime	The number of seconds this server has been running.
electionMetrics	This pertains to the functioning of a *replica set* and is covered in more detail in Chapter 13, *Deploying a Replica Set*.
freeMonitoring	Covered in the next sub-section.
locks	Returns a detailed breakdown on each lock currently existing in the database.
network	Information on database network usage.

op* and repl	Details on replication and the *oplog*. This is covered in more detail in Chapter 13, *Deploying a Replica Set.*
security	Gives information on security configuration and status.
sharding	Details on the *sharded cluster(s)*. This is covered in more detail in Chapter 15, *Deploying a Sharded Cluster.*
transactions	Information on the status of transactions: currently active, committed (that is, successful), aborted (that is, failed), and so on.
wiredTiger	Gives information on the status of the WiredTiger engine driving this database. This information is only available if your mongod instance is using the WiredTiger (default) storage engine. As of this writing, this category returns 425 individual statistics!
mem	Returns information on MongoDB memory usage.
metrics	General metrics in all areas of database operations. This set of statistics is *extremely* useful in establishing a performance benchmark.

This table provides statistics that might indicate problems with your database:

`db.serverStatus()` **statistic**	Notes
`asserts.msg`	The number of warning messages that have been issued. An excessive number of messages could indicate problems with the database. Consult the database log for more information. The location of the database log is configured in the `mongod.conf` file.
`connections.current`	The number of users (including applications and `mongo` shell connections) currently connected.
`connections.available`	The number of remaining connections. A value of less than `1000` indicates serious connection problems. Check to see if applications are keeping connections open needlessly. Also, check to see if any applications are tying up the database in endless loops.
`extra_info.page_faults`	A large value here indicates potential problems with memory and large datasets. Memory issues can be resolved through improvements in application program code, or by reducing the size of individual documents through a process of refining your database design. Large dataset issues can be handled through *sharding*, covered in `Chapter 15`, *Deploying a Sharded Cluster*.
`metrics.commands.<command>.failed`	If this number is high, you need to look for the applications that are issuing the `<command>` that failed and determine wherein lies the error.
`metrics.getLastError`	A JSON document giving details of the last error.
`metrics.operation.writeConflicts`	If this value is high, it indicates a bottleneck whereby write operations are preventing queries from being executed.
`metrics.cursor.timedOut`	Represents the number of times a *cursor* (that is, the return value from a find operation) timed out. A high number might indicate an application issue or a bottleneck where writes are blocking read operations.

As of MongoDB 4.2, the Storage Node Watchdog is in the Community (that is, free) as well as Enterprise editions of MongoDB. This feature allows MongoDB to automatically shut down the monitored `mongod` instance where its underlying OS filesystem becomes unresponsive. This feature can be activated by setting a MongoDB server parameter either using `db.adminCommand( { setParameter: 1,watchdogPeriodSeconds: <value> } )` or by adding `set.Parameter.watchdogPeriodSeconds` to the `mongod.conf` file. See `https://docs.mongodb.com/master/administration/monitoring/#storage-node-watchdog`.

 MongoDB provides a separate utility called `mongostat` that provides real-time database monitoring. For more information, see the documentation on `mongostat` (`https://docs.mongodb.com/manual/reference/program/mongostat/#bin.mongostat`).

 As of MongoDB 4.0, a feature known as free monitoring is available. This feature automatically monitors execution times, memory, and CPU utilization, as well as statistics on the number of operations performed. The feature is enabled using the `db.enableFreeMonitoring()` command. Here is the documentation reference for more information: `https://docs.mongodb.com/manual/administration/free-monitoring/#free-monitoring`.

And that concludes our coverage of essential MongoDB administration techniques.

Summary

In this chapter, you learned how to get into the `mongo` shell using a set of command-line options. You also learned how to use the `.mongorc.js` file to customize the shell. You learned how to perform basic database operations, such as conducting queries, and inserting, updating, and deleting documents. You learned the basics of how to construct a `query filter` using query operators, and how to control the output by creating a *projection*. You also learned that the same filter created for a query can also be applied to update and delete operations. In the process of learning about performing queries, you were shown various examples of common ad hoc reports required by management.

You now know that there is a set of db.collection.XXX shell methods to perform operations on single documents, including findOne(), insertOne(), updateOne(), replaceOne(), and deleteOne(). There is a parallel set of db.collection.XXX shell methods that operate on one or more documents, including find(), insertMany(), updateMany(), and deleteMany(). You can use the xxxOne() set of methods to ensure only a single document is affected, which might be useful when performing critical updates. You also learned how these tools can be used to spot irregularities in the data, and how to make adjustments.

You then learned how to perform backup and restore using the mongodump and mongorestore tools. You learned about the various command-line switches available for these utilities and were presented with a number of points to consider in the interest of maintaining data integrity, especially when operating upon servers in a replica set. Finally, you were given an overview and some practical techniques on how to monitor database performance. As you learned, the two major methods are db.collection.stats() and db.serverStatus(). In addition to a summary of the major statistics available to monitor database operations, you also learned about which parameters to look to that might indicate impaired database performance.

In the next section of the book, you will learn how to build a dynamic database-driven web application using Python, based upon MongoDB. The next chapter will address database design for a typical small enterprise, *Sweets Complete, Inc.*, a fictitious company offering confections for sale online.

Section 2: Building a Database-Driven Web Application

2

Each chapter in this section covers a series of practical hands-on tasks that teach you how to perform routine DevOps tasks involving MongoDB. The program code examples provided are in the popular Python programming language. Staying with the initial scenario involving the fictitious company *Sweets Complete, Inc.*, this section takes you through a progression of major concepts ultimately leading to a dynamic database-driven website.

This section contains the following chapters:

- `Chapter 4`, *Fundamentals of Database Design*
- `Chapter 5`, *Mission-Critical MongoDB Database Tasks*
- `Chapter 6`, *Using AJAX and REST to Build a Database-Driven Website*

4
Fundamentals of Database Design

This chapter introduces a fictitious company called *Sweets Complete Inc.*, and presents a database model based upon their needs. Special focus is placed upon moving from a set of customer requirements to a working MongoDB database design. The basic technique you will learn in this chapter is to first model the proposed document using JSON syntax, which can be tested in the `mongo` shell. You will then learn how to adapt the newly designed document to a Python entity class that will be put to use in the next chapter.

The topics that we will cover in this chapter are as follows:

- Reviewing customer requirements
- Building MongoDB document structures
- Developing the corresponding Python module

Technical requirements

The minimum recommended hardware required for this chapter are as follows:

- Desktop PC or laptop
- 2 GB free disk space
- 4 GB of RAM
- 500 KBPS or faster internet connection.

The software requirements for this chapter are as follows:

- **OS** (**Linux or macOS**): Docker, Docker Compose, Git (optional)
- **OS** (**Windows**): Docker for Windows and Git for Windows
- Python 3.x and the PyMongo driver (already installed in the Docker container used for the book)

The installation of the required software and how to restore the code repository for the book is explained in Chapter 2, *Setting Up MongoDB 4.x*. To run the code examples in this chapter, open a Terminal window (or Command Prompt) and enter these commands:

```
cd /path/to/repo
docker-compose build
docker-compose up -d
docker exec -it learn-mongo-server-1 /bin/bash
```

When you are finished practicing the commands covered in this chapter, be sure to stop Docker as follows:

```
docker-compose down
```

The code used in this chapter can be found in the book's GitHub repository at https://github.com/PacktPublishing/Learn-MongoDB-4.x/tree/master/chapters/04.

Reviewing customer requirements

It is very important to note that this is not a book on software design or application architecture. Rather, our focus is on developing a useful set of classes that allow you to gain database access regardless of which Python framework you are using. In this section, we cover how to come up with a good database design based on customer requirements.

It is extremely important to interview the stakeholders when developing a new application or adding a new feature to an existing application. Please bear in mind that the stakeholders do not just include management; the people that you interview should also include those who will actually be using the new feature or application! For a good discussion on this topic, go to http://www.agilemodeling.com/essays/stakeholders.htm for an article on agile modeling.

Introducing Sweets Complete Inc.

Sweets Complete Inc. is a fictitious company representing small enterprises that sell products online. They sell specialized confections, including cake, candy, and donuts, to an international clientele. People who use their website fall into one of four groups, each with different needs:

- **Casual website visitors**: This group consists of people who happened upon the site through word of mouth, search engines, and links from associated sites. They generally want to get information about the products on offer, and will hopefully be enticed to make a purchase and become a member.
- **Members**: Members are regular individual customers and partners who sell their products (for example, 7-Eleven). They do not need to be sold on the products. They mainly want to set up recurring purchases and check on the status of their orders.
- **Administrative staff**: Staff need to set up member accounts, manage the list of products, and generate sales reports for management.
- **Management**: Often, Sweets Complete employees who work in a managerial role do not need direct access to the data, relying instead upon reports issued by their staff; however, increasingly, managers want direct access and quick reports.

Sweets Complete will be our example company for this part of the book. Let's now dive into what we need to know to produce a solid database design.

What data comes out of the application?

As you review customer requirements, one of the first things you need to ascertain is *what information is the application expected to produce*? This is referred to as the *output* and can take many forms and encompass many different technologies. Some examples of the output include the following:

- Printed reports
- Charts and graphs
- A list of products on a web page
- Data produced in response to a **representational state transfer** (**REST**) request
- Output appearing on the screen of an *app* running on a smart device (for example, a tablet, or mobile phone)

The data required to satisfy these various forms of response will often be grouped in a certain logical manner, which will lead you to the database document structures definition. Different groups of users have different requirements for output:

Group	Desired Output
Casual website visitors	Ability to search lists of products Purchase confirmation
Members	Purchase history with totals Product availability information
Administrative assistants	Product inventory information Updating costs and prices
Management	Monthly sales reports broken down by product and geographic area Analyzing costs versus sales Setting prices

 REST itself is not a protocol. For a good discussion of what it is, see https://en.wikipedia.org/wiki/Representational_state_transfer. For a list of official standards and protocols associated with REST see: http://standards.rest/.

What data goes into the application?

Continuing your review, the next piece of information you might consider is what information goes into the application. As with outputs, inputs to your application can take many forms, including the following:

- HTML forms
- Data obtained from external web services (for example, Google Maps)
- Streaming data (for example, the real-time location of warehouse workers)

It is useful to examine what form of input to expect from members of the different groups outlined in the previous section. Here is an example of what activities each group might perform to produce inputs:

Group	Input-Producing Activities
Casual website visitors	All *clicks* are registered as product-interest data. Making purchases, generating purchase data.
Members	Reading up on products, giving management an idea of which products are trending upward. Making purchases, generating purchase data.
Administrative assistants	Entering product information. Updating inventory information as new stock arrives.
Management	Placing orders for the raw materials needed to produce products. Defining product lines. Creating sales promotions.

Once you have determined the outputs (described in the previous subsection) that the application is expected to produce, you then have an idea of what data needs to go into the system. At first glance, it would seem logical to conclude that for every piece of output data, there needs to be a corresponding input source. For example, if one of the printed reports includes a list of product titles, somehow that information needs to be entered into the system.

In some cases, however, this theory does not hold true: your application might massage input data to produce the desired output. Not all inputs will appear directly as outputs and vice versa. For example, if the printed report is expected to produce a total for each customer purchase, this total does not need to be entered manually by an administrative assistant. Instead, the totals are calculated from the product price and quantity purchased, along with any other fees and taxes that might apply.

One simple technique is to visualize customer requirements, as illustrated in the following diagram:

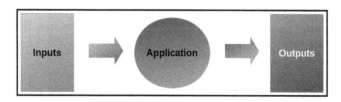

Imagine placing all the inputs on the left and all the outputs on the right. The difference between the two is produced by the application software.

What needs to be stored?

So, at last, we arrive at the big question: what needs to be stored in our database? You might be tempted to go with the simple answer: *everything*. A lot depends on how many unique individual data items we deal with and how many of each can we expect. Your initial tendency might be to take the easy route and just store everything: storage is cheap, and a MongoDB installation can be scaled up easily through the process of *sharding*.

 Sharding is covered in more detail in `Chapter 15`, *Deploying a Sharded Cluster*.

It is in your interest to economize, however, if for no other reason than to avoid having the database engine clogged up with massive amounts of unneeded data. Also, no matter how cheap it is, there will be only so many times management will grant your request for a purchase order to buy more storage! Furthermore, the less data your application needs to sort through, the more efficient it will be.

So, to answer the question, you should store any data items that cannot be generated from existing information. This means that the cost of a product and the quantity of items that are purchased needs to be stored. The extended price of a purchase does not need to be stored, however, as it can be easily generated from the quantity and price. Other items that need to be stored are identifying information, such as transaction numbers, purchase dates, and so forth.

A quick word on software design

Although we are not trying to teach you software design, it might be worth noting that Martin Fowler, a very influential thinker in the IT world, had this to say:

> *I've concluded that the majority of the time you need to take an evolutionary approach to software design, which explains my involvement in the agile community. I've also concluded that patterns are one of the most effective ways to organize and communicate ideas about design.*

> *- Martin Fowler,* `https://martinfowler.com/design.html`, *second paragraph*

In this book, we will not go over software design patterns, but encourage you to have an awareness of these patterns. In an absolutely shameless plug, here are two good books on that subject from Packt Publishing:

- *Architectural Patterns*, by Pethuru Raj, Anupama Raman, and Harihara Subramanian (https://www.packtpub.com/application-development/architectural-patterns)
- *Learning Python Design Patterns: Second Edition*, by Chetan Giridhar (https://www.packtpub.com/application-development/learning-python-design-patterns-second-edition)

You could also, of course, read Martin Fowler's famous book, *Patterns of Enterprise Application Architecture* (https://martinfowler.com/books/eaa.html).

Building MongoDB document structures

For the purposes of this book, we will not focus on actual accounting database storage: most small enterprises either outsource their accounting to a separate firm or use a canned software accounting package, which has its own database and storage capabilities. Instead, we will focus on defining the following collections:

- Products
- Customers
- Purchases

In addition, we will include one more collection called common, which holds the defaults for drop-down menus, product categories, units of measurement, social-media codes (for example, FB is the code for Facebook), and so forth.

Defining the products document structure

Following the preceding suggestions, after reviewing the customer requirements and interviewing stakeholders, you might discover that each of the groups defined previously needs the following information about products:

Group	Information Needed
Casual Website Visitors	Product photo, category, title, price, and description
Members	Same as website visitors + unit of measurement and units on hand
Administrative Assistants	Same as members + cost per unit
Management	Same as administrative assistants

 It is the job of your application to determine the website visitor's role. Your application must also determine, based on the role, what level of access to the database should be granted and what information this user can view.

A decision that you need to make for this example is how to store information about the photo. One obvious approach is to simply store a URL that points to the photo. Another possibility is to store the photo itself directly in the database, using Base64 encoding. Yet a third approach would be to store the actual photo file in MongoDB using GridFS. For our illustration, in this section, we will assume that product photo information is stored as either a URL or a Base64-encoded string. The next chapter, Chapter 5, *Mission-Critical MongoDB Database Tasks*, will discuss how to deliver an image directly from the database. Chapter 10, *Working with Complex Documents across Collections*, discusses how to use GridFS.

The information that should be stored for each product could be summarized in the following JSON document:

```
product =
{
    "productPhoto"  :  "URL or base 64 encoded PNG",
    "skuNumber"     :  "4 letters from title + 4 digit number",
    "category"      :  "drawn from common.categories",
    "title"         :  "Alpha-numeric + spaces",
    "description"   :  "Detailed description of the product",
    "price"         :  "floating point number",
    "unit"          :  "drawn from common.unit",
    "costPerUnit"   :  "floating point number",
    "unitsOnHand"   :  "integer"
}
```

For the actual technical specification for Base64 encoding, see RFC 4648
(`https://tools.ietf.org/html/rfc4648`).

Primary key considerations

For those of you used to working with RDBMS, you will notice that a critical field is
missing: the famous *primary key*. In MongoDB, when a document is inserted into the
database, a unique key is generated and given the reserved field name `_id`. Here is an
example drawn from the products collection:

```
"_id" : ObjectId("5c5f8e011a2656b4af405319")
```

It is an instance of BSON data type `ObjectId`, and is generated as follows:

- 4-byte UNIX timestamp
- 5-byte random value
- 3-byte counter (starting with a random value)

This means that, in MongoDB, there is no need to define a unique key; however, it is often
convenient to store a unique key of your own creation that is somehow tied to the data
entered into the collection. For our purposes, we will create a unique field, `productKey`,
which is a condensed version of the product title.

It's interesting to note that because the `ObjectId` contains a UNIX
timestamp, it's possible to use it to determine the date and time a
document was created. You can use the `ObjectId.getTimestamp()`
method for this purpose (`https://docs.mongodb.com/manual/reference/`
`method/ObjectId.getTimestamp/#objectid-gettimestamp`).

The following is an example of the first product entered into the collection, along with our
completed document structure:

```
db.products.findOne();
{
    "_id" : ObjectId("5c7ded398c3a9b9b9577b7e9"),
    "productKey"    : "apple_pie",
    "productPhoto"  : "iVBORw0KGgoAAAANSUhEUgAAASwAAADCCAYAAADzRP8zA ... ",
    "skuNumber"     : "APPL501",
    "category"      : "pie",
    "title"         : "Apple Pie",
    "description"   : "Donec sed dui hendrerit iaculis vestibulum ... ",
```

```
        "price"        : "4.99",
        "unit"         : "tin",
        "costPerUnit"  : "4.491",
        "unitsOnHand"  : 693
   }
```

 Please note that we truncated the contents of `productPhoto` and `description` to conserve book space.
The reference for `ObjectId` is https://docs.mongodb.com/manual/reference/bson-types/#objectid.

Defining the purchases document structure

Following the preceding suggestions, after reviewing the customer requirements and interviewing stakeholders, you might discover that each of the groups defined previously needs the following information about purchases:

Group	Information Needed
Casual Website Visitors	Purchase history, including the date, products purchased, and the total price paid
Members	Same as website visitors
Administrative Assistants	Same as members + customer contact information
Management	Same as administrative assistants + customer geospatial and address information in order to generate sales reports by geographic region

You will note from the preceding table that documents in the purchases collection need three distinct pieces of information, which are summarized in the next table:

Information Category	Distinct Pieces of Information
Purchase Data	Purchase date
Customer Data	Name, address, city, state or province, postal code, country, and geospatial information
Products Purchased	For each product: quantity purchased, SKU number, category, title, and price

As always, there are decisions that you must make regarding how much customer and product information you will store. In legacy RDBMS circles, you run the risk of violating the cardinal rule: no duplication of data. Accordingly, assuming that you are accustomed to the RDBMS way of thinking, the tendency would be to establish a *relationship* between the products, purchases, and customers collections (that is, a SQL JOIN). Alternatively, you might decide to embed the associated customer and products directly into each purchase document.

In `Chapter 7`, *Advanced MongoDB Database Design*, we will cover how to embed documents. For the purposes of this chapter, however, in order to keep things simple (for now), we will simply copy the appropriate pieces of customer and product data directly into the purchase document without the benefit of embedded objects.

Furthermore, in order to avoid having to create a *join table*, joining multiple products to a single purchase, we will simply define an embedded array of products. In this part of the book, we will use very simple operators to access embedded array information. In `Chapter 8`, *Using Documents with Embedded Lists and Objects*, we will go into further detail on how to actually manipulate embedded array information.

Another decision that needs to be made is whether or not to define a unique identifying field. As we mentioned in the previous section, MongoDB already creates a unique identifier, the mysterious _id field, which is an `ObjectId` instance. For our purposes, we add a new identifying field, `transactionId`, which is useful for accounting purposes. Although we can extract information on the date and time from the _id field, for the purposes of this illustration, we will use the date to form our new identifier, followed by two letters representing customer initials, followed by a random four-digit number.

Finally, we need to decide whether or not to include the extended price in each purchase document. An argument in favor of this approach is that by precalculating this information and storing it, we experience faster speed when it's time to generate sales reports. Also, by storing this value at the time the purchase document is created, we avoid possible errors that might come from overly complicated reporting programming code. For the purposes of this example, we have decided to precalculate the extended price and store it in each purchase document.

Therefore, the final structure for `purchase` would appear as shown here from a `find()` query:

```
purchase = {
    "_id" : ObjectId("5c8098d90f583b515e4d3f41"),
    "transactionId"  : "20181008VC5473",
    "dateOfPurchase" : "2018-10-08 11:03:37",
    "extendedPrice"  : 303.36,
    "customerKey"    : "VASHCARS8772",
    "firstName"      : "Vashti",
    "lastName"       : "Carson",
    "phoneNumber"    : "+1-148-236-8772",
    "email"          : "vcarson148@Swisscom.com",
    "streetAddressOfBuilding" : "5161 Green Mound Ride",
    "city"           : "West Haldimand (Port Dover)",
    "stateProvince"  : "ON",
    "locality"       : "Ontario",
```

```
    "country"        : "CA",
    "postalCode"     : "N0A",
    "latitude"       : "42.9403",
    "longitude"      : "-79.945",
    "productsPurchased" : [
        {
            "productKey"    : "glow_in_the_dark_donut",
            "qtyPurchased"  : 384,
            "skuNumber"     : "GLOW437",
            "category"      : "donut",
            "title"         : "Glow In The Dark Donut",
            "price"         : "0.79"
        }
    ]
}
```

Coordinate pairs such as *latitude* and *longitude* lend themselves to geospatial queries. Accordingly, it might be convenient to represent them as MongoDB GeoJSON objects. This subject is covered in more detail in Chapter 9, *Handling Complex Queries in MongoDB*.

Defining the common document structure

When creating an application based upon a traditional RDBMS, you would normally define a plethora of tiny tables that are consulted when presenting the website user with HTML select options, radio buttons, or checkboxes. The only other alternative would be to hardcode such defaults and options into configuration files.

When using MongoDB, you can take advantage of two radical differences from a legacy RDBMS:

- Each document in a collection *does not* have to have the same schema.
- Documents can contain arrays.

The following table summarizes the options needs for Sweets Complete:

Item	Key	Value
gender	M	Male
	F	Female
	X	Other
social media	FB	Facebook
	TW	Twitter

As you can see, for these two items, a dictionary is defined with key–value pairs. For other items, summarized in the following table, there are lists for which there is no key:

Item	Values
categories	Cake, chocolate, pie, cookie, donut
unit	Box, tin, piece, item

To add flexibility, we can simply add each of the aforementioned items as a separate document, consisting of a single list. Later, when we need to display options or to confirm form submissions against the options, we perform a lookup based on the item. So, finally, here is how the JavaScript to populate the common collection might appear:

```
db.common.insertMany( [
{ "key" : "gender",
  "data" : { "M" : "male", "F" : "female", "X" : "Other" }
},
{ "key" : "socialMedia",
  "data" : {"FB":"facebook", "GO":"google", "LI":"linkedin",
            "LN":"line", "SK":"skype", "TW":"twitter"}
},
{ "key" : "categories",
  "data" : ["cake","chocolate","pie","cookie","donut"]
},
{ "key" : "unit",
  "data" : ["box","tin","piece","item"]
}]);
```

We will now turn our attention to developing application code that takes advantage of these data structures.

Developing a corresponding Python module

We are now ready to start defining a Python *module* containing a class definition equivalent to each MongoDB document described previously. Following the same order as the previous section, let's go collection by collection, starting with products. First, we need to determine what every single entity class needs and put this into a base class.

The reason why we create *entity* classes is to give us a programmatic way to work with database documents. Eventually, starting with the next chapter, we will show you how to perform important database operations that will make direct use of these classes.

Determining common needs in a base class

We first define a directory structure that will contain information pertinent to
`sweetscomplete`. Under this, we create a subdirectory entity where our new base module
will be placed. We then create an empty `__init__.py` file at each directory level to
indicate the presence of a Python module. In the entity directory, we create a
`base.py` file to contain our new base class. Eventually, we will also define entity classes for
customers, products, and purchases.

The resulting directory structure appears as follows:

```
/path/to/repo/chapters/04/src/
└──── sweetscomplete
     ├──── entity
     │     ├──── base.py
     │     ├──── common.py
     │     ├──── customer.py
     │     ├──── __init__.py
     │     ├──── product.py
     │     └──── purchase.py
     └──── __init__.py
```

> When using Python version 3.3 and above, you no longer need to create a
> `__init__.py` file to indicate the presence of a module or package.

In the `Base` class, we import the JSON module and define a property field that defaults to
an empty dictionary. Subclasses inheriting from the `Base` class will contain precise field
definitions specific to customers, products, and purchases:

```python
# sweetscomplete.entity.base
import json
class Base(dict) :
    fields = dict()
```

Next, we define the core `set()` and `get()` methods that are used to assign values to
properties or retrieve those values:

```python
def set(self, key, value) :
    self[key] = value
def get(self, key) :
    if key in self :
        return self[key]
    else :
        return None
```

We can then define two methods `setFromDict()` and `setFromJson()` that populate properties from a Python dictionary or a JSON document, respectively. You will note that these methods rely upon the current value of fields:

```
def setFromDict(self, doc) :
    for key, value in self.fields.items() :
        if key in doc : self[key] = doc[key]
def setFromJson(self, jsonDoc) :
    doc = json.loads(jsonDoc)
    self.setFromDict(doc)
```

We make use of these two methods in the __init__ constructor method. We first check to see whether a parameter has been provided to the constructor. If so, we initialize the fields to their defaults. Following this, if the document argument that is provided is of a `dict` type, we use `setFromDict()`; otherwise, if the document provided is a string, we use `setFromJson()`. If no document is provided to the constructor, then an empty object instance is created. We are then free to use the `set()` method described earlier to populate the selected properties:

```
def __init__(self, doc = None) :
    if doc :
        for key, value in self.fields.items() : self[key] = value
        if isinstance(doc, dict) :
            self.setFromDict(doc)
        elif isinstance(doc, str) :
            self.setFromJson(doc)
```

Finally, just in case we need to produce a JSON document from the entity, we add a `toJson()` method:

```
def toJson(self, filtered = []) :
    jsonDoc = dict()
    for k,v in self.items() :
        if k in filtered :
            pass
        elif k == '_id' :
            jsonDoc[k] = str(v)
        else :
            jsonDoc[k] = v
    return json.dumps(jsonDoc)
```

We are now ready to look at the entity class `sweetscomplete.entity.product.Product`.

Defining classes for the products collection

The most important thing to do at this point is to define the `fields` property. You will note that we also assign defaults in this dictionary structure. Here is how the core class might appear:

```
# sweetscomplete.entity.product
from sweetscomplete.entity.base import Base
class Product(Base) :
    fields = dict({
        'productKey'    : '',
        'productPhoto'  : '',
        'skuNumber'     : '',
        'category'      : '',
        'title'         : '',
        'description'   : '',
        'price'         : 0.00,
        'unit'          : '',
        'costPerUnit'   : 0.00,
        'unitsOnHand'   : 0.00
    })
```

We then add a few useful methods to set and retrieve the unique key, and also the product title:

```
def getKey(self) :
    return self['productKey']
def setKey(self, key) :
    self['productKey'] = key
def getTitle(self) :
    return self['title']
```

Let's look at a demo script for `product`.

Demo script for product

In the `src` directory (the directory that contains the `sweetscomplete` directory), we define a demonstration script that tests our new class. We call the script `/path/to/repo/chapters/04/demo_sweetscomplete_product_entity.py`. In the first few lines of code, we tell Python where to find the source code and import classes needed for the demo, as shown here:

```
# sweetscomplete.entity.product.Product
# tell Python where to find module source code
import os,sys
```

```
sys.path.append(os.path.realpath('src'))
import pprint
from datetime import date
from sweetscomplete.entity.product import Product
```

Next, we define a demo key and the initial block of data in the form of a dictionary:

```
key = 'TEST' + date.today().isoformat().replace('-', '')
doc = dict({
    'productKey'   : key,
    'productPhoto': 'TEST',
    'skuNumber'    : 'TEST0000',
    'category'     : 'test',
    'title'        : 'Test',
    'description'  : 'test',
    'price'        : 2.22,
    'unit'         : 'test',
    'costPerUnit'  : 1.11,
    'unitsOnHand'  : 333
})
```

Finally, we use the entity class, providing the sample dictionary as an argument:

```
product = Product(doc)
print("\nProduct Entity Initialized from Dictionary")
print('Title: '    + product.getTitle())
print('Category: ' + product.get('category'))
print(product.toJson())
```

We execute it using the following command:

```
python /path/to/repo/chapters/04/demo_sweetscomplete_product_entity.py
```

From this, we get the following output:

```
File Edit View Search Terminal Help
Product Entity Initialized from Dictionary
Title: Test
Category: test
{"productKey": "TEST20190307", "productPhoto": "TEST", "skuNumber": "TEST0000",
"category": "test", "title": "Test", "description": "test", "price": 2.22, "unit
": "test", "costPerUnit": 1.11, "unitsOnHand": 333}


-----------------
(program exited with code: 0)
Press return to continue
```

Next, let's look at a unit test for the product entity.

Unit test for the product entity

Rather than creating demo scripts, it is considered a best practice to create a unit test. First, you need to import `unittest` and set up a test class that inherits from `unittest.TestCase:`. Create a file named `test_product.py` and place it in a directory structure named `test/sweetscomplete/entity`. As with our demo script mentioned previously, we use the `os` and `sys` modules to tell Python where to find the source code. We also import `json`, `unittest`, and our `Product` class:

```python
# sweetscomplete.entity.product.Product test
import os,sys
sys.path.append(os.path.realpath("../../../src"))
import json
import unittest
from sweetscomplete.entity.product import Product
class TestProduct(unittest.TestCase) :
    productFromDict = None
```

Next, you define a dictionary with fields that are populated with test data. The
`productFromDict` property is then initialized in the test method, `setUp()`:

```
testDict = dict({
    'productKey'   : '00000000',
    'productPhoto': 'TEST',
    'skuNumber'    : 'TEST0000',
    'category'     : 'test',
    'title'        : 'Test',
    'description'  : 'test',
    'price'        : 1.11,
    'unit'         : 'test',
    'costPerUnit'  : 2.22,
    'unitsOnHand'  : 333
})

def setUp(self) :
    self.productFromDict = Product(self.testDict)
```

You can now define methods that test whether the `setKey()`, `set()`, and `get()` methods
return expected values. This also incidentally tests whether or not the object was correctly
populated (that is, it indirectly tests the constructor):

```
def test_product_from_dict(self) :
    expected = '00000000'
    actual   = self.productFromDict.getKey()
    self.assertEqual(expected, actual)

def test_product_from_dict_get_and_set(self) :
    self.productFromDict.set('skuNumber', '99999999')
    expected = '99999999'
    actual   = self.productFromDict.get('skuNumber')
    self.assertEqual(expected, actual)
```

Finally, you add the following lines to the end of the file to allow command-line execution:

```
def main() :
    unittest.main()

if __name__ == "__main__":
    main()
```

Here is the result from the test script located at
`/path/to/repo/chapters/04/test/sweetscomplete/entity/test_product.py:`

```
File  Edit  View  Search  Terminal  Help
. .
- - - - - - - - - - - - - - - - - - - - - - - - - - - - - - -
Ran 2 tests in 0.000s

OK

- - - - - - - - - - - - - - - -
(program exited with code: 0)
Press return to continue
```

 There are many other tests that we could show. For example, we could test to see whether an entity initialized from a Python dictionary is the same as one initialized from a JSON document with the same values. We could also test to make sure that an entity is initialized to the defaults where neither a dictionary nor a string is passed. However, for the sake of brevity, we have only included the basic tests shown in this chapter.

Defining classes for the remaining collections

The remaining classes to be defined follow the same logic as described in the previous section. Again, the main challenge is to correctly define the fields that you plan to use. For this purpose, we simply refer back to the original data structures described earlier in the chapter. The key classes that we need include classes to represent the `purchases` and `customers` collections. Some example Python entity classes for these can be found as follows:

- The `Purchases` collection:
  `/path/to/repo/chapters/04/src/sweetscomplete/entity/purchase.py`
- The `Customers` collection:
  `/path/to/repo/chapters/04/src/sweetscomplete/entity/customer.py`

The `common` collection is a simple sequence of documents, each of which has a key and an associated list or dictionary. The Python class associated with this collection can be found in the `purchases` collection,
at `/path/to/repo/chapters/04/src/sweetscomplete/entity/common.py`.

Summary

In this chapter, you learned how to break down customer requirements by determining what needs to come out of the proposed application. From this breakdown, you were able to deduce what needs to go into the application by way of form submissions, data obtained from web service calls, and so on. Finally, you were given guidance on how to determine what needs to be stored.

The next section showed you how to create MongoDB document structures in JSON format after analyzing the needs of *Sweets Complete Inc*. From there, you went on to learn how to convert the JSON documents into Python classes. In each case, the Python entity classes included a sample demo script showing you how the entity class can be used. In addition, you were shown how to develop an appropriate unit test for the new class.

In the next chapter, you will learn how to apply your Python knowledge to the task of performing critical database tasks, such as adding, editing, and deleting documents, as well as basic queries.

5
Mission-Critical MongoDB Database Tasks

This chapter builds upon the work from the previous chapter, Chapter 4, *Fundamentals of Database Design*. The reason why the tasks in this chapter are considered *mission-critical* is because they will teach you how to perform vital database operations that will need to be performed every day. These tasks include reading, writing, updating, and deleting documents from a database.

First, you will be presented with a series of common tasks based upon the Sweets Complete dataset. You'll then learn how to define *domain* classes, which perform critical database access services, including performing queries and adding, editing, and deleting documents. These capabilities are then applied to practical tasks, including updating product information and generating a sales report that details total purchases by month and by year.

In the process, you will be shown how to formulate queries in JavaScript, and then translate that query into its equivalent using Python. Finally, you'll learn how to integrate MongoDB into a simple web infrastructure in order to have a customer log in, make a purchase, see the results, and confirm or cancel the purchase.

In this chapter, we will cover the following topics:

- Creating a Connection class
- Defining domain service classes
- Generating a product sales report
- Updating product information
- Simple web infrastructure
- Adding a new purchase
- Canceling a purchase

Let's get started!

Technical requirements

The following is the minimum recommended hardware for this chapter:

- Desktop PC or laptop
- 2 GB free disk space
- 4 GB of RAM
- 500 Kbps or faster internet connection

The following are the software requirements for this chapter:

- OS (Linux or Mac): Docker, Docker Compose, Git (optional)
- OS (Windows): Docker for Windows and Git for Windows
- Python 3.x, the PyMongo driver, and Apache (already installed in the Docker container used for this book)

Installation of the required software and how to restore the code repository for this book is explained in Chapter 2, *Setting Up MongoDB 4.x*. To run the code examples in this chapter, open a Terminal window (Command Prompt) and enter these commands:

```
cd /path/to/repo
docker-compose build
docker-compose up -d
docker exec -it learn-mongo-server-1 /bin/bash
```

To run the demo website described in this chapter, follow the setup instructions given in Chapter 2, *Setting Up MongoDB 4.x*. Once the Docker container for Chapters 3 through 12 has been configured and started, there are two more things you need to do:

1. Add the following entry to the local hosts file on your computer:

    ```
    172.16.0.11    learning.mongodb.local
    ```

     The local hosts file is located at
    `C:\windows\system32\drivers\etc\hosts` on Windows and
    `/etc/hosts` for Linux and Mac.

2. Open a browser on your local computer (outside of the Docker container) and go to this URL:

    ```
    http://learning.mongodb.local/
    ```

3. Click on the link for Chapter 5.

When you are finished working with the examples covered in this chapter, return to the Terminal window (Command Prompt) and stop Docker, as follows:

```
cd /path/to/repo
docker-compose down
```

Now, let's start creating a connection class.

The code used in this chapter can be found in this book's GitHub repository at `https://github.com/PacktPublishing/Learn-MongoDB-4.x/tree/master/chapters/05`, and also at `https://github.com/PacktPublishing/Learn-MongoDB-4.x/tree/master/www/chapter_05`.

Creating a Connection class

It is extremely useful to have a generic connection class available that can be used throughout the application. This class needs to have the following capabilities:

- Accept a database connection string, which can be provided by a separate config file.
- Create a connection to a database.
- Create a connection to a collection.

First, we must discuss the role of the `pymongo.MongoClient` class.

pymongo.MongoClient

`pymongo.MongoClient` is actually an alias for `pymongo.mongo_client.MongoClient`. This class accepts a URI style connection string as an argument or individual parameters, as shown here. The arguments in square brackets `[]` are optional:

```
from pymongo.mongo_client import MongoClient
client =
MongoClient(<host>[,<port>,<document_class>,<tz_aware>,<connect>,**kwargs])
```

The main parameters are summarized in this table:

Parameter	Notes
host	Can be *localhost*, a *DNS address* (for example, db.unlikelysource.com), or an *IP address*.
port	If not specified, defaults to 27017; otherwise, specify the MongoDB listening port.
document_class	Defaults to *dict*. This parameter controls what gets returned when you run a database query. This parameter ties the entity classes discussed in the previous chapter into MongoDB.
tz_aware	If set to *True*, any *datetime* instances reflect the server's time zone.
connect	If set to *True*, it causes the client to immediately attempt a database connection; otherwise, the connection is only made when results are required.

**kwargs stands for a series of optional keyword arguments (*key* = *value* pairs) that can be specified. As an example, if you wish to supply the username and password authentication keyword arguments, you would create the client instance as follows:

```
client = pymongo.mongo_client.MongoClient(host='localhost', port=27017,
    document_class=dict, tz_aware=False, connect=True,
    username='fred', password='very_secret_password');
```

The other keyword arguments are listed here:

- **Performance**: maxPoolSize, minPoolSize, maxIdleTimeMS, socketTimeoutMS, connectTimeoutMS, serverSelectionTimeoutMS, waitQueueTimeoutMS, waitQueueMultiple, heartbeatFrequencyMS, appname, driver, event_listeners, compressors, zlibCompressionLevel, and retryWrites
- **Read Concern**: readConcernLevel
- **Write Concern**: w, wTimeoutMS, journal, and fsync
- **Replica Set Options**: replicaSet, readPreference, readPreferenceTags, maxStalenessSeconds
- **Authentication**: username, password, authSource, authMechanism, and authMechanismProperties
- **SSL**: ssl_certfile, ssl_keyfile, ssl_pem_passphrase, ssl_cert_reqs, ssl_ca_certs, ssl_crlfile, and ssl_match_hostname

The full list of `pymongo.mongo_client.MongoClient` options can be found here: `https://api.mongodb.com/python/current/api/pymongo/mongo_client.html#module-pymongo.mongo_client`.
For more information on MongoClient high availability, go to `https://api.mongodb.com/Python/current/examples/high_availability.html`.

Instead of specifying a set of parameters, you can also create a pymongo client instance using a URI style connection string in place of the `host` argument. For more information, go to `https://docs.mongodb.com/manual/reference/connection-string/#connection-string-uri-format`.
For specifics on *read concerns*, go to `https://docs.mongodb.com/manual/reference/read-concern/#read-concern`.
The following is a link to the docs on *write concerns*: `https://docs.mongodb.com/manual/reference/write-concern/#write-concern`.

Creating the connection

Although the pymongo client is extremely easy to use, it is much more convenient to wrap its functionality into a `Connection` class. We start by importing the `MongoClient` class. We then define two properties: one to represent the database and another to represent the client. Most importantly, take note of the third argument when creating a `MongoClient` instance – the name of the entity class associated with this connection:

```
# db.mongodb.connection
from pymongo import MongoClient
class Connection :
    db     = None
    client = None
    def __init__(self, host = 'localhost', port = 27017, \
                 result_class = dict) :
        self.client = MongoClient(host, port, result_class)
```

Next, we define a number of useful methods to grant outside access to the client and database connection instances:

```
    def getClient(self) :
        return self.client
    def getDatabase(self, database) :
        self.db = self.client[database]
        return self.db
```

Finally, we add one more method that allows us to gain access to both the database and a specific collection:

```
def getCollection(self, collection) :
    if collection.find(".") :
        database, collection = collection.split(".",1)
        self.getDatabase(database)
    return self.db[collection]
```

Here is a demo script that uses the `Connection` class to gain access to the `products` collection in the `sweetscomplete` database. Note that after the `import` statements, we create a `Connection` instance by specifying the hostname, port, and result class:

```
# /path/to/repo/chapters/05/demo_db_connection_sweetscomplete_product.py
from db.mongodb.connection import Connection
from sweetscomplete.entity.product import Product
conn = Connection('localhost', 27017, Product)
prodCollect = conn.getCollection("sweetscomplete.products")
```

After that, we use the pymongo collection's `find_one()` method to retrieve a single result:

```
result = prodCollect.find_one()
print("\nResult from Query:")
print('Class:    ' + str(type(result)))
print('Key:      ' + result.getKey())
print('Title:    ' + result.getTitle())
print('Category: ' + result.get('category'))
print('JSON:     ' + result.toJson(['productPhoto']))
```

As shown in the following output, the result is a `sweetscomplete.entity.product.Product` instance. We are using the methods we defined in the entity class we discussed in Chapter 4, *Fundamentals of Database Design*. You can also see that the `toJson()` method includes a filter that's used here to preclude `productPhoto` from the output:

```
File  Edit  View  Search  Terminal  Help
Result from Query:
Class:    <class 'sweetscomplete.entity.product.Product'>
Key:      apple_pie
Title:    Apple Pie
Category: pie
JSON:     {"_id": "5c80b6c2d5e85963e1714dc9", "productKey": "apple_pie", "skuNum
ber": "APPL501", "category": "pie", "title": "Apple Pie", "description": "Cras s
it amet eros congue, rhoncus sem sed, consequat arcu. Nam gravida libero ac male
suada cursus. Ut quis massa sit amet enim faucibus suscipit ac viverra elit.", "
price": "4.99", "unit": "tin", "costPerUnit": 4.4910000000000005, "unitsOnHand":
 736.0}


------------------
(program exited with code: 0)
Press return to continue
```

Now that you have an idea of what a `Connection` class looks like, we'll have a look at developing a *domain service* class.

Defining domain service classes

If your main interest is only writing short demo scripts, you could just use the `Connection` class discussed in the previous section. However, it is a best practice to wrap actual database access into a *domain* or *service* component. The *domain* is where the *business logic* is placed. In this context, *business logic* would include database access methods that allow us to perform critical database tasks such as adding to the database, making updates, removing documents, and performing queries.

For the purposes of the book, we rely on the of the **Action-Domain-Responder** (**ADR**) software design pattern (`https://github.com/pmjones/adr`). It is a newer design pattern that provides much-needed refinements to the more traditional **Model-View-Controller** (**MVC**) design pattern (`https://web.archive.org/web/20120729161926/http://st-www.cs.illinois.edu/users/smarch/st-docs/mvc.html`).

We will start our discussion with a domain service that represents the `products` collection.

Products collection domain service

As with our entity classes, we start by defining a base class for all the domain services. What the base class provides for us is a property that represents instances of `pymongo.database.Database` and `db.mongodb.connection.Connection`.

The constructor initializes these two properties using the connection to retrieve the database instance:

```
# sweetscomplete.domain.base
class Base :
    db = None
    conn = None
    def __init__(self, conn, database) :
        self.conn = conn
        self.db   = conn.getDatabase(database)
```

We then define a domain service class called
`sweetscomplete.domain.product.ProductService` that inherits from the base class:

```
# sweetscomplete.domain.product
from sweetscomplete.domain.base import Base
from sweetscomplete.entity.product import Product
class ProductService(Base) :
    # useful methods definitions go here
```

We are then free to add useful methods as needed. These methods are wrappers for core
`pymongo` collection methods including the following:

- `find()`
- `find_one()`
- `insert_one()`
- `insert_many()`
- `update_one()`
- `update_many()`
- `delete_one()`
- `delete_many()`

If you dig through the documentation of these methods (`https://api.mongodb.com/`
`python/current/api/pymongo/collection.html`), it will not surprise you to learn that they
mimic the functionality of the equivalent *mongo* shell methods described in `Chapter 3`,
Essential MongoDB Administration Techniques. As an example of a method useful for classes
consuming the service, consider `fetchByKey()`, shown here. It performs a simple query,
using `find_one()`, and returns a single `Product` instance:

```
def fetchByKey(self, key) :
    query  = dict({"productKey" : key})
    return self.db.products.find_one(query)
```

In a similar manner, we could return an iteration of `Product` instances that only contain
`productKey` and `title`:

```
def fetchAllKeysAndTitles(self) :
    projection = dict({"productKey":1,"title":1,"_id":0})
    for doc in self.db.products.find(None, projection) :
        yield doc
```

And, of course, where would we be without the ability to add and delete? Examples of such methods are shown here:

```
def addOne(self, productEntity) :
    return self.db.products.insert_one(productEntity)

def deleteOne(self, query) :
    return self.db.products.delete_one(query)
```

Finally, here is a demo script that makes use of the domain service. After performing the appropriate imports, you can see that we create a connection by specifying a `Product` as the result class. With the `Connection` instance, we are then able to create a `ProductService` instance:

```
# /path/to/repo/chapters/05
# demo_sweetscomplete_product_read_add_edit_delete.py
import os,sys
sys.path.append(os.path.realpath("src"))
import pprint
import db.mongodb.connection

from datetime import date
from sweetscomplete.entity.product import Product
from sweetscomplete.domain.product import ProductService

conn = db.mongodb.connection.Connection('localhost',27017,Product)
service = ProductService(conn, 'sweetscomplete')
```

Next, we initialize a JSON document with sample data. As you may recall, our entity class is able to accept either a `dict` or a JSON document as a constructor argument:

```
key = 'test' + date.today().isoformat().replace('-', '')
doc = '''\
{
    "productKey"   : "%key%",
    "productPhoto": "TEST",
    "skuNumber"    : "TEST0000",
    "category"     : "test",
    "title"        : "Test",
    "description"  : "test",
    "price"        : 1.11,
    "unit"         : "test",
    "costPerUnit"  : 2.22,
    "unitsOnHand"  : 333
}
'''.replace('%key%',key)
```

We can then add the new product, find it using the demo key, delete it, and display a list of all product titles:

```python
print("\nAdding a Single Test Product\n")
product = Product(doc)
if service.addOne(product) :
    print("\nProduct " + key + " added successfully\n")

print("\nFetch Product by Key\n")
doc = service.fetchByKey(key)
if doc :
    print(doc.toJson())

query = dict({"productKey" : key})
if service.deleteOne(query) :
    print("\nProduct " + key + " deleted successfully\n")

print("\nList of Product Titles and Keys\n")
for doc in service.fetchAllKeysAndTitles(6) :
    print(doc['title'] + '  [' + doc['productKey'] + ']')
```

Here is the output from the demo script:

```
File Edit View Search Terminal Help
Adding a Single Test Product

Product test20190313 added successfully

Fetch Product by Key
{"_id": "5c88ba970317e31daf50e5e3", "productKey": "test20190313", "productPhoto"
: "TEST", "skuNumber": "TEST0000", "category": "test", "title": "Test", "descrip
tion": "test", "price": "1.11", "unit": "test", "costPerUnit": "2.22", "unitsOnH
and": 333}

Product test20190313 deleted successfully

List of Product Titles and Keys
Apple Pie  [apple_pie]
Birthday Cake  [birthday_cake]
Black And Tan Truffel  [black_and_tan_truffel]
Cake With Chocolate Frosting  [cake_with_chocolate_frosting]
Carrot Cake  [carrot_cake]
Cherry Pie  [cherry_pie]


--------------------
(program exited with code: 0)
Press return to continue
```

Next, we'll examine a more practical application: producing a sales report.

Generating a product sales report

Management has come to you with a request for a sales report for 2018, by country, broken down by month. You decide on a simple strategy: perform a query on the `purchases` collection that returns the total extended price of each purchase, along with the purchase date and country. Before getting into the actual code, we will look at the core `pymongo` collection command; that is, `find()`.

 Please note that you could use the MongoDB Aggregation Framework to provide the requested breakdown. Aggregation is a complex topic, however, and is covered in `Chapter 9`, *Handling Complex Queries in MongoDB*.

Using the pymongo.collection.Collection.find() class

The core class used for report queries is `pymongo.collection.Collection.find()`.

Here is the generic syntax:

```
find(filter=None, projection=None, skip=0, limit=0, \
     no_cursor_timeout=False, cursor_type=CursorType.NON_TAILABLE, \
     sort=None, ... /* other options not shown */)
```

The `find()` arguments match their mongo shell equivalents. The more widely used parameters are summarized here:

Parameter	Default	Notes
filter	None	A dictionary consisting of key/value pairs, where the key represents the field to search and the value represents the search criteria. If set to None, all documents in the collection are selected.
projection	None	A dictionary consisting of key/value pairs, where the key represents the field to show or suppress from the output. To show a field, set the value to 1. To suppress the field, set the value to 0. If set to None, all fields in the selected documents are returned.
skip	0	During results iteration, this parameter causes MongoDB to skip this many documents before starting to return query results. This is the equivalent to the SQL OFFSET keyword.
limit	0	This parameter causes MongoDB to return this many documents in the query result. This is equivalent to the SQL LIMIT keyword.

`sort`	None	Lets you specify which fields are used to order the results. This parameter is a dictionary of key/value pairs. The key is the field to influence the sort. The value is 1 for ascending and -1 for descending.
`collation`	None	You can supply a `pymongo.collation.Collation` instance that allows you to influence how search and sort operate. This is especially useful if you are collecting information on a website that services an international clientele.
`hint`	None	Allows you to recommend which index MongoDB should use when performing the query.
`max_time_ms`	None	Sets a time limit in milliseconds for how long a query is allowed to run.
`show_record_id`	False	If set to True, adds the internal records identifier to the output in the form of a `recordId` field.
`session`	None	If you supply a `pymongo.client_session.ClientSession` instance, you are able to initiate transaction support, which allows a set of database operations to be treated as a single atomic block. The session also gives you control over read and write concerns, which are important when operating on a replica set.

In addition to these parameters, other parameters that are supported but not discussed here include `no_cursor_timeout`, `cursor_type`, `allow_partial_results`, `oplog_replay`, `batch_size`, `max`, `min`, `return_key`.

All `find()` parameters are optional. Without any parameters, the result is a dump of the entire collection. Please refer to the following links for more information:

`pymongo.collection.Collection.find()`: https://docs.mongodb. com/manual/reference/method/db.collection.find/#db.collection. find

`pymongo.client_session.ClientSession`: https://api.mongodb. com/python/current/api/pymongo/client_session.html#pymongo. client_session.ClientSession

`pymongo.collation.Collation`: https://api.mongodb.com/python/ current/api/pymongo/collation.html#pymongo.collation.Collation

Modeling the query in JavaScript

A really great technique that you can use to formulate complex queries is to model the query using the *mongo* shell or use MongoDB Compass (covered in `Chapter 9`, *Handling Complex Queries in MongoDB*). That way, you are able to get an idea of what data is returned, which might lead to further refinements. You can then adapt the query to Python and the `pymongo.collection.Collection.find()` method.

In the *mongo* shell, we use the `sweetscomplete` database. After that, we can formulate our query document in the form of a JavaScript variable query. Next, we define the projection, which controls which fields appear in the output. We can then execute this query:

```
query = {"dateOfPurchase":{"$regex":/^2018/}};
projection = {"dateOfPurchase":1,"extendedPrice":1,"country":1,"_id":0};
db.purchases.find(query, projection). \
            sort({"country":1,"dateOfPurchase":1});
```

We will achieve this result:

```
File  Edit  View  Search  Terminal  Help
> query = {"dateOfPurchase":{"$regex":/^2018/}};
{ "dateOfPurchase" : { "$regex" : /^2018/ } }
> projection = {"dateOfPurchase":1,"extendedPrice":1,"country":1,"_id":0};
{ "dateOfPurchase" : 1, "extendedPrice" : 1, "country" : 1, "_id" : 0 }
> db.purchases.find(query, projection).sort({"country":1,"dateOfPurchase":1});
{ "dateOfPurchase" : "2018-02-05 11:48:08", "extendedPrice" : 691.89, "country" : "AD" }
{ "dateOfPurchase" : "2018-01-24 11:48:08", "extendedPrice" : 47645.450000000004, "country" : "AU" }
{ "dateOfPurchase" : "2018-02-22 11:48:08", "extendedPrice" : 35914.950000000004, "country" : "AU" }
{ "dateOfPurchase" : "2018-03-07 11:48:08", "extendedPrice" : 3818.7400000000002, "country" : "AU" }
{ "dateOfPurchase" : "2018-06-15 11:48:07", "extendedPrice" : 7982.530000000001, "country" : "AU" }
{ "dateOfPurchase" : "2018-07-25 11:48:08", "extendedPrice" : 12594.65, "country" : "AU" }
{ "dateOfPurchase" : "2018-08-09 11:48:07", "extendedPrice" : 3213.56, "country" : "AU" }
{ "dateOfPurchase" : "2018-10-07 11:48:08", "extendedPrice" : 5690.9400000000005, "country" : "AU" }
{ "dateOfPurchase" : "2018-10-31 11:48:08", "extendedPrice" : 7601.73, "country" : "AU" }
{ "dateOfPurchase" : "2018-11-13 11:48:08", "extendedPrice" : 1157.13, "country" : "AU" }
{ "dateOfPurchase" : "2018-11-18 11:48:08", "extendedPrice" : 3603.5200000000004, "country" : "AU" }
{ "dateOfPurchase" : "2018-11-21 11:48:07", "extendedPrice" : 11762.3, "country" : "AU" }
{ "dateOfPurchase" : "2018-11-29 11:48:08", "extendedPrice" : 6029.0599999999995, "country" : "AU" }
{ "dateOfPurchase" : "2018-12-03 11:48:08", "extendedPrice" : 15239.01, "country" : "AU" }
{ "dateOfPurchase" : "2018-04-04 11:48:08", "extendedPrice" : 9981.17, "country" : "BD" }
{ "dateOfPurchase" : "2018-08-25 11:48:07", "extendedPrice" : 3769.5600000000004, "country" : "BD" }
{ "dateOfPurchase" : "2018-07-09 11:48:08", "extendedPrice" : 11117.289999999999, "country" : "BM" }
{ "dateOfPurchase" : "2018-07-21 11:48:08", "extendedPrice" : 10638.38, "country" : "BM" }
{ "dateOfPurchase" : "2018-08-30 11:48:07", "extendedPrice" : 59.85, "country" : "BM" }
{ "dateOfPurchase" : "2018-01-21 11:48:08", "extendedPrice" : 2935.09, "country" : "CA" }
Type "it" for more
>
```

Having successfully modeled the query using the *mongo* shell, we will now turn our attention back to the `purchases` domain service class, where we adapt the JavaScript query to Python code.

Adding the appropriate methods to the domain service class

There are two ways to go about modeling queries in a domain service class. One approach is to create a method that explicitly produces the results for a report. In this approach, the complex logic needed to produce the report goes into the domain service class. This approach, which we call the *Custom Query Method* approach, is useful in situations where it is called frequently, and where you wish to hide the complexity of the logic from any calling program.

An alternative approach is to simply create a wrapper for existing pymongo functionality. We call this the *Generic Query Method* approach. This approach is useful if you wish to extend the flexibility of your domain service class, but places the burden on the calling program to formulate the appropriate query and logic to process the results.

We'll start by discussing the *Generic Query Method* approach.

Generic Query Method approach

First, we define the generic method in the domain service class. Note that we supply defaults for the `find()` method arguments; that is, `skip`, `limit`, `no_cursor_timeout`, and `cursor_type`:

```
# sweetscomplete.domain.purchase
from sweetscomplete.domain.base import Base
from sweetscomplete.entity.purchase import Purchase
from bson.objectid import ObjectId
from pymongo.cursor import CursorType
class PurchaseService(Base) :
    def genericFetch(self, query, proj = dict(), order = None) :
        for doc in self.db.purchases.find(query, proj,
            0, 0, False, CursorType.NON_TAILABLE, order) :
            yield doc
    # other methods not shown
```

As you can see from the following code, the onus for results processing is on the calling program. As with other demo scripts, we first establish imports and set up the connection:

```
# /path/to/repo/chapters/05/demo_sweetscomplete_purchase_sales_report.py
import os,sys
sys.path.append(os.path.realpath("src"))
import db.mongodb.connection
import re
from bson.regex import Regex
```

```
from sweetscomplete.entity.purchase import Purchase
from sweetscomplete.domain.purchase import PurchaseService
conn = db.mongodb.connection.Connection('localhost',27017,Purchase)
service = PurchaseService(conn, 'sweetscomplete')
```

We then define the variables that will be used in the script, including the query and projection dictionaries:

```
year    = 2018
pattern = '^' + str(year)
regex   = Regex(pattern)
query   = dict({'dateOfPurchase':{'$regex':regex}})
proj    = dict({'dateOfPurchase':1,'extendedPrice':1,'country':1,'_id':0})
order   = [('country', 1),('dateOfPurchase', 1)]
results = {}
```

Here is the logic that produces final results totals in the form of a dictionary – {country:{month:total}}:

```
for doc in service.genericFetch(query, proj, order) :
    country = doc['country']
    dop     = doc['dateOfPurchase']
    month   = str(dop[5:7])
    if country in results :
        temp = results[country]
        if month in temp :
            price = temp[month] + doc['extendedPrice']
        else :
            price = doc['extendedPrice']
        temp.update({month:price})
    else :
        results.update({country:{month:doc['extendedPrice']}})
```

And finally, here is the logic used to produce the final report's output:

```
print("\nSales Report for " + str(year))
import locale
grandTotal = 0.00
for country, monthList in results.items() :
    print("\nCountry: " + country)
    print("\tMonth\tSales");
    for month, price in monthList.items() :
        grandTotal += price
        showPrice = locale.format_string("%8.*f", (2, price), True)
        print("\t" + month + "\t" + showPrice)
showTotal = locale.format_string("%12.*f", (2, grandTotal), True)
print("\nGrand Total:\t" + showTotal)
```

The following screenshot shows the partial result:

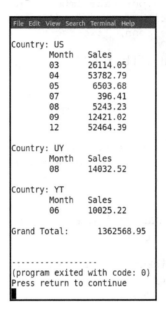

```
File Edit View Search Terminal Help

Country: US
        Month    Sales
        03       26114.05
        04       53782.79
        05        6503.68
        07         396.41
        08        5243.23
        09       12421.02
        12       52464.39

Country: UY
        Month    Sales
        08       14032.52

Country: YT
        Month    Sales
        06       10025.22

Grand Total:        1362568.95


-------------------
(program exited with code: 0)
Press return to continue
```

Now, let's look at the Custom Query Method approach.

Custom Query Method approach

In this approach, the complex logic shown previously in the calling program is transferred to the domain service class in a custom method called `salesReportByCountry()` instead. As with the logic shown previously, first, we initialize the variables:

```
def salesReportByCountry(self, year) :
    from bson.regex import Regex
    pattern = '^' + str(year)
    regex   = Regex(pattern)
    query   = dict({'dateOfPurchase':{'$regex':regex}})
    proj    = dict({'dateOfPurchase':1,'extendedPrice':1, \
                    'country':1,'_id':0})
    order   = [('country', 1),('dateOfPurchase', 1)]
    results = {}
```

We then implement exactly the same logic shown in the calling program prior:

```
for doc in self.genericFetch(query, proj, order) :
    country = doc['country']
    dop     = doc['dateOfPurchase']
```

```
month    = str(dop[5:7])
if country in results :
    temp = results[country]
    if month in temp :
        price = temp[month] + doc['extendedPrice']
    else :
        price = doc['extendedPrice']
    temp.update({month:price})
else :
    results.update({country:{month:doc['extendedPrice']}})
return results
```

The calling program is now greatly simplified:

```
# /path/to/repo/chapters/05
# demo_sweetscomplete_purchase_sales_report_custom.py
# imports and setting up the connection are the same as shown above

# running a fetch based on date
year     = 2018
results = service.salesReportByCountry(year)
print("\nSales Report for " + str(year))
import locale
grandTotal = 0.00
for country, monthList in results.items() :
    print("\nCountry: " + country)
    print("\tMonth\tSales");
    for month, price in monthList.items() :
        grandTotal += price
        showPrice = locale.format_string("%8.*f", (2, price), True)
        print("\t" + month + "\t" + showPrice)

showTotal = locale.format_string("%12.*f", (2, grandTotal), True)
print("\nGrand Total:\t" + showTotal)
```

The results are not shown here as they are identical to the results provided in the previous subsection. Now, we'll learn how to develop classes and methods to perform update operations.

Updating product information

The core classes used for updates are `update_one()` and `update_many()`. The difference between the two is that the latter potentially affects multiple documents. Here is the generic syntax. The parameters in square brackets `[]` are optional:

```
update_one(<filter>, <update>, [<upsert>, \
    <bypass_document_validation>, <collation>, \
    <array_filters>, <session>])
```

The arguments match their *mongo* shell equivalents and are summarized here:

Parameter	Default	Notes
`filter`	--	A query dictionary that is identical to that used for the `find()` method.
`update`	--	A `$set : { key / value pairs }` dictionary, where the key corresponds to the field and the value is the updated information.
`upsert`	False	If set to True, a new document is added if the query filter does not find a match.
`bypass_document_validation`	False	If set to True, the write operation is instructed not to validate. This improves performance at the cost of the potential loss of data integrity.
`collation`	None	You can supply a `pymongo.collation.Collation` instance, which allows you to influence how search and sort operate. This is especially useful if you are collecting information on a website that services an international clientele.
`array_filters`	None	Gives you control over which updates are applied to the elements of an embedded array.
`session`	None	If you supply a `pymongo.client_session.ClientSession` instance, you are able to initiate transaction support, which allows a set of database operations to be treated as a single atomic block. The session also gives you control over read and write concerns, which are important when operating on a replica set.

When defining an update method in a domain service class, don't forget that the update document needs a `$set` update operator as its key.

The return value after an update is a `pymongo.results.UpdateResult` object, from which we need to extract information. The most pertinent piece of information is `modified_count`. If this value is 0, it means the update failed. Here is how the update method might appear in the `ProductService` class:

```
def editOneByKey(self, prodKey, data) :
    query = dict({"productKey" : prodKey})
    updateDoc = dict({"$set" : data})
    result = self.db.products.update_one(query, updateDoc)
    return self.db.products.find_one(query) \
            if (result.modified_count > 0) else False
```

We can then modify the `ProductService` demo script shown earlier by adding the following code:

```
updateDoc = {
    'productPhoto' :'REVISED PHOTO',
    'price'        : 2.22
};
result = service.editOneByKey(key, updateDoc)
if not result :
    print("\nUnable to find this product key: " + key + "\n")
else :
    print("\nProduct " + key + " updated successfully\n")
    print(result.toJson())
```

Here is the modified output:

```
File  Edit  View  Search  Terminal  Help
Adding a Single Test Product

Product test20190313 added successfully

Fetch Product by Key
{"_id": "5c88c5ef0317e321c5f5f057", "productKey": "test20190313", "productPhoto"
: "TEST", "skuNumber": "TEST0000", "category": "test", "title": "Test", "descrip
tion": "test", "price": 1.11, "unit": "test", "costPerUnit": 2.22, "unitsOnHand"
: 333}

Product test20190313 updated successfully

{"_id": "5c88c5ef0317e321c5f5f057", "productKey": "test20190313", "productPhoto"
: "REVISED PHOTO", "skuNumber": "TEST0000", "category": "test", "title": "Test",
 "description": "test", "price": 2.22, "unit": "test", "costPerUnit": 2.22, "uni
tsOnHand": 333}

Product test20190313 deleted successfully


-------------------
(program exited with code: 0)
Press return to continue
```

As you can see, the original values for `productPhoto` and price were TEST and 1.11. After the update, the same fields now show REVISED PHOTO and 2.22. The code shown previously is located at `/path/to/repo/chapters/05/demo_sweetscomplete_product_read_add_edit_delete.py`.

Please refer to the following links for more information:
`pymongo.results.UpdateResult`: https://api.mongodb.com/python/current/api/pymongo/results.html#pymongo.results.UpdateResult.
`pymongo.collection.Collection.update_one()`: https://api.mongodb.com/python/current/api/pymongo/collection.html#pymongo.collection.Collection.update_one
`pymongo.collection.Collection.update_many()`: https://api.mongodb.com/python/current/api/pymongo/collection.html#pymongo.collection.Collection.update_many

Now, let's learn how to handle a purchase.

Handling a customer purchase

At first glance, adding a purchase seems like a trivial process. Looking deeper, however, it becomes clear that we need to somehow bring together information about the customer making the purchase, as well as the products to be purchased.

Altogether, the process of adding a new purchase entails the following:

1. Authenticate the customer.
2. Select products and quantities to be purchased.
3. Add the purchase to the database.

Before we can accomplish all of this, however, we need to introduce three new classes that provide web support: `web.session.Session`, `web.auth.Authenticate`, and `web.responder.Html`.

Web environment support classes

Later in this book, starting with `Chapter 7`, *Advanced MongoDB Database Design*, when we introduce our next company, *Book Someplace, Inc.*, we'll cover MongoDB integration into the popular Python framework known as *Django* (`https://www.djangoproject.com/`). For the small company Sweets Complete, however, all that is required is an extremely simple web environment. The first class to investigate is `Session`.

The web.session.Session class

As you might be aware, the HTTP protocol (see `https://tools.ietf.org/html/rfc2616`), upon which the World Wide Web is based, is *stateless* in nature. What that means is that each web request is treated as if it's the first request ever made. Accordingly, the protocol itself provides no way to retain information between requests. Therefore, here, we will define a classic authentication mechanism that uses information stored in an HTTP *cookie* to provide a reference to more information stored on the server.

> We will not cover web classes in detail as they are beyond the scope of this book (that is, they do not address MongoDB-driven application development). If you are interested, the source code for the strictly web-based classes referenced in this chapter can be found in `/path/to/repo/chapters/05/src/web` in the source code repository associated with this book.

The `web.session.Session` class (`/path/to/repo/chapters/05/src/web/session.py`) provides this support by way of the methods discussed here:

Method	Description
`genToken()`	Generates a random token that serves to uniquely identify the session.
`getToken()`	Returns the generated token. If it's not available, it reads the token from the token cookie.
`getTokenFromCookie()`	Uses `http.cookie.SimpleCookie` to read a cookie called `token`, if it exists.
`buildSessFilename()`	Builds a filename that consists of the base directory (supplied during class construction), the token, and a `.sess` extension.
`writeSessInfo()`	Writes the information supplied to the session file in JSON format.
`readSessInfo()`	Retrieves information from the session file and converts it from JSON into `dict()`.

setSessInfoByKey()	Uses readSessInfo() to retrieve the contents of the session file. Updates the dictionary with the new information based upon the key supplied. The method then executes writeSessInfo().
getSessInfoByKey()	Performs readSessInfo() to retrieve the contents of the session file. Returns the value of the key supplied, or *False* if the key doesn't exist.
cleanOldSessFiles()	This is called upon object construction to remove old session files.

The Session class is then used by the SimpleAuth class (described next) to store information when a user successfully logs in.

The web.auth.SimpleAuth class

As the name implies, this class is used to perform user authentication. The class constructor accepts the domain service class and a base directory as arguments. The base directory is then used to create a Session instance that the SimpleAuth class consumes. You can also see a getSession() method, which grants access to the internally stored Session instance. Likewise, the getToken() method is used to retrieve the token generated by Session.

The workhorse of this class is authByEmail(). It uses the domain service class to perform a lookup based on the customer's email address, used as part of the web login process (described later in this chapter). The method uses bcrypt.checkpw() to check the provided plaintext, UTF-8 encoded password, against the stored hashed password.

The authenticate() method brings everything together. It accepts an email address and plaintext password. The first thing to do is run authByEmail(). If successful, this method retrieves a Customer instance, the session class is used to generate a token, and the customer key is stored in the session. Otherwise, the method returns False. You see that the username argument, in this usage, is the user email.

What happens on the next request cycle after a successful login? As you may recall, successful authentication creates a token. The token needs to be set in a cookie. This action is performed by another class, discussed next. On the next request cycle, the user's browser presents the token cookie to the application. There needs to be a getIdentity() method that presents the customer key, stored in the session, to the domain service class to retrieve customer information.

The full Python code can be found at /path/to/repo/chapters/05/src/web/auth.py. The last class we'll cover here is used to respond to a web request. It contains methods that are useful when publishing an HTML web page. It also handles setting the token cookie mentioned previously.

The web.responder.Html class

The class constructor accepts the filename of the specified HTML document and serves as a template. As this class deals strictly with HTML (as compared to JSON output, for example), the `Content-Type` header is set to text/HTML.

The next two methods populate the headers and inserts lists. Headers represent HTTP headers that need to be sent upon output. The class constructor, shown previously, adds the first header to the list: the `Content-Type` header represents a list of placeholders in the HTML template document, which are subsequently replaced by a given value.

Finally, the `render()` method pulls everything together. First, the headers are assembled, separated by a carriage-return linefeed (`\r\n`), as mandated by the HTTP protocol. Next, the inserts are processed: the placeholders in the seed HTML template are replaced with their respective values. The rendered HTML is then returned for output by the calling program.

> The full Python code can be found at
> `/path/to/repo/chapters/05/src/web/responder.py`.

This concludes the discussion of support web classes. We will now turn our attention to the `PurchaseService` domain service class. Here, it's important to address the `pymongo.collection` methods, which can add documents to the database.

The pymongo.collection.Collection.insert_one() and insert_many() class

Adding a document to the database involves using either the `insert_one()` or `insert_many()` method from the `pymongo.collection.Collection` class. The generic syntax for `insert_one()` is as follows. The parameters enclosed in square brackets `[]` are optional:

```
insert_one(<document>, [<bypass_document_validation>, <session>])
```

The generic syntax for `insert_many()` is also shown here. The parameters enclosed in square brackets `[]` are optional:

```
insert_many(<iteration of documents>, [<ordered>, \
            <bypass_document_validation>, <session>])
```

The more important parameters are summarized here:

Parameter	Default	Notes
document(s)	--	For insert_one(), this is a single document. For insert_many(), this is an iteration of documents.
bypass_document_validation	False	If set to True, the write operation is instructed not to validate. This improves performance at the cost of the potential loss of data integrity.
session	None	If you supply a pymongo.client_session.ClientSession instance, you are able to initiate transaction support, which allows a set of database operations to be treated as a single atomic block. The session also gives you control over read and write concerns, which are important when operating on a replica set.
ordered	True	This parameter is only available for the insert_many() method. If set to True, the documents are inserted in the literal order presented in the iteration. In case of an error, all remaining inserts are not processed. If set to False, all documents inserts are attempted independently, regardless of the status of other inserts in the iteration. In this case, parallel processing is possible.

Please refer to the following links for more information:

pymongo.collection.Collection.insert_many(): https://api.mongodb.com/python/current/api/pymongo/collection.html#pymongo.collection.Collection.insert_many

pymongo.collection.Collection.insert_one(): https://api.mongodb.com/python/current/api/pymongo/collection.html#pymongo.collection.Collection.insert_one

Now, let's learn how to use this method in the PurchaseService domain service.

Inserting documents into the purchase domain service

In the sweetscomplete.domain.purchase.PurchaseService class, we need to define a method that adds a document to the database. As with the other domain service classes described earlier in this chapter, we import the Base domain service class, as well as the Purchase entity class. In addition, we need access to the ObjectId and CursorType classes, as shown here:

```
from sweetscomplete.domain.base import Base
from sweetscomplete.entity.purchase import Purchase
from bson.objectid import ObjectId
from pymongo.cursor import CursorType
class PurchaseService(Base) :
```

As you may recall from a discussion earlier in this chapter, a decision needs to be made about the extended price of a purchase. In this implementation, we calculate the extended price just prior to storage, as shown here. The addOne() method then returns the new ObjectId, or a boolean False value if the insertion failed:

```
def addOne(self, purchaseEntity) :
    extPrice = 0.00
    for key, purchProd in purchaseEntity['productsPurchased'].items() :
        extPrice += purchProd['qtyPurchased'] * purchProd['price']
    purchaseEntity['extendedPrice'] = extPrice
    result = self.db.purchases.insert_one(purchaseEntity)
    return result.inserted_id \
        if (isinstance(result.inserted_id, ObjectId)) else False
```

As a follow-up to a purchase, in this class, we also define a custom fetchLastPurchaseForCust() method that does a lookup of purchases by customer key in descending order. It then returns the first document, which represents the latest purchase:

```
def fetchLastPurchaseForCust(self, custKey) :
    query    = dict({"customerKey" : custKey})
    projection = None
    return self.db.purchases.find_one(
        query, projection, 0, 0, False,
        CursorType.NON_TAILABLE, [('dateOfPurchase', -1)])
```

Now, we will turn our attention to the web, where we'll do the following:

- Authenticate the customer.
- Present a list of products to be purchased.
- Process the purchase.

Web application for product purchase

In this section, we'll discuss how to incorporate the domain services class with the web support classes to produce a simple web application. To accomplish this, we'll create three Python scripts to handle different aspects of the three main phases. Each Python web script makes use of one (or more) of the domain service classes. The flow for this discussion is illustrated in the following diagram:

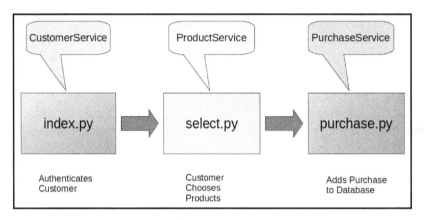

Before we dive into the coding details, let's have a quick look at configuring Apache for simple Python **Common Gateway Interface** (**CGI**) script processing.

Configuring Apache for Python CGI

Using the CGI process is a quick and simple way to get your Python code on the web. For the purposes of this illustration, we'll use Apache as an example. Please bear in mind that 40% of internet web servers are now running *Nginx*; however, due to space considerations, we will only present the Apache configuration here.

First, you need to install the Apache `mod_wsgi` module. Although this can be done through the normal package management for your operating system, it's recommended to install it using `pip3`. If you see a warning about missing operating system dependencies, make sure these are installed first before proceeding with this command:

```
pip3 install mod-wsgi
```

IMPORTANT: If you are using a Docker container, as outlined at the beginning of this chapter, this command is not needed as `mod-wsgi` is already installed in the provided Docker container. However, if you are setting up a Python web application on a customer server that is not using Django, you need to run this command to install `mod-wsgi`. The `mod-wsgi` installation documentation can be found at `https://modwsgi.readthedocs.io/en/develop/user-guides/quick-installation-guide.html`.

You then need to create an Apache configuration for CGI. Here is a sample configuration file. You can see the full version at `/path/to/repo/docker/learning.mongodb.local.conf`:

```
AddHandler cgi-script .py
<Directory "/path/to/repo/www/chapter_05">
    Options +FollowSymLinks +ExecCGI
    DirectoryIndex index.py
    AllowOverride All
    Require all granted
</Directory>
```

At the beginning of any Python script you plan to run using CGI, you need to place an operating system directive indicating the path to the executable needed to run the script. In the case of Linux, the directive might appear as follows:

`#!/usr/bin/python3`

Finally, make sure the web server user has the rights to read the script and that the script is marked as executable. Now, let's have a look at creating a web authentication infrastructure.

Simple customer authentication

When a customer wishes to make a purchase on the Sweets Complete website, they need to log in. For the purposes of this chapter, we'll present a very simple login system that accepts a username and password from an HTML form and verifies it against the customer's collection.

In `Chapter 7`, *Advanced MongoDB Database Design*, you'll learn how to tie MongoDB to Django and use Django's built-in authentication system with MongoDB as the authentication source.

We will start with a Python script called
`/path/to/repo/www/chapter_05/index.py` that's designed to run via the CGI process.
This script creates an instance of our customer domain service and uses the
`web.auth.Authenticate` class described earlier in this section. The very first line is
extremely important when running Python via CGI. It allows the web server to pass off
processing to the currently installed version of Python.

You'll notice that we tell Python where the source code is located and initialize important
variables:

```python
#!/usr/bin/python3
import os,sys
src_path = os.path.realpath("../../chapters/05/src")
sys.path.append(src_path)
success         = False
message         = "<b>Please Enter Appropriate Login Info</b>"
sess_storage    = os.path.realpath("../data")
```

Next, we enable error handling using the `cgitb` library, which displays errors in an easily
readable format, complete with backtrace and suggestions. We also create an instance of the
`web.responder` class, described earlier in this section:

```python
import cgitb
cgitb.enable(display=1, logdir=os.path.realpath("../data"))
from web.responder import Html
html_out = Html('templates/index.html')
```

Next, we use the `cgi` library to parse the form posting, looking for username and password
fields:

```python
import cgi
form = cgi.FieldStorage()
if 'username' in form and 'password' in form :
```

If these are present in the post data, we retrieve the values and set up a customer domain
service instance:

```python
username = form['username'].value
password = form['password'].value
from web.auth import SimpleAuth
from db.mongodb.connection import Connection
from sweetscomplete.domain.customer import CustomerService
from sweetscomplete.entity.customer import Customer
conn = Connection('localhost', 27017, Customer)
cust_service = CustomerService(conn, 'sweetscomplete')
```

We can then use the `SimpleAuth` class to perform authentication, as described earlier in this section. You can see that if authentication is successful, we add a header to the output responder class, which then causes a cookie containing the authentication token to be set. This is extremely important for the next request, where we want the user to be identified:

```
auth = SimpleAuth(cust_service, sess_storage)
if  auth.authenticate(username, password) :
    success = True
    html_out.addHeader('Set-Cookie: token=' + \
        auth.getToken() + ';path=/')
    message = "<b>HOORAY!  Successful Login.</b>"
else :
    message = "<b>SORRY!  Unable to Login.</b>"
```

The template used for this script is `/path/to/repo/www/templates/index.html`. Here is the body:

```
<div style="width: 60%;">
    <h1>Simple Login</h1>
    <hr>
    <div style="width:100%;float:left;">
    <a href="/chapter_05/select.py">PRODUCTS</a></div>
    <form name="simple_auth" method="post" action="/chapter_05/index.py">
        <div class="labels">Email Address:</div>
        <div class="inputs"><input type="email" name="username" /></div>
        <div class="labels">Password:</div>
        <div class="inputs"><input type="password" name="password" /></div>
        <div class="labels"><input type="submit" value="Login" /></div>
    </form>
    <hr>
    <br>%message%
</div>
```

From a web browser on your local computer, enter the following URL:

```
http://learning.mongodb.local/chapter_05/index.py
```

The rendered output appears as follows:

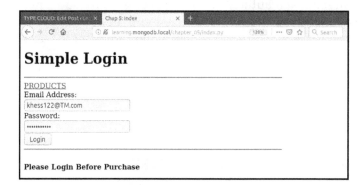

After clicking the **Login** button, a success message will appear. If you examine the `/path/to/repo/ww/data` folder, you will see that a new session file has been created. The contents of the file might appear as follows:

```
File Edit View Search Terminal Help
root@server1:/# ls -l /repo/www/data
total 4
-rw-r--r-- 1 www-data www-data 27 May  4 04:54 5cd00efc47474e01b029ea9aac6255f6.sess
root@server1:/# cat /repo/www/data/5cd00efc47474e01b029ea9aac6255f6.sess
{"custKey": "KURTHESS8667"}root@server1:/#
root@server1:/#
```

The next subsection deals with presenting a list of products for the customer to purchase.

Selecting products from a web page

Now, we'll turn our attention to the main page, following a successful login. The first bit of coding that needs to be done is in the `sweetscomplete.domain.product.ProductService` class. We need to add a method that returns product information in a format suitable for web consumption. The objective is to be able to display product titles and prices. Once selected, the web form returns the selected product key and quantity to purchase.

Here is the new `fetchAllKeysAndTitlesForSelect()` method. It uses the
`pymongo.collection.Collection.find()` method and supplies a projection argument
that includes only the `productKey`, `title`, and `price` fields. The return value is stored as
a Python dictionary in the `keysAndTitles` class variable:

```
# sweetscomplete.domain.product.ProductService
def fetchAllKeysAndTitlesForSelect(self) :
    if not self.keysAndTitles :
        projection = dict({"productKey":1,"title":1,"price":1,"_id":0})
        for doc in self.db.products.find(None, projection) :
            self.keysAndTitles[doc.getKey()] = doc.get('title') + \
                ' [ $' + str(doc.get('price')) + ' ]'
    return self.keysAndTitles
```

Next, we make an addition to the `web.responder.Html` class described earlier in this
chapter. The `buildSelect()` method accepts a parameter's name, which is used to
populate the name attribute of the HTML select element. The second argument comes
directly from the product domain service method we just described; that
is, `fetchAllKeysAndTitlesForSelect()`. The return value of the new `buildSelect()`
method is an HTML select element. The product key is returned when a product is selected:

```
def buildSelect(self, name, dropdown) :
    output = '<select name="' + name + '">' + "\n"
    for key, value in dropdown.items() :
        output += '<option value="' + key + '">' + value
        output += '</option>' + "\n"
    output += '</select>' + "\n"
    return output
```

We can now examine the Python `/path/to/repo/www/chapter_05/select.py` web
script. As with the login script described earlier, we include an opening tag that tells the
operating system how to execute the script. We follow that by including our source code in
the Python system path and enabling errors to be displayed:

```
#!/usr/bin/python3
import os,sys
src_path = os.path.realpath('../../chapters/05/src')
sys.path.append(src_path)
import cgitb
cgitb.enable(display=1, logdir='../data')
```

Next, we perform imports and set up two database connections – one to the `customers`
collection and another to `products`:

```
from web.responder import Html
from web.auth import SimpleAuth
```

```
from db.mongodb.connection import Connection
from sweetscomplete.domain.customer import CustomerService
from sweetscomplete.entity.customer import Customer
from sweetscomplete.domain.product import ProductService
from sweetscomplete.entity.product import Product

cust_conn    = Connection('localhost', 27017, Customer)
cust_service = CustomerService(cust_conn, 'sweetscomplete')
prod_conn    = Connection('localhost', 27017, Product)
prod_service = ProductService(prod_conn, 'sweetscomplete')
```

We then initialize the key variables and set up our output responder class:

```
sess_storage   = os.path.realpath("../data")
auth           = SimpleAuth(cust_service, sess_storage)
cust           = auth.getIdentity()
html_out       = Html('templates/select.html')
html_out.addInsert('%message%', '<br>')
```

Now for the fun part! First, we use the authentication class to retrieve the user's identity. As you may recall from our discussion earlier, if the user has successfully authenticated, a session is created and propagated via a cookie. An unauthenticated website visitor sees nothing.

For customers, we pull our list of product keys and titles from the products domain service. Here, we make an assumption that a typical purchase has a maximum of 6 line items. In the *real world*, of course, we would most likely use a JavaScript frontend that allows customers to add additional purchase fields at will. That discussion is reserved for the next chapter, Chapter 6, *Using AJAX and REST to Build a Database-Driven Website*. For the purposes of this chapter, we'll keep things simple and limit the selection to 6. Finally, the output is rendered:

```
cust = auth.getIdentity()
if cust :
    html_out.addInsert('%name%', cust.getFullName())
    keysAndTitles = prod_service.fetchAllKeysAndTitlesForSelect()
    for x in range(0, 6) :
        htmlSelect = html_out.buildSelect('dropdown' + str(x+1), \
                                          keysAndTitles)
        html_out.addInsert('%dropdown' + str(x+1) + '%', htmlSelect)
else :
    html_out.addInsert('%name%', 'guest')

print(html_out.render())
```

The HTML template we used is `/path/to/repo/www/templates/index.html`, the body of which is shown here:

```html
<h1>Welcome to Sweets Complete</h1>
You are logged in as: %name%
<a href="/chapter_05/index.py">LOGIN</a>
<form name="products" method="post" action="/chapter_05/purchase.py">
    Choose Product(s) to Purchase:
    <br>%dropdown1%   Quantity:
        <input type="number" name="qty1" value="1"/>
    <br>%dropdown2%   Quantity:
        <input type="number" name="qty2" value="0"/>
    <!-- other inputs not shown -->
    <br><input type="submit" name="purchase" value="Purchase" />
</form>
```

Here is how the output appears:

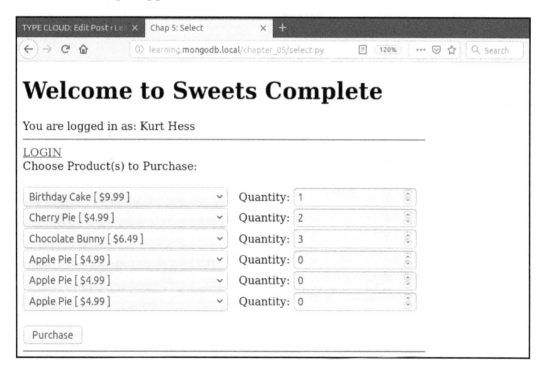

Obviously, you would use much cleverer styling to improve the appearance! For the purposes of this book, however, we are using only basic HTML and minimal CSS. The last thing we need to do is purchase selections and process them.

Capturing and processing purchase information from a web form

In order to provide the customer with information on their last purchase, in the domain service class for purchases, we need to add the following method. The technique for finding the last purchase involves including a `sort` parameter in the `pymongo.collection.Collection.find_one()` method, which tells the driver to sort by `dateOfPurchase` in descending order:

```
def fetchLastPurchaseForCust(self, custKey) :
    query  = dict({"customerKey" : custKey})
    projection = None
    return self.db.purchases.find_one(query, projection, 0, 0, \
        False, CursorType.NON_TAILABLE, [('dateOfPurchase', -1)])
```

Returning to the HTML responder class, we add another custom method that pulls information out of the `Purchase` object that's presented as an argument. Note that `productsPurchased` is itself a dictionary of `ProductPurchase` objects. Accordingly, a `for` loop is needed for extraction:

```
def buildLastPurchase(self, purchInfo) :
    output = ''
    output += '<h3>Purchase Info</h3>' + "\n"
    output += '<table width="50%">' + "\n"
    output += '<tr><th>Date</th><td>'
    output += purchInfo.get('dateOfPurchase')+'</td></tr>' + "\n"
    output += '<tr><th>Total</th><td>'
    output += str(purchInfo.get('extendedPrice'))+'</td></tr>' + "\n"
    output += '</table>' + "\n"
    output += '<h3>Product(s)</h3>' + "\n"
    output += '<table width="60%">' + "\n"
    output += '<tr><td><i>Title</i></td>'
    output += '<td><i>Quantity</i></td>'
    output += '<td><i>Price</i></td></tr>' + "\n"
    for key, prodInfo in purchInfo.get('productsPurchased').items() :
        output += '<tr>'
        output += '<td>' + prodInfo['title'] + '</td>'
        output += '<td>' + str(prodInfo['qtyPurchased']) + '</td>'
        output += '<td>' + str(prodInfo['price']) + '</td>'
        output += '<tr>' + "\n"
    output += '</table>' + "\n"
    return output
```

The `/path/to/repo/www/chapter_05/purchase.py` web script starts much like the others shown previously:

```
#!/usr/bin/python3
import os,sys
src_path = os.path.realpath("../../chapters/05/src")
sys.path.append(src_path)
import cgitb
cgitb.enable(display=1, logdir=".")
```

The imports are more complicated. In order to build a `Purchase` instance, we need access to all three domain service classes for customers, products, and purchases:

```
from web.responder import Html
from web.auth import SimpleAuth
from db.mongodb.connection import Connection
from sweetscomplete.domain.customer import CustomerService
from sweetscomplete.entity.customer import Customer
from sweetscomplete.domain.product import ProductService
from sweetscomplete.entity.product import Product
from sweetscomplete.entity.product_purchased import ProductPurchased
from sweetscomplete.domain.purchase import PurchaseService
from sweetscomplete.entity.purchase import Purchase
```

Then, we need to build instances of these services, as shown here. Note that, in each case, we are creating a connection to a *mongod* instance running on the localhost. Each connection instance represents a connection to a different MongoDB collection: `customers`, `products`, and `purchases`. The third argument causes the PyMongo driver to return results in the form of the corresponding Python entity classes we defined previously, each associated with a specific collection. Here is the code that creates these connections:

```
cust_conn     = Connection('localhost', 27017, Customer)
cust_service  = CustomerService(cust_conn, 'sweetscomplete')
prod_conn     = Connection('localhost', 27017, Product)
prod_service  = ProductService(prod_conn, 'sweetscomplete')
purch_conn    = Connection('localhost', 27017, Purchase)
purch_service = PurchaseService(purch_conn, 'sweetscomplete')
```

Next, we set up session storage, create a `web.authenticate.SimpleAuth` instance, and check the customer's identity. An instance of our output HTML responder class is also created. Note that if there is no identity, we redirect back home:

```
sess_storage = os.path.realpath("../data")
auth         = SimpleAuth(cust_service, sess_storage)
cust         = auth.getIdentity()
html_out     = Html('templates/purchase.html')
if not cust :
    print("Location: /chapter_05/index.py\r\n")
    print()
    quit()
```

We now scan the CGI input to see if a field called `purchase`, which is the name of the submit button, has been posted:

```
import cgi
form = cgi.FieldStorage()
if 'purchase' in form :
```

If so, we are now ready to start building the `Purchase` instance. We start by populating `dateOfPurchase` with today's date and time. We get customer information from the identity provided by the authenticator. This information is used to form the `transactionId`:

```
import datetime
import random
today = datetime.datetime.today()
dateOfPurchase = today.strftime('%Y-%m-%d %H:%M:%S')
ymd    = today.strftime('%Y%m%d')
fname = cust.get('firstName')
lname = cust.get('lastName')
transactionId  = ymd + fname[0:4] + lname[0:4] + \
                 str(random.randint(0, 9999))
```

The `Purchase` instance is created and populated with `Customer` information first:

```
purchase = Purchase()
purchase.set('transactionId',  transactionId)
purchase.set('dateOfPurchase', dateOfPurchase)
purchase.set('customerKey',    cust.getKey())
purchase.set('firstName',      cust.get('firstName'))
purchase.set('lastName',       cust.get('lastName'))
purchase.set('phoneNumber',    cust.get('phoneNumber'))
purchase.set('email',          cust.get('email'))
purchase.set('streetAddressOfBuilding',
             cust.get('streetAddressOfBuilding'))
```

```
purchase.set('city',          cust.get('city'))
purchase.set('stateProvince', cust.get('stateProvince'))
purchase.set('locality',      cust.get('locality'))
purchase.set('country',       cust.get('country'))
purchase.set('postalCode',    cust.get('postalCode'))
purchase.set('latitude',      cust.get('latitude'))
purchase.set('longitude',     cust.get('longitude'))
```

We then loop through the purchase information that's posted, only posting items where the quantity is greater than zero:

```
productsPurchased = dict()
for x in range(1,7) :
    qtyKey  = 'qty' + str(x)
    if qtyKey in form :
        qty = int(form[qtyKey].value)
        if  qty > 0 :
            dropKey = 'dropdown' + str(x)
            prodKey = form[dropKey].value
            prodInfo = prod_service.fetchByKey(prodKey)
            if prodInfo :
                prodPurch = ProductPurchased(prodInfo.toJson())
                prodPurch.set('qtyPurchased', qty)
                productsPurchased[prodKey] = prodPurch
```

The newly created dictionary of `ProductPurchased` instances is inserted into the `Purchase` object and saved to the database:

```
purchase.set('productsPurchased', productsPurchased)
purch_service.addOne(purchase)
```

Finally, some output is produced using the HTML output responder class:

```
if cust :
    html_out.addInsert('%message%', '<b>Thanks for your purchase!</b>')
    html_out.addInsert('%name%', cust.getFullName())
    mostRecent = html_out.buildLastPurchase( \\
        purch_service.fetchLastPurchaseForCust(cust.getKey()))
    html_out.addInsert('%purchase%', mostRecent)
else :
    html_out.addInsert('%name%', 'guest')
print(html_out.render())
```

The body of the associated web
template, `/path/to/repo/www/chapter_05/templates/purchase.html`, is shown
here:

```
<div style="width: 60%;">
    <h1>Welcome to Sweets Complete</h1>
    You are logged in as: %name%
    <a href="/chapter_05/select.py">SELECT</a>
    <h2>Here is a your most recent purchase:</h2>
    <hr>
    %purchase%
    <hr>
    <br>%message%
</div>
```

And finally, here is the output:

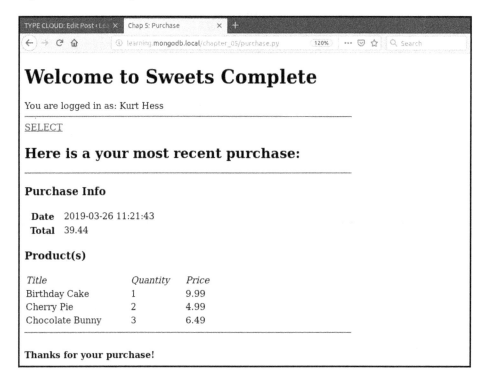

Now, let's turn our attention to the cancellation process.

Canceling a purchase

There are two ways to handle canceling a purchase. One technique is to hold purchase information in a temporary file, or in session storage, until the customer is ready to confirm this. Once confirmed, the purchase information can be stored permanently in the database after payment confirmation. Another technique we will discuss here is storing the purchase information immediately and then providing the customer with a cancel option. If a purchase is canceled, we simply delete the document from the database.

Before we approach the changes that need to be made to the `/path/to/repo/www/chapter_05/purchase.py` web script, we must first discuss the `pymongo.collection.Collection.delete_*` methods.

The pymongo.collection.Collection.delete_one() and delete_many() methods

Oddly, both the `delete_one()` and `delete_many()` methods have a lot in common with `find()`: they both accept a filter dictionary, which is identical to a query document. Here is the syntax summary for these commands:

```
delete_one(filter, collation=None, session=None)
delete_many(filter, collation=None, session=None)
```

The main difference between these two commands is how many documents they are capable of deleting: one or many. Here is a brief summary of the command syntax:

Parameter	Default	Notes
filter	None	A dictionary consisting of key/value pairs, where the key represents the field to search and the value represents the search criteria. If set to None, all the documents in the collection are selected.
collation	None	You can supply a `pymongo.collation.Collation` instance, which allows you to influence how search and sort operate. This is especially useful if you are collecting information on a website that services an international clientele.
session	None	If you supply a `pymongo.client_session.ClientSession` instance, you are able to initiate transaction support, which allows a set of database operations to be treated as a single atomic block. The session also gives you control over read and write concerns, which are important when operating on a replica set.

Please refer to the following links for more information:
pymongo.collection.Collection.delete_one(): https://api.
mongodb.com/python/current/api/pymongo/collection.html#pymongo.
collection.Collection.delete_one
pymongo.collection.Collection.delete_many(): https://api.
mongodb.com/python/current/api/pymongo/collection.html#pymongo.
collection.Collection.delete_many

Integrating purchase cancellation into the web flow

The first order of business is to add a `Cancel` button to the last page in the web flow described earlier. We add this to the
`/path/to/repo/www/chapter_05/templates/purchase.html` HTML template:

```
<form action="/chapter_05/purchase.py" method="post">
<input type="submit" name="cancel" value="Cancel" /></form>
```

We then modify `purchase.py` to scan for this extra input in the form post data. The main changes involve changing the message inserts to account for a cancellation. We also scan for the presence of `cancel` in the form:

```
if cust :
    html_out.addInsert('%name%', cust.getFullName())
    lastPurch  = purch_service.fetchLastPurchaseForCust(cust.getKey())
    if not lastPurch :
        html_out.addInsert('%purchase%', 'No recent purchase found')
    else :
        mostRecent = html_out.buildLastPurchase(lastPurch)
        html_out.addInsert('%purchase%', mostRecent)
        if 'cancel' in form :
            if purch_service.deleteByKey(lastPurch.getKey()) :
                html_out.addInsert('%message%', \
                    '<b>Purchase cancelled.</b>')
            else :
                html_out.addInsert('%message%', \
                    '<b>Sorry! Unable to cancel purchase.</b>')
        else :
            html_out.addInsert('%message%', \
                '<b>Thanks for your purchase!</b>')
else :
    html_out.addInsert('%name%', 'guest')

print(html_out.render())
```

The resulting output reverts to the purchase prior to the recently canceled one:

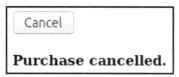

This brings us to the end of this chapter. Let's summarize what we have covered in this chapter.

Summary

In this chapter, you learned how to create a set of database classes that can then be used throughout your application. The Connection class provides access to a pymongo.mongo_client.MongoClient instance, from which you can access a database and collection. You then learned how to define domain service classes, each of which provides methods that are pertinent to a MongoDB collection. In many cases, these classes provide wrappers for various pymongo.collection.Collection.* methods, including find(), find_one(), update_one(), update_many(), insert_one(), insert_many(), delete_one(), and delete_many().

You were then shown how to model a query first in the *mongo* shell using JavaScript and then how to adapt it to Python using the domain services classes. First, you learned how to generate a product sales report. Then, you were shown how to use the domain service classes to update a product.

Finally, the last two sections introduced a set of classes that provide support for a web application. In the example covered in this chapter, you learned how to authenticate a customer using CustomerService. You then learned how to use ProductService to generate a list of products available for purchase. Finally, you learned how to use all three domain service classes to build a Purchase instance that was then saved to the database. After that, we provided a short section on canceling a purchase.

The next chapter moves you to the next level of web development by showing you how to use a responder class to produce **JavaScript Object Notation (JSON)** output, and a web environment that can accept **Asynchronous JavaScript and XML (AJAX)** queries or **Representation State Transfer (REST)** requests.

6
Using AJAX and REST to Build a Database-Driven Website

In this chapter, you will move on to the next level, away from a standard web application and into what is needed to have the application respond to AJAX queries and **REpresentational State Transfer** (**REST**) requests. The tasks presented in this chapter build on the example presented in the previous chapter, Chapter 5, *Mission-Critical MongoDB Database Tasks*.

In this chapter, we will first introduce you to the concept of pagination, and how it might be applied to a purchase history. You will then learn how to incorporate jQuery DataTables into a list of products for purchase. You will also be shown how to serve graphics directly from the database. Finally, you will learn how to handle a request coming from a fictitious external partner using REST.

In this chapter, we will cover the following topics:

- Paginating a customer's purchase history
- Writing code to accept an AJAX request from jQuery DataTables
- Serving graphics directly from the database
- Responding to REST requests

Technical requirements

The minimum recommended hardware is as follows:

- A desktop PC or laptop
- 2 GB of free disk space
- 4 GB of RAM
- 500 KBPS or faster internet connection.

The software requirements are as follows:

- **OS (Linux or macOS)**: Docker, Docker Compose, and Git (optional)
- **OS (Windows)**: Docker for Windows and Git for Windows
- Python 3.x, a PyMongo driver, and Apache (already installed in the Docker container used for this book)

The installation of the required software and how to restore the code repository for this book is explained in Chapter 2, *Setting Up MongoDB 4.x*. To run the Python code examples in this chapter from the command line, open a Terminal window (Command Prompt) and enter these commands:

```
cd /path/to/repo
docker-compose build
docker-compose up -d
docker exec -it learn-mongo-server-1 /bin/bash
```

To run the demo website described in this chapter, follow the setup instructions given in Chapter 2, *Setting Up MongoDB 4.x*. Once the Docker container for chapters 3 through 12 has been configured and started, there are two more things you need to do:

1. Add the following entry to the local hosts file on your computer:

 The local hosts file is located at C:\windows\system32\drivers\etc\hosts on Windows, and for Linux and macOS, at /etc/hosts.

```
172.16.0.11    learning.mongodb.local
```

2. Open a browser on your local computer (outside of the Docker container) to this URL:

```
http://learning.mongodb.local/
```

3. Click on the link for chapter 6.

When you are finished working with the examples covered in this chapter, return to the Terminal window (Command Prompt) and stop Docker, as follows:

```
cd /path/to/repo
docker-compose down
```

The code used in this chapter can be found in this book's GitHub repository at `https://github.com/PacktPublishing/Learn-MongoDB-4.x/tree/master/chapters/06`, as well as `https://github.com/PacktPublishing/Learn-MongoDB-4.x/tree/master/www/chapter_06`.

Paginating MongoDB database queries

Fortunately, gone are the bad old days where developers would just display all the results of a database query, regardless of the number of lines. The problem with this approach is that website's visitors were forced to scroll endlessly to find what they were looking for among the results. In order to provide a better experience for your website visitors, you can apply **pagination** to your query results. The idea is simple: on any given web page, only display a preset number of lines. At the bottom of the page, the website visitors see a navigation control that allows them to move between pages of data.

To implement pagination, you need to set a parameter reflecting how many items you want to appear on a theoretical page. A common setting is 20 items per page. The objective is to scan through the query result set 20 results at a time, with options to go forward and backward.

Modifying the purchases domain service

Returning to the `pymongo.collection.Collection.find()` method, you might recall two options we have not yet used: `skip` and `limit`. These two options allow us to achieve pagination. In the `sweetscomplete.domain.purchase.Purchase` domain service, we add a new method, `fetchPaginatedByCustKey()`. In the new method, we accept a customer key as an argument in order to build a query:

```
# sweetscomplete.domain.purchase.Purchase
    def fetchPaginatedByCustKey(self, custKey, skip, limit) :
        query = {'customerKey' : custKey}
        for doc in self.db.purchases.find(query, None, skip, limit) :
            yield doc
```

We also accept arguments for `skip` and `limit`, which is crucial for pagination to occur properly.

Adding pagination to the responder class

The next step is to add a method to the `web.responder.html.Html` class that accepts the output produced by `fetchPaginatedByCustKey()` and returns HTML. We also need to accept the current page number and a base URL to create links to the previous and next pages. Note that if the previous page number drops below 0, we simply reset it to 0.

The method might appear as follows:

```
# web.responder.html.Html
def buildPurchaseHistory(self, purchHist, pageNum, baseUrl) :
    import locale
    nextPage = pageNum + 1
    prevPage = pageNum - 1
    if prevPage < 0 : prevPage = 0
```

 We do not define an upper limit for the pagination in this illustration, so the link for the next page might go past the end of the query result. In our example, in this case, a page with no results is displayed. If an upper limit is needed, create a domain service method that returns the value of `pymongo.collection.count_documents()`. You could then multiply the lines per page by the page number and check to see whether the result is less than the total number of documents in the query.

Next, we begin developing the output by defining some styles and displaying a banner:

```
output = ''
output += '<style>' + "\n"
output += '.col1 { width:20%;float:left;font-size:10pt; }' + "\n"
output += '.col2 { width:10%;float:left;font-size:10pt; }' + "\n"
output += '.col3 { width:70%;float:left;font-size:10pt; }' + "\n"
output += '.full { width:100%;float:left;font-size:10pt; }' + "\n"
output += '</style>' + "\n"
output += '<h3>Purchase History</h3>' + "\n"
output += '<div class="col1"><b>Date</b></div>'
output += '<div class="col2"><b>Total</b></div>'
output += '<div class="col3"><b>Products</b></div>' + "\n"
```

The outer loop iterates through the results of the query to the `purchases` collection. The inner loop goes through the products purchased. The final output is appended to a string, represented by the `output` variable, as shown here:

```
for purchase in purchHist :
    output += '<div class="col1">' + \
            purchase.get('dateOfPurchase') + '</div>'
    output += '<div class="col2">'
    output += locale.format('%.*f', \
            (2, purchase.get('extendedPrice')), True)
    output += '</div>'
    output += '<div class="col3">'
    output += '<div class="col3"><i>Title</i></div>'
    output += '<div class="col2"><i>Qty</i></div>'
    output += '<div class="col1"><i>Price</i></div>' + "\n"
    for key, prodInfo in purchase.get('productsPurchased').items()

        output += '<div class="col3">'
        output += prodInfo['title'] + '</div>'
        output += '<div class="col2">'
        output += locale.format('%4d', \
                prodInfo['qtyPurchased']) + '</div>'
        output += '<div class="col1">'
        output += locale.format('%.*f', \
                (2, prodInfo['price']), True) + '</div>'
    output += '</div>' + "\n"
```

Finally, we present the **navigation controls** in the form of simple links with a `page` parameter:

```
# output control
output += '<div class="full">'
output += '<a href="' + baseUrl + '?page='
output += str(prevPage) + '">Previous</a>'
output += '  | ' + str(pageNum)
output += ' |  '
output += '<a href="' + baseUrl + '?page='
output += str(nextPage) + '">Next</a>'
output += '</div>'
return output
```

Obviously, the output could use refinements in the form of JavaScript functions to show or hide details on the products purchased. Background colors and styling can improve the appearance as well. We forego that to simplify the discussion and retain focus on our main goal: demonstrating how to get a Python web script to produce paginated output generated by MongoDB.

Consolidating configuration

Before we get to the action script, we will introduce the concept of storing configuration in one place. This is a useful technique as it allows you to define a dictionary of values that change from site to site. This fosters reusability and makes maintenance easier as all such parameters are in one place. For this purpose, here is the path, `/path/to/repo/chapters/06/src/config/config.py`, in which a class, `Config`, is defined. It has a `config` property and a single method, `getConfig()`:

```
# config.config
import os
class Config :
    config = dict({
        'db' : {
            'host'     : 'localhost',
            'port'     : 27017,
            'database' : 'sweetscomplete'
        },
        'pagination' : {
            'lines' : 4,
            'baseUrl' : '/chapter_06/history.py'
        },
        'session' : {
            'storage' : os.path.realpath('../../www/data')
        }
    })
    def getConfig(self, key = None) :
        if key :
            return self.config[key]
        return self.config
```

We then use this class and its configuration in the new script, described next.

> All the Python CGI web action scripts described in this chapter must give the web server user the `execute file` permission!

Adding pagination to a web application

We will now return to the web application developed in Chapter 5, *Mission-Critical MongoDB Database Tasks*. To keep things simple, we add a new web action script, `/path/to/repo/www/chapter_06/history.py`, which presents a paginated list of a customer's purchase history using the classes and methods described previously.

As we are following the **action-domain-responder** software design pattern, mentioned briefly in the previous chapter, you need to understand that the web scripts represent the *action* component. These scripts intercept web requests, consult with the domain service, and use the responder to produce output. Refer to `https://github.com/pmjones/adr`.

As with the other scripts, we import the classes described in the previous chapter, the `web.responder.html.Html` class, as well as `web.auth.SimpleAuth`. The `cgitb` Python module is used to display errors and should be removed from a live server:

```python
#!/usr/bin/python
import os,sys
src_path = os.path.realpath("../../chapters/06/src")
sys.path.append(src_path)
import cgitb
cgitb.enable(display=1, logdir=".")
from config.config import Config
from web.auth import SimpleAuth
from web.responder.html import Html
from db.mongodb.connection import Connection
from sweetscomplete.domain.customer import CustomerService
from sweetscomplete.entity.customer import Customer
from sweetscomplete.domain.purchase import PurchaseService
from sweetscomplete.entity.purchase import Purchase
```

Next, we create a `CustomerService` instance used for authentication. Of course, `PurchaseService` is used to return the customer's purchase history. Note how we use the new `Config` class to supply parameters instead of hardcoding them in the following code:

```python
config        = Config()
db_config     = config.getConfig('db')
cust_conn     = Connection(db_config['host'], db_config['port'], Customer)
cust_service  = CustomerService(cust_conn, db_config['database'])
purch_conn    = Connection(db_config['host'], db_config['port'], Purchase)
purch_service = PurchaseService(purch_conn, db_config['database'])
sess_config   = config.getConfig('session')
auth          = SimpleAuth(cust_service, sess_config['storage'])
cust          = auth.getIdentity()
response      = Html('templates/history.html')
```

We use `cgi` to pull the page number. As you will recall, the web responder class uses this page number in order to generate links for the previous and next pages. You might note that as a security measure, we force the page number to `int`. This prevents attackers from attempting a cross-site scripting attack (`https://www.owasp.org/index.php/Cross-site_Scripting_(XSS)`):

```
import cgi
info = cgi.FieldStorage()
if 'page' in info :
    pageNum = int(info['page'].value)
else :
    pageNum = 0
response.addInsert('%message%', \
    'You are currenty viewing page: ' + str(pageNum))
```

As you can see, the web script code is quite simple, mainly because all the work is being done by the `purchases` domain service and responder classes. The only real tricky bit is to calculate the `skip` value by multiplying the page number by the `limit` value:

```
if cust :
    response.addInsert('%name%', cust.getFullName())
    page_config = config.getConfig('pagination')
    limit = page_config['lines']
    skip = int(pageNum * limit)
    purchHist  = purch_service.fetchPaginatedByCustKey(
        cust.getKey(), skip, limit)
    if not purchHist :
        response.addInsert('%history%', 'No recent purchase found')
    else :
        historyHtml = response.buildPurchaseHistory(
            purchHist, pageNum, page_config['baseUrl'])
        response.addInsert('%history%', historyHtml)
else :
    response.addInsert('%name%', 'guest')
    response.addInsert('%history%', \
        'Need to login to view purchase history')
```

 Please note that in the code examples, we use the backslash convention (\\) to indicate that the code presented should be on one single contiguous line, where the length of the line exceeds the width of the page.

As with the other web scripts presented so far, we create a simple HTML template to be used with the script. That template is located in `/path/to/repo/www/chapter_06/templates/history.html`, and the body of it appears as follows:

```html
<div style="width: 60%;">
    <h1>Welcome to Sweets Complete</h1>
    You are logged in as: %name%
    <hr>
    <a href="/chapter_06/index.py">LOGIN</a> |
    <a href="/chapter_06/select.py">SELECT</a> |
    <a href="/chapter_06/purchase.py">RECENT</a>
    <hr>
    %history%
    <hr>
    <br>%message%
</div>
```

Here is the output for a fictitious customer, Yadira Conway (login email: `yconway172@Turkcell.com`):

Welcome to Sweets Complete

You are logged in as: Yadira Conway

LOGIN | SELECT | RECENT

Purchase History

Date	Total	Products		
2019-05-08	1236.57	*Title*	*Qty*	*Price*
11:48:07		Chocolate Chip Cookie	51	0.88
		Star And Sprinkles Donut	403	0.59
2017-12-03	4054.61	*Title*	*Qty*	*Price*
11:48:07		Layer Cake	190	2.99
		Birthday Cake	349	9.99
2017-02-20	1484.24	*Title*	*Qty*	*Price*
11:48:07		Cherry Pie	231	4.99
		Peanut Truffle	95	3.49
2019-01-22	28191.86	*Title*	*Qty*	*Price*
11:48:08		Cake With Chocolate Frosting	984	8.99
		Chocolate Bunny	918	6.49
		Apple Pie	536	4.99
		Dozen Donuts	379	6.99

Previous | 0 | Next
You are currenty viewing page: 0

In the pagination config, we set the number of lines to 4 to better illustrate how the pagination works. From the preceding screen, we then click **Next**, the page parameter in the URL goes to 1, and we see the remaining set of products:

Welcome to Sweets Complete

You are logged in as: Yadira Conway

LOGIN | SELECT | RECENT

Purchase History

Date	Total	Products	Qty	Price
2018-10-05 11:48:08	849.41	*Title*	*Qty*	*Price*
		Star And Sprinkles Donut	492	0.59
		Chocolate Bar	187	2.99
2019-03-16 11:48:08	1279.66	*Title*	*Qty*	*Price*
		Chocolate Chip Cookie	637	0.88
		Carrot Cake	90	7.99
2018-11-06 11:48:08	4852.28	*Title*	*Qty*	*Price*
		Schweineoehrchen Cookie	701	0.49
		Pumpkin Pie	789	4.99
		Star And Sprinkles Donut	317	0.59

Previous | 1 | Next
You are currenty viewing page: 1

In the next section, you will learn a slightly different way of designing your web application by introducing **jQuery DataTables**, which makes an AJAX request to the web application and expects a JSON response.

Handling AJAX requests

AJAX stands for **Asynchronous JavaScript and XML**. It is not a protocol in and of itself, but is rather a mixture of technologies that includes HTML and JavaScript. This mixture of technologies represents a major paradigm shift away from the traditional approach where a person sitting in front of a desktop PC opens their browser and makes a request of a website, which in turn delivers static HTML content.

Here is a diagram of the traditional request-response sequence:

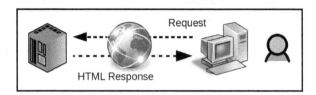

AJAX, on the other hand, turns the tables on the traditional request-response sequence illustrated in the preceding diagram. Instead, what happens is that the initial request delivers HTML content, including one or more JavaScript applications. The JavaScript applications run locally on the person's browser or smart device. From that point on, in most cases, it is the JavaScript applications that make small-scale asynchronous requests in the background.

Here is a diagram of an AJAX request-response sequence:

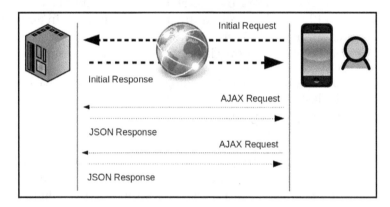

 It should be noted that most applications today use **JavaScript Object Notation (JSON)** as the data exchange format, rather than the older and more verbose **eXtensible Markup Language (XML)**. Refer to `https://www.json.org/json-en.html`.

There are two primary advantages to this approach:

- It avoids a full-page refresh of the HTML, CSS, and graphic images. JavaScript is able to repaint a small portion of the screen, leaving the rest intact.
- Because the background AJAX requests can operate asynchronously, the overall user experience is much improved. Overall performance tends to become balanced and lengthy waits for a server response are avoided.

In order to illustrate how this might play out using MongoDB, let's turn our attention back to the product selection page from the basic web application for Sweets Complete described in the previous chapter. Wouldn't it be nice to have a neatly styled, searchable, and paginated table of products for purchase, rather than the current, rather unsightly clump of HTML select elements? This is where jQuery DataTables comes into play.

A brief overview of jQuery and DataTables

jQuery is one of the most popular JavaScript libraries. It includes classes and functions that allow you to quickly traverse and manipulate a web page, without having to force a full-page refresh every time something changes. A minimal version is available to reduce the amount of overhead. It is actively maintained and is supported by a large variety of browsers.

If you look at the actual statistics on `https://github.com/jquery/jquery`, you will see that the library has almost 300 contributors, over 53,500 star ratings, and has been *forked* over 19,400 times (that means 19,400 other projects are using this library). In terms of the market share of all the websites surveyed, **W3Techs** places jQuery at over 75% as of July 2020. The next highest comparable JavaScript library is Bootstrap, which comes in at 21%.

The other thing to note when looking at the GitHub source code is the date of the last *commit*. If the last commit was very recent, it means the project is being actively maintained.

DataTables is a jQuery plugin, requiring jQuery in order to operate. It creates a widget in which there are rows of data displayed. The widget is fully configurable in terms of size, fonts, styles, and so forth. It can include a search box, supports pagination, and lets you set what columns can be used as sort criteria.

Documentation references:

- **jQuery:** `https://jquery.com/`.
- **jQuery DataTables:** `https://www.datatables.net/`
- **W3Techs:** `https://w3techs.com/technologies/history_overview/javascript_library/all/y`

Incorporating jQuery and DataTables

In order to incorporate jQuery and DataTables into the simple web application developed in the previous chapter, we turn our attention to `/path/to/repo/www/chapter_06/templates/select.html`. This is the web template associated with the product selection process. In the template, we need to do the following:

- Load a minimal version of jQuery.
- Load the DataTables plugin.
- Load a stylesheet used by DataTables.
- Add JavaScript, using jQuery, to initialize the data table and process table elements that are outside the DOM.

 DOM refers to the **Document Object Model**, which is used extensively by the JavaScript language. Its original intention was to model an HTML web page as a tree of objects. Although that is still its most popular use, variations of DOM now allow you to model XML documents, among others. The current standard is a W3C recommendation – the latest being version 4, introduced in April 2020 (`https://www.w3.org/DOM/DOMTR`).

In the HEAD portion of the HTML page, we include links to a `datatables` stylesheet:

```
<link rel="stylesheet" type="text/css" \\
    href="https://cdn.datatables.net/1.10.19/css/jquery.dataTables.css">
```

In addition, we need to create a skeleton template that DataTables uses to create the widget. There are two important things to note here. The first is that the entire data table is enclosed in an HTML form. The second is that unless you perform some manipulation in the JavaScript code used to initialize the data table, the number of columns must match the number being sent to the data table following an AJAX request:

```
<form id="form_products" method="post" action="/chapter_06/purchase.py">
    <p>Choose Product(s) to Purchase:</p>
    <table id="data_table" class="display">
        <thead>
            <tr>
                <th>Title</th>
                <th>Price</th>
                <th>Qty</th>
            </tr>
        </thead>
        <tbody>
            <tr>
                <td>Row 1 Data 1</td>
```

```
                    <td>Row 2 Data 2</td>
                    <td>Row 3 Data 3</td>
            </tr>
        </tbody>
    </table>
    <br><input type="submit" name="purchase" value="Purchase" />
</form>
```

Finally, we add JavaScript to initialize the data table. This is where we designate the data source as an AJAX request. The `%ajax_url%` parameter is substituted for an actual URL by the web action `select.py` script. In this illustration, we set the number of rows displayed by the data table to 6 in order to see the effect on pagination:

```
<script charset="utf8"
src="https://code.jquery.com/jquery-3.3.1.min.js"></script>
<script charset="utf8"
src="https://cdn.datatables.net/1.10.19/js/jquery.dataTables.js"></script>
<script type="text/javascript" charset="utf8">
$(document).ready( function () {
    table = $('#data_table').dataTable({
        'ajax'       : {'url':'%ajax_url%','type': 'GET'},
        'pageLength' : 6,
        'autoWidth'  : false
    });
});</script>
```

jQuery can be directly downloaded and included in a directory off the website's document root. The advantage to this approach is that your website is using locally available code, over which you have complete control. Also, the version is *frozen*, which means you don't have to worry about an update breaking something. The disadvantage is that you also have to perform periodic updates in order to apply security patches. Alternatively, you can simply include a link to the jQuery **Content Distribution Network (CDN)**, which gives you the latest release of your chosen version. For the purposes of this chapter, we will use the jQuery CDN.

For the purposes of this example, we will not dive deeply into jQuery DataTables configuration. The DataTables website is well-documented, and you can find plenty of examples there. The documentation on the DataTables options is located at `https://www.datatables.net/manual/options`.

Finally, we include JavaScript to ensure that all the data table elements are returned when the form is posted. Otherwise, the only data returned is the page currently showing on the screen. What this script does is add data table inputs that are not in the current DOM (that is, not showing up on the screen) as *hidden* `form` elements:

```
<script type="text/javascript" charset="utf8">
$('#form_products').on('submit', function(e){
    var form = this;
    var params = table.$('input').serializeArray();
    $.each(params, function(){
      if(!$.contains(document, this)){
          $(form).append(
              $('<input>').attr('type', 'hidden')
                          .attr('name', this.name).val(this.value)
          );}
}); }); </script>
```

Unfortunately, however, the hidden elements end up serialized as *lists*, which we need to deal with when form results are processed. Next, we will define a JSON responder class.

Defining a JSON responder class

As you have seen earlier in this chapter, we used a responder class that produced HTML. For the purposes of answering AJAX requests, we develop a responder class that produces JSON. For this purpose, we create a file, `/path/to/repo/chapters/06/src/web/responder/json.py`, in which we define a `JsonResponder` class.

As we plan to return JSON, in the `__init__()` method, we can add a `Content-Type` header set to `application/json`:

```
# web.responder.json
import json
class JsonResponder :
    headers   = []
    data      = []
    def __init__(self) :
        self.addHeader('Content-Type: application/json')
```

As with the `Html` responder class, we define a method that allows us to add additional headers. Instead of the `addInsert()` method used in the `Html` responder class, we define a method, `addData()`, which simply adds to an internal dictionary:

```
def addHeader(self, header) :
    self.headers.extend([header])
def addData(self, key, value) :
    self.data[key] = value
```

We also add a custom method that returns data in a format acceptable to DataTables. The `title` and `price` fields are straight out of the database query. You might note that the `qty` field, on the other hand, is actually HTML that ends up getting rendered inside the data table as a form input field. We append the product key to the `qty_` prefix and extract it later. Also, note that we append each entry for the products as a straight list, which is what the data table is expecting:

```
def addProductKeyTitlePrice(self, key, data) :
    temp = []
    for product in data :
        title = product.get('title')
        price = product.get('price')
        qty   = '<input class="qty" name="qty_'
        qty  += product.get('productKey')
        qty  += '" value="0" type="number" />'
        temp.append([title,price,qty])
    self.data[key] = temp
```

The internal dictionary, `data`, is then returned in JSON format in the `render()` method:

```
def render(self) :
    output = ''
    for item in self.headers :
        output += item + "\r\n"
    output += "\r\n"
    output += json.dumps(self.data)
    return output
```

Let's now learn how to deliver data from MongoDB.

Delivering data from MongoDB

To deliver data from MongoDB, we modify
the sweetscomplete.domain.product.ProductService domain service class. As per
the action-domain-responder pattern, the domain service has no knowledge of what format
the output takes. The only concern of the new method is to make sure it delivers a list of
sweetscomplete.entity.product.Product entities that contain the information
needed. Formatting is handled by the responder class.

The new method, fetchAllKeysTitlesPricesForAjax(), returns a list of products that
includes productKey, title, and price:

```
# sweetscomplete.domain.product.ProductService
def fetchAllKeysTitlesPricesForAjax(self, skip = 0, limit = 0) :
    projection = dict({"productKey":1,"title":1,"price":1,"_id":0})
    result = []
    for doc in self.db.products.find(None, projection, skip, limit) :
        result.append(doc)
    return result
```

In addition, we need a very simple method that only returns product keys:

```
def fetchKeys(self) :
    result = dict()
    projection = dict({"productKey":1,"_id":0})
    for doc in self.db.products.find(None, projection) :
        result[doc.getKey()] = doc.getKey()
    return result
```

We will use the preceding method later to reconstruct the names of data table form
elements associated with the quantity purchased.

Creating an AJAX handler web action script

The next step is to create a script that jQuery DataTables can call, and subsequently returns
the requested data in JSON format. This action script
(/path/to/repo/www/chapter_06/ajax.py) makes use of both the products domain
service as well as the new JSON responder class discussed in the previous sections.

As with the other web scripts we've described so far, we add an initial tag that tells the web server's CGI process how to run the script. We then perform the appropriate imports:

```python
#!/usr/bin/python
import os,sys
src_path = os.path.realpath('../../chapters/06/src')
sys.path.append(src_path)
from config.config import Config
from web.responder.json import JsonResponder
from web.auth import SimpleAuth
from db.mongodb.connection import Connection
from sweetscomplete.domain.product import ProductService
from sweetscomplete.entity.product import Product
```

Next, we use the new `Config` class to create the `Connection`, `Product`, and `ProductService` instances:

```python
config       = Config()
db_config    = config.getConfig('db')
prod_conn    = Connection(db_config['host'], db_config['port'], Product)
prod_service = ProductService(prod_conn, db_config['database'])
```

Finally, we run a query off `ProductService` and use the result to populate the data property of the `JsonResponder` class. We then output the rendered output:

```python
response = JsonResponder()
data     = prod_service.fetchAllKeysTitlesPricesForAjax()
response.addProductKeyTitlePrice('data', data)
print(response.render())
```

In order to test whether or not this script works, you can run it directly from your browser. Assuming you have followed the directions for creating a test environment, use this URL:
`http://learning.mongodb.local/chapter_06/ajax.py`

Here is the current output:

Next, we will learn how to modify the product select web action script.

Modifying the product select web action script

In order to activate the data table, we now need to make some modifications to the original script that presents a list of products to be purchased. As you recall from the last chapter, we simply presented six HTML select elements, with separate inputs for the quantities purchased. I think we can all agree this looks pretty bad! We now replace that logic with a single data table.

There are three main differences between /path/to/repo/www/chapter_05/purchase.py and the one we're examining for this illustration. The first is that we use the new Config class to provide a common configuration. The second major difference is that the responder class has been moved one level deeper. Instead of from web.responder.Html, the purchase script now defines this class as web.responder.html.Html:

```
from config.config import Config
from web.responder.html import HtmlResponder
```

The third difference is that instead of using the responder class `addInsert()` method to substitute six HTML select elements, we only need to provide a URL for the data table AJAX source, represented in the template as `%ajax_url%`:

```
response = HtmlResponder('templates/select.html')
response.addInsert('%message%', '<br>')
response.addInsert('%ajax_url%', config.getConfig('ajax_url'))
```

The remainder of the script remains the same, as described in the previous chapter. The output, however, is radically different, as shown here using the `http://learning.mongodb.local/chapter_06/select.py` URL:

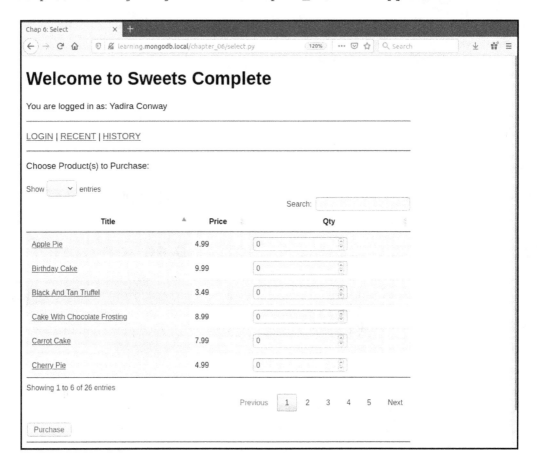

In the next section, we will learn how to modify the product purchase web action script.

Modifying the product purchase web action script

We can now turn our attention to the web script that processes purchases: /path/to/repo/www/chapter_06/purchase.py. This script was first introduced in Chapter 5, *Mission-Critical MongoDB Database Tasks*. As with the modified select.py script, described in the previous section of this chapter, we use the new Config service, as well as having moved the responder class, as shown:

```
from web.responder.html import HtmlResponder
html_out = HtmlResponder('templates/purchase.html')
```

The main difference in this version is how a list of the products purchased is built. As before, we initialize an empty dictionary, productsPurchased. We then add a dictionary obtained from the products domain service that contains product keys:

```
prodKeys = prod_service.fetchKeys()
productsPurchased = dict()
```

We loop through the list of product keys and recreate the quantity purchased form element name sent to the data table: qtyKey = 'qty_' + key. We then check to see whether this key is present in the form's HTTP POST data. If so, we retrieve that value, returned in the form of a cgi.MiniFieldStorage instance.

In this block of code, we loop through return values from the product keys database lookup. In the loop, we recreate the form of the field name by prepending the qty_ prefix. We then check to see whether this is present in the form posting. If so, it's retrieved into a variable, item:

```
for key, value in prodKeys.items() :
    qtyKey = 'qty_' + key
    if qtyKey in form :
        item = form[qtyKey]
```

As you learned from the discussion on how to recover all the data table elements not present in the DOM, they are appended as list items to the form. Accordingly, we need to check to see whether the item is an instance of list. If so, we retrieve element 0. We then force the quantity to the int data type, which has the effect of removing any potential XSS markup:

```
if isinstance(item, list) :
    item = item[0]
qty = int(item.value)
```

Finally, once we have the quantity, we check to make sure the value is greater than 0. If so, we look up the product information using the product key and the `products` domain service, and create a `ProductPurchased` instance:

```
if  qty > 0 :
        prodInfo = prod_service.fetchByKey(key)
        if prodInfo :
            prodPurch = ProductPurchased(prodInfo.toJson())
            prodPurch.set('qtyPurchased', qty)
            productsPurchased[key] = prodPurch
```

The remaining code from the original `purchase.py` script is exactly as presented in the previous chapter. The `ProductPurchased` instance is added to the `Purchase` instance, and then saved to the database. Also, all the output logic remains the same. The output from a completed purchase is not shown as it is exactly the same as presented in the previous chapter.

Next, we will have a look at how to serve binary data directly from MongoDB.

Serving binary data directly from MongoDB

When browsing through the sample data for products, you might have noticed a field, `productPhoto`, with an exceptionally large amount of seemingly gibberish data. Here is a screenshot from the first product in the collection showing part of the first lines of data for this field:

```
File  Edit  View  Search  Terminal  Help
> db.products.findOne({}, {"productPhoto":1,"_id":0});
{
        "productPhoto" : "iVBORw0KGgoAAAANSUhEUgAAASwAAADCCAYAAADzRP8zAAAACXBIWXMAAAsTAAALEwEAmpwYAAAABmJ
LR0QA/wD/AP+gvaeTAABn5ElEQVR42u1dB1gUV9feJGrsil1RsIKCggqCoIJYwIIdQbBAUHsYKcUFLH3XhNj77Ek9t6NURNjYkxPvvTe
83/l/Pe970DsMjM7W1lk7v0cJxGdZdmZdmZ+c+cp736HTacsb1ArMyzGoyq8usGrMi2mnRlra05UwLwDSZ2WlmfzIjvf2P2QNmGczkKa6dJW
9rSVn4uF2YLmP0iAik5+xezAcye006btrSlLUcvT2aPjIGpzZ1iFNisLkkV3aq0HRKf7k5+5+N0xV8sKn7NZmaltNOnLW1py1GrMrMVUDVVs
iTDm4cST8/Wkl/PVlHf3+wjv7vo/X0X0X0N9pvW+vSlJ7aadSWtroTliLuGdFWW//v3vrmJ4vv2C7r0k3oo9P9DC8zyznaaa6Q27r0zhEtxx2ePOu1h
YD+znw2Lb0PPPP0f6MLK/diq1pSlt2XM0h6ve+AFadwxbpzb0oApX8b+XXM3X9nxx39Vxxq0Q0TR0TQB1fmcW8Lv8fd/j24dTbwVGQ1/AA+F27d+uBD333
HN5KoVFXniekgeF0b/eWJwLbgP6tBR+/5iZt3Z6z6UtbdlygRx6VQChz24vAZVYsSQL07/J5KvVy1JikNFI9Y9N9K0DG/ux/LQ8RRw9pJJ+/zuR2
ZdtFOsLW1py5ZrkgA+u1YnSAKW4EEhXxxUe4kUvMO/KGLj8fdxz81+zJ3QXvDKEm+N1Gl9LW9rSLpxreWbBL7HV9dS2bh5JC5V9e9CVBE
/zyzXtol0Ja2tKXkRVXXXXe1FHmf2sk8hDl1aruwnNPVVw+9NV9V1Q1w1au3i3klUuWIZ4d0bOeMPPRLou2tKUtYSeGRXXQ+jZqyL6L2tKP+
zvG5yGF2tre.PjOb6rpVFj77J2aR2mXSIrYK70JSG/moEcyu62SalsuXLUn9ew9XSkS2j6PfHayTDOHvZZ7cWkr9PbeFY/qMHVC0ZyltFb
KQrxmztbocGkEekEIDc+/OzWnXqgT66Z0VDgUpY/vx4Urq3jLax+Pi26LRkvLa0VSiWB7N9uhydKg0QK13yReravgLtXXP39xbylltsx00
2aK85s2sCndszkR6ey6A/3nect4Wkfnx0K/HxImytpl10bWnr2VwQz1umE1X4QClo3NCVGxgxtT8e3JeEqC2KKAezbB8soY0J3J3cn0tkMcL
q1ShDPXu1JwObrRxJf7BQ0d6gBXCcMbqrmEH/ATNf7dJqS1vP1sKmfk+nb40JC/ak1Zlx9IR5R5TEpBDDVa9eqaEqUj1vNauVpTmpPrPPsZgb
```

This data represents Base64-encoded binary data, which, when decoded, is a photo of the product. In this section, we will show you how to cause a web script to display the product photo directly out of the database.

It is important to note that the technique described in this section has advantages and disadvantages over just uploading the photos directly to the images folder of your website's document root. The disadvantage is that this technique produces a larger HTML page, and the photo doesn't render as fast. The advantage is that you have more control over the photos as the database can be secured, and the only way an attacker can access the photo directly would be to do a screen scrape followed by a Base64 decode. This is often enough to discourage casual website visitors from attempting to rip off your photos.

Producing a Base64 image tag

There are two approaches to rendering a Base64-encoded image directly from the database. One technique is to create a standalone Python script that accepts a product key as a parameter, looks up the product, Base64 decodes the data, and outputs an appropriate `Content-Type` header for that image, along with the actual binary image data.

The second approach is to output the image using the following format:

```
<img src="data:image/xxx;base64,yyy" />
```

Here, `xxx` is the image type (for example, `image/png`) and `yyy` is the actual Base64 string. Although this approach puts the burden on the browser to perform the decoding, it minimizes the impact on the database and web server. For the purposes of illustration, we will use the second approach.

Accordingly, we return to our HTML `responder` class and add a method that produces the appropriate HTML image tag:

```
# web.responder.html.Html
mimeTypes    = dict({'png':'image/png','jpg':'image/jpeg'})
mimeDefault = 'jpg'
def buildBase64Image(self, name, mimeType, contents) :
    tag = '<img name="' + name + '" src="data:'
    if mimeType in self.mimeTypes :
        tag += self.mimeTypes[mimeType]
    else :
        tag += self.mimeTypes[mimeDefault]
    tag += ';base64,' + contents + '" />'
    return tag
```

Next, we will create an HTML template for the product details.

Creating an HTML template for the product details

We can now create an HTML template that gives us the product details. The body is shown here:

```
<h2>%title%</h2>
<table width="80%">
<tr>
    <td align="top">%photo%</td>
    <td align="top">
        <table>
            <tr><th>Product
Key</th><td> </td><td>%productKey%</td></tr>
            <tr><th>SKU
Number</th><td> </td><td>%skuNumber%</td></tr>
<tr><th>Category</th><td> </td><td>%category%</td></tr>
            <tr><th>Price</th><td> </td><td>%price%</td></tr>
            <tr><th>Unit</th><td> </td><td>%unit%</td></tr>
            <tr><th>On
Hand</th><td> </td><td>%unitsOnHand%</td></tr>
        </table>
    </td>
</tr>
</table>
<p>%description%</p>
```

Let's now create a display script.

Creating a display script

The next thing we do is define a standalone Python action script, `/path/to/repo/www/chapter_6/details.py`, which uses the product domain service and the graphic responder to render the photo and details on a specific product. As with the other scripts, we include a hashtag that instructs the web server's CGI process on how to handle the script. After that is a set of `import` statements:

```
#!/usr/bin/python
# tell python where to find source code
import os,sys
src_path = os.path.realpath('../../chapters/06/src')
sys.path.append(src_path)
import cgitb
cgitb.enable(display=1, logdir='../data')
```

```
from config.config import Config
from web.responder.html import HtmlResponder
from db.mongodb.connection import Connection
from sweetscomplete.domain.product import ProductService
from sweetscomplete.entity.product import Product
```

Next, we set up the database connection and product service using the `Config` class. We also initialize `HtmlResponder`:

```
config       = Config()
db_config    = config.getConfig('db')
prod_conn    = Connection(db_config['host'], db_config['port'], Product)
prod_service = ProductService(prod_conn, db_config['database'])
message      = 'SORRY! Unable to find information on this product'
response     = HtmlResponder('templates/details.html')
```

We can now check to see whether a product key was included in the URL. If so, we perform a lookup using the domain service:

```
import cgi
form = cgi.FieldStorage()
if 'product' in form :
    key = form['product'].value
    product = prod_service.fetchByKey(key)
```

If a product is found for this key, we use the responder's `addInsert()` method to perform replacements in the template:

```
if product :
    title   = product.get('title')
    message = 'Details on ' + title
    imgTag  = response.buildBase64Image('photo', 'png', \
        product.get('productPhoto'))
    response.addInsert('%photo%', imgTag)
    response.addInsert('%title%', title)
    response.addInsert('%productKey%', product.get('productKey'))
    response.addInsert('%skuNumber%', product.get('skuNumber'))
    response.addInsert('%category%', product.get('category'))
    response.addInsert('%price%', product.get('price'))
    response.addInsert('%unit%', product.get('unit'))
    response.addInsert('%unitsOnHand%', product.get('unitsOnHand'))
    response.addInsert('%description%', product.get('description'))
    response.addInsert('%message%', message)
    print(response.render())
```

Finally, we bail out to the main page if a product is not found or is not in the URL:

```
    else :
        print("Location: /chapter_06/index.py\r\n")
        print()
        quit()
else :
    print("Location: /chapter_06/index.py\r\n")
    print()
    quit()
```

Next, we will modify the AJAX response.

Modifying the AJAX response

At this point, it's time to revisit `web.responder.JsonResponder`. The objective is to modify the output sent to the data table in such a manner that all product titles become links to the detail page for that product. Accordingly, in `addProductKeyTitlePrice()`, notice how `title` is now wrapped in the following:

```
<a href="/chapter_06/details.py?product=PRODKEY">TITLE</a>
```

In the new method, shown here, we loop through `incoming` and extract information from the `product` instances. We build a list, `temp`, which is ultimately added to the internal dictionary, `data`. Here is the modified block of code:

```
def addProductKeyTitlePrice(self, key, incoming) :
    temp = []
    for product in incoming :
        prodKey = product.get('productKey')
        title   = '<a href="/chapter_06/details.py?product='
        title  += prodKey + '">' + product.get('title') + '</a>'
        price   = product.get('price')
        qty     = '<input class="qty" name="qty_' + prodKey
        qty    += '" value="0" type="number" />'
        temp.append([title,price,qty])
    self.data[key] = temp
```

Next, let's revisit the product selection action script.

Revisiting the product selection action script

Now, when we log in and choose the **SELECT** option, the title field in the data table is now a link. If we click on a link, we are taken to the new product details page, which displays the Base64-encoded photo, along with other pertinent details. The example shown here is the **Apple Pie** product:

If you then view the page source, you will see that the HTML `img` tag contains Base64 data. We have truncated the screenshot to conserve space in the book:

```
<h2>Birthday Cake</h2>
<table width="80%">
<tr>
    <td align="top"><img name="photo" src="data:image/png;base64,iVBORw0KGgoAAAANSUhEUgAAAP8AAAE1CAYAAAr2jQ3AAAA
    <td align="top">
        <table>
            <tr><th>Product Key</th><td> </td><td>birthday_cake</td></tr>
            <tr><th>SKU Number</th><td> </td><td>BIRT104</td></tr>
            <tr><th>Category</th><td> </td><td>cake</td></tr>
            <tr><th>Price</th><td> </td><td>9.99</td></tr>
            <tr><th>Unit</th><td> </td><td>box</td></tr>
            <tr><th>On Hand</th><td> </td><td>144.0</td></tr>
        </table>
    </td>
</tr>
</table>
<p>In a magna pretium, laoreet neque eget, consectetur justo. Vivamus pharetra sapien at sem ultrices semper. Ut
```

In the next section, we will turn our attention to a different way of accessing the web application, called **REST**.

Responding to REST requests

In this section, we will assume a fictional scenario where Sweets Complete starts a partner channel. The partners need access to basic product information, including product keys, titles, and prices. For maximum flexibility, for the purposes of this illustration, let's say that Sweets Complete has decided to develop a simple **Application Programming Interface (API)**.

All the partner needs to do is to make a REST request of the Sweets Complete website, and the partner can obtain information on a single product or all products. Before we get into the exact details, however, it's important to digress for a moment and examine what REST is.

Understanding REST

REST was introduced in a doctoral thesis by Roy Fielding, best known for his work on **HyperText Markup Language (HTML)** and the extremely popular Apache web server. His thesis described a network of resources identified by a **Uniform Resource Identifier (URI)**. As a user accesses these resources, the current **state** (that is, the contents of the page) is *transferred* to them.

The first implementation of this scheme was the World Wide Web. So, in a sense, by simply opening up a browser on a PC and making a request for a web page, you are making a REST request! More recently, the ideas behind Fielding's original thesis have been greatly expanded such that we now refer to **RESTful web services**, which can include software (or firmware) running on literally any device connected to the internet, making use of **HyperText Transfer Protocol (HTTP)**.

Please refer to the following links for more information:

- Documentation reference on REST: `http://standards.rest/`
- Roy Fielding: `https://en.wikipedia.org/wiki/Roy_Fielding`
- HTML: `https://www.w3.org/TR/html52/`
- Apache: `http://httpd.apache.org/`
- URI: `https://tools.ietf.org/rfc/rfc3986`
- HTTP: `https://tools.ietf.org/html/rfc2616`

Understanding HTTP

In order to develop a RESTful web service, it's important to have a basic understanding of how HTTP works. In this section, we will not go into packet-level detail, but we will cover the general concepts you need to know in order to create an effective REST service.

HTTP status codes

RFC 2616 defines a set of **status codes** set when the RESTful web service responds to a REST request. The status code gives the client making the request a quick idea of the request's success or failure. The default response in most situations is status code 200, which simply means OK. More details on expected responses depend on the nature of the request (addressed in the next sub-section). Here is a brief summary of the *family* of status codes:

Code family	General nature
1xx	Informational
2xx	Success
3xx	Redirect
4xx	Client error
5xx	Server error

In cases where the RESTful web service you are defining needs to perform a redirect, it's a good idea to set a status code of either 301 (moved permanently) or 307 (temporary redirect), depending on the nature of the redirect.

Use the 4xx family of status codes in situations where the request is bad for some reason. If the requester is not authenticated or authorized, you can set status code 401 (unauthorized) or 403 (forbidden). If the URI a client uses to make the request points to a resource that does not exist, set status code 404 (not found) or 410 (gone). Also, if the request just doesn't make sense (which is often a sign of an attack in progress!), you can set status code 400 (bad request) or 406 (not acceptable).

Finally, the 5xx family is used when there is an error within your RESTful web application. Status code 500 (internal server error) is the most widely used in such situations. If a resource that your RESTful application needs is not available (for example, the MongoDB server is down), you can set a status code of 501 (not implemented) or 503 (service unavailable).

For more information, see *RFC 2616, Section 10*: `https://tools.ietf.org/html/rfc2616#section-10`

HTTP methods

Possibly the most important aspect of HTTP that you need to know is that each HTTP request identifies an **HTTP method** (also called an **HTTP verb**). The default method for most REST clients is GET. Another method you might have already encountered is POST, often used with HTML forms. The following table summarizes the methods most often handled by RESTful web services:

Method	Description/Status code returned
GET	Requests information from the application. Think of GET as a database query. In most RESTful web services, if the GET request is accompanied by an ID, only the document matching that ID is returned. Otherwise, all documents are returned. If the request succeeds, typically status code 200 (OK) is returned. If only a subset of the requested information is returned, your RESTful application can set a status code of 206 (partial content).
POST	A POST request contains a block of data. The associated RESTful web service typically performs a database insert. Return status code 201 (created) if the insert succeeded. If the operation succeeded but nothing was created, your application can return status code 204 (no content). Otherwise, the default success code 200 (OK) is acceptable.
PUT	The outcome of a PUT request depends on whether the identified document already exists. If it exists, PUT ends up performing an update on the existing document. If the document doesn't exist, PUT is normally interpreted as an insert. As with POST, success status codes could be 200, 201, or 204.
DELETE	As the name implies, this method requests that the RESTful web application delete the identified document. Success status codes usually default to 200 (OK). 202 (accepted) can be used if the request has been accepted but not yet performed. 204 (no content) is used if the action was successfully performed, but there is no other information to be provided.

There are a number of other methods used primarily for the purposes of the transmission. These include HEAD, TRACE, OPTIONS, and CONNECT. For more information on these other methods, refer to *RFC 2616, Section 9* (`https://tools.ietf.org/html/rfc2616#section-9`).

HTTPS is a secure form of HTTP, and is entirely based upon HTTP. For all intents and purposes, the same application can make a request to a URI that starts with HTTP or HTTPS without having to rewrite anything.

 The most widely used version of HTTP is version 1.1. HTTP/2 was published as an RFC in May 2015. It is now widely available and is considerably faster and more secure than HTTP/1.1. For more information, refer to `https://en.wikipedia.org/wiki/HTTP/2`.

To begin our discussion of how to develop a RESTful web application, we need to define a class that responds to REST requests.

JSON standards for a REST response

Although not mandatory, the common practice is to return a JSON document in response to a REST request. It is worth noting, at this juncture, that many developers are opting to respond with a formalized JSON document structure based on an emerging standard. We will digress briefly to discuss some of the more commonly used standards.

The first question that doubtlessly comes to mind is *why bother with standards for JSON?* This is a very good question, for which there is a simple answer: if your RESTful application delivers a standards-based response, it's easier for REST clients to handle, easier to maintain, and extends the application's usefulness over time. The next question is: *is there an "official" standard for JSON responses?* Unfortunately, the answer to the latter question is no! There are, however, *emerging* standards, the more popular of which we summarize in this section.

JSend

JSend (`https://github.com/omniti-labs/jsend`) is an extremely simple standard that covers three main categories of response: **success**, **fail**, and **error**. Each type further defines mandatory keys. For example, all three types require a `status` key. *Success* and *fail* require a `data` key. The *error* type requires a key called `message`.

Here is an example of a successful response to a `GET` request:

```
{
    status : "success",
    data : {
        "posts" : [
            { "id" : "1111AAAA", "productKey" : "1111AAAA", "title" :
```

```
        "Chocolate Bunny" },
                      { "id" : "2222BBBB", "productKey" : "2222BBBB", "title" :
        "Chocolate Donut" }
                    ]
            }
        }
```

Next, we will look at JSON:API.

JSON:API

JSON:API (`https://jsonapi.org/format/1.1/`) has actually been published but has not yet been ratified as of the time of writing. The formatting and available tags are more complex and diverse than with JSend. Reading through the documentation on JSON:API, you can see that the guidelines are much more rigid and precise than JSend.

JSON:API has the following accepted `Content-Type` header value:

```
Content-Type: application/vnd.api+json
```

This means that if the `Accept` header in a client request specifies this content type, your application is expected to respond with a JSON:API-formatted JSON response. If the `Content-Type` header of a client request uses the value shown, your REST application needs to understand this formatting.

Top-level keys in a JSON:API-formatted request or response must include at least one of the following: `meta`, `data`, or `errors`. In addition, the `jsonapi`, `links`, and `included` top-level keys are available for use. The resulting document can end up being quite complex. The advantage of this format is that the response becomes *self-describing*. The client has an idea where the request originated and where to go for related information.

Here is an example:

```
{
  "links": {
    "self": "http://sweetscomplete.com/products/AAAA1111"
  },
  "data": {
    "type": "product",
    "id": "AAAA1111",
    "attributes": {
      "title": "Chocolate Bunny"
    },
    "relationships": {
      "category": {
```

```
        "links": {
          "related": "http://sweetscomplet.com/products/category/chocolate"
        }
      }
    }
  }
}
```

In the preceding example, `links` tells the client the URI used to retrieve this information, `data` is the information requested, `relationships` tells the client how to obtain more information of this nature.

hal+json

Finally, in our brief roundup of emerging JSON standards, we have **HAL**, which stands for **Hypertext Application Language**. Although filed as an **Internet Engineering Task Force (IETF) Request For Comment (RFC)**, the last such submission was for version 8, which expired in May, 2016 (`https://tools.ietf.org/html/draft-kelly-json-hal-08`). As with JSON:API, HAL has its own accepted `Content-Type` header:

```
Content-Type: application/hal+json
```

The proposal is much like JSON:API. In contrast to JSON:API, however, the specification for `hal+json` has fewer mandatory (that is, *required*) keys, and more recommended (that is, *should include*) ones. As with JSON:API, `hal+json` includes provisions for a self-referencing JSON response, such that the client can retrieve the same information and learn where to go for further information. Furthermore, again as with JSON:API, there is provision for multiple responses and embedded documents within the response.

Here is an example based on the one shown previously in the JSON:API discussion:

```
{
    "_links" : {
        "self" : { "href" : "http://sweetscomplete.com/products/AAAA1111"
},
        "find" : { "href" :
"http://sweetscomplete.com/products/category/chocolate" }
    },
    "type" : "product",
    "id"   : "AAAA1111",
    "title": "Chocolate Bunny"
}
```

Now, we will turn our attention to the actual responder class.

Strategies for creating a REST responder

As with previously defined classes, we will define a responder class, `RestResponder`, which resides under `web.responder.rest`. One of our first tasks is to determine the HTTP request method and the content type accepted by the client making the request. We can do this easily by examining the headers in the incoming request.

Our new REST responder activates a *strategy* that returns the response using the desired JSON formatting. So, before diving into the responder class, let's digress and examine the base class and three strategy classes, which reside under `web.responder.strategy`.

Please refer to the following links for more information:

- **HTTP request method**: `https://tools.ietf.org/html/rfc2616#section-5.1.1`
- **Strategy pattern**: `https://en.wikipedia.org/wiki/Strategy_pattern`

Response strategy base class

All of the response strategy classes return data. We need to decide what else should be returned. Here are the keys we return in the JSON response for the purposes of this discussion:

Key	Description
status	HTTP status code
data or error	Data if the operation is successful, and there is data to be returned Error if a problem was encountered
link	The original page requested

In the base class, we need to import the `json` module and define a few properties needed:

```
# web.responder.strategy.base
import json
class Base :
    headers   = []
    data    = {}
    error   = ''
    status = 200
    link    = ''
```

Next, even though HTTP status codes are available in the
`http.HTTPStatus` class (`https://docs.python.org/3/library/http.html#http.`
`HTTPStatus`), the codes are configured as an *enum*, which proves awkward to use for our
purposes. Accordingly, we define the HTTP status codes we use as a dictionary:

```
http_status  = {
    200 :  'OK',
    400 :  'BAD REQUEST',
    404 :  'NOT FOUND',
    500 :  'INTERNAL SERVER ERROR',
    501 :  'NOT IMPLEMENTED'
}
```

Next, we define a constructor method that accepts the data to be transmitted, any error
messages, HTTP status code, and a link to this request:

```
def __init__(self, data, error, status, link) :
    self.data  = data
    self.error = error
    self.link  = link
    if status in self.http_status :
        self.status = status
    else :
        self.status = 400
        self.error  = self.error
        self.error += ', Supported status codes: 200,400,404,500,501'
```

Finally, we define a method, `addHeader()`, which simply adds to the list of headers, as
with the other responder classes. Also, we add a new method, `buildOutput()`, called by
the `render()` method implemented by the strategy classes that inherit from the base class.
In the `buildOutput()` method, we set the status code using a separate header `status`. We
also set the content type to that supplied by the strategy.

Thus, for example, `HalJsonStrategy` would specify `application/hal+json`. The
`payload` argument includes the original data produced wrapped in the additional keys, as
required by the content type:

```
def addHeader(self, header) :
    self.headers.extend([header])
def buildOutput(self, content_type, payload) :
    self.addHeader('Status: ' + str(self.status) + ' ' + \
                   self.http_status[self.status])
    self.addHeader('Content-Type: ' + content_type)
    output = ''
    for item in self.headers :
        output += item + "\r\n"
```

```
output += "\r\n"
output += json.dumps(payload)
return output
```

Next, we create strategy classes to match the three JSON formats outlined previously.

JSend strategy

All of our strategy classes include a `content_type` property, set to the content type for that strategy:

```
# web.responder.strategy.jsend
from web.responder.strategy.base import Base
class JsendStrategy(Base) :
    content_type = 'application/json'
```

The most critical method is `render()`. If `self.error` is set, the JSend top-level key, `status`, is set to `error`. If `self.data` is set, on the other hand, the `status` key changes to `success`. The data stored in `self.data` is included in the final payload. The final return value is the product of the `buildOutput()` method from the base class:

```
def render(self) :
    if self.error :
        payload = { 'status' : 'error', 'message' : self.error }
    if self.data :
        count = 1
        post  = {}
        for key, value in self.data.items() :
            post.update({ key : value })
        if count == 1 :
            payload = { 'status' : 'success',
                        'link'   : self.link,
                        'data'   : { 'post' : post } }
        else :
            payload = { 'status' : 'success',
                        'link'   : self.link,
                        'data'   : { 'posts' : post } }
    return self.buildOutput(self.content_type, payload)
```

Next, let's look at the JSON:API strategy.

The JSON:API strategy

A very similar process occurs within the strategy class representing the JSON:API format:

```
# web.responder.strategy.json_api
from web.responder.strategy.base import Base
class JsonApiStrategy(Base) :
    content_type = 'application/vnd.api+json'
```

Again, if `self.error` is set, an `error` key contains the error message. If there is data in `self.data`, it is wrapped in the appropriate syntax for this format. The base class `buildOutput()` method is called to assemble the final response:

```
def render(self) :
    payload = dict({'links':{'self':self.link }})
    if self.error :
        payload['error'] = self.error
    if self.data :
        count = 1
        data = {}
        for key, value in self.data.items() :
            if 'category' in value :
                prodType = value['category']
            else :
                prodType = 'N/A'
            data.update({count:{'type':prodType,'id':key,\
                        'attributes':value}})
            count += 1
        payload['data'] = data
    return self.buildOutput(self.content_type, payload)
```

Next, lets look at the `hal+json` strategy.

The hal+json strategy

Finally, the `HalJsonStrategy` class is configured in a similar manner:

```
# web.responder.strategy.hal_json
from web.responder.strategy.base import Base
class HalJsonStrategy(Base) :
    content_type = 'application/hal+json'
```

You might notice that the `render()` method for this format is even more concise as data values are directly placed at the top level:

```
def render(self) :
    payload = dict({'_links':{'self':{'href':self.link }}})
    if self.error :
        payload['error'] = self.error
    if self.data :
        for key, value in self.data.items() :
            payload.update({ key : value })
    return self.buildOutput(self.content_type, payload)
```

Let's now create a REST responder.

Creating a REST responder

Since all the JSON formatting is done by the strategy classes, the job of the responder class becomes much simpler: all this class needs to do is to accept the incoming environment information, data, error message, and HTTP status code. All of this information is passed to the responder by the action script. We start by importing the strategies and initializing the key properties:

```
# web.responder.rest
from web.responder.strategy.jsend import JsendStrategy
from web.responder.strategy.json_api import JsonApiStrategy
from web.responder.strategy.hal_json import HalJsonStrategy
class RestResponder :
    json_strategy = None
    header_accept = None
```

The most important property definition determines what formatting types are accepted. This property is in the form of a dictionary and is used to match the information in the incoming HTTP `Accept` header (`https://tools.ietf.org/html/rfc2616#section-14.1`) against the supported MIME types:

```
accepted_types = {
    'json'     : { 'mime_type' : 'application/json',
                   'strategy' : 'JsendStrategy' },
    'json_api' : { 'mime_type' : 'application/vnd.api+json',
                   'strategy' : 'JsonApiStrategy' },
    'hal_json' : { 'mime_type' : 'application/hal+json',
                   'strategy' : 'HalJsonStrategy' }
}
```

In the constructor, from the supplied environment, we pull out the link and Accept header. If no Accept header is sent by the agent making the request, we default to `application/json`:

```
def __init__(self, environ, data, error, status) :
    if 'REQUEST_URI' in environ :
        link = environ['REQUEST_URI']
    if 'HTTP_ACCEPT' in environ :
        self.header_accept = environ['HTTP_ACCEPT']
    else :
        self.header_accept = 'application/json'
```

We can now determine which strategy class to invoke by simply looping through the list of accepted types, and testing to see whether the `mime_type` key is present in the Accept header. Note that we default to JSend if the Accept header is not present, or if the contents of this header do not match the types we support:

```
strategy = 'json'
for key, info in self.accepted_types.items() :
    if info['mime_type'] in self.header_accept :
        strategy = key
        break
if strategy == 'hal_json' :
    self.json_strategy = HalJsonStrategy(data, error, status, link)
elif strategy == 'json_api' :
    self.json_strategy = JsonApiStrategy(data, error, status, link)
else :
    self.json_strategy = JsendStrategy(data, error, status, link)
```

The `render()` method of the responder class is trivial. All it needs to do is to return the value of the `render()` method of the chosen strategy class!

```
def render(self) :
    return self.json_strategy.render()
```

Now, we can look at what domain services are needed to handle REST requests.

Modifying the products domain service

We have mentioned a number of times that the simple web application defined in this chapter follows the action-domain-responder design pattern. Accordingly, the domain service just needs to keep doing the same job it has been doing all along. This class has no awareness of how the request is made, nor is it concerned with the output format. The action script handles the incoming request, and the responder class handles the output.

The domain service we need to deal with is
sweetscomplete.domain.product.ProductService. We need to define a method that
returns a dictionary of all products. The final result is a dictionary of products with
productKey as the key and a partially populated Product entity as the value. The entity
needs to include productKey, title, and price:

```
# sweetscomplete.domain.product.ProductService
def fetchAllKeyCategoryTitlePriceForRest(self, skip = 0, limit = 0) :
    projection = dict({"productKey":1,"category":1,\
                       "title":1,"price":1,"_id":0})
    result = dict({})
    for doc in self.db.products.find(None, projection, skip, limit) :
        result[doc.getKey()] = doc
    return result
```

Likewise, for this illustration, the domain service needs a method that performs a
MongoDB findOne() operation, returning a single Product entity instance based on the
product key:

```
def fetchOneKeyCategoryTitlePriceForRest(self, prod_key) :
    query      = dict({"productKey":prod_key})
    projection = dict({"productKey":1,"category":1,\
                       "title":1,"price":1,"_id":0})
    doc        = self.db.products.find_one(query, projection)
    result     = { doc.getKey() : doc }
    return result
```

Finally, we define an action script to intercept the incoming REST request, and make use of
the RESTful domain service and responder classes.

REST request action handler

REST requests are handled via the web server's CGI gateway by means of a
script, /path/to/repo/www/chapter_06/rest.py. Because this is a web script, we need
the opening tag that identifies the location of the Python executable to the web server's CGI
process. The script also needs to have its permissions set so that it is executable. After the
opening tag is the expected series of import statements:

```
#!/usr/bin/python
import os,sys
src_path = os.path.realpath('../../chapters/06/src')
sys.path.append(src_path)
import cgi
from collections import Counter
```

```
from config.config import Config
from web.responder.rest import RestResponder
from db.mongodb.connection import Connection
from sweetscomplete.domain.product import ProductService
from sweetscomplete.entity.product import Product
```

Next, as with the other action scripts described in this chapter and the previous one, we use the `Config` class to create a database connection and define a `ProductService` instance:

```
config       = Config()
db_config    = config.getConfig('db')
prod_conn    = Connection(db_config['host'], db_config['port'], Product)
prod_service = ProductService(prod_conn, db_config['database'])
```

We can now define variables that are acquired from the environment and domain service query:

```
data         = None
error        = None
link         = '/'
status       = 200
params       = cgi.FieldStorage()
```

In this part of the action script, we determine the HTTP method. In this example, we will only define processing for a GET request. If the request is anything other than GET, we set the status code to 501, along with an error message. If the request is GET, we then look for a parameter named key in the incoming parameters captured by the Python cgi library.

If a parameter named key is present, we call `fetchOneKeyCategoryTitlePriceForRest()` from the domain service. Otherwise, we call `fetchAllKeyCategoryTitlePriceForRest()`:

```
if 'REQUEST_METHOD' in os.environ :
    method = os.environ['REQUEST_METHOD']
    if method == 'GET' :
        if 'key' in params :
            data = prod_service.fetchOneKeyCategoryTitlePriceForRest( \
                params['key'].value)
        else :
            data = prod_service.fetchAllKeyCategoryTitlePriceForRest()
```

If there are no results, we set the status code to 404 (not found) with an error message. Also, if there is no request method at all in the environment (which might mean that the script is called from the command line), we set the status code to 400 (bad request). Note that the default status code is 200 (OK), defined previously.

The block of code shown here shows how to handle situations with no data, an unsupported HTTP method, or a bad request:

```
    if not data :
        error = 'No results'
        status = 404
    else :
        error = 'Set HTTP method to GET'
        status = 501
else :
    error = 'Unable to determine request method'
    status = 400
```

The response is quite simple. We create a `RestResponder` class instance and pass it the environment, data, error, and status code. We then print the return value of `render()`, which, as you may recall, is actually the `render()` method of the chosen strategy:

```
response = RestResponder(os.environ, data, error, status)
print(response.render())
```

Here is the output from a REST client (Firefox plugin) where the Accept header indicates JSON:API formatting:

Finally, here is the output when a `DELETE` request is made:

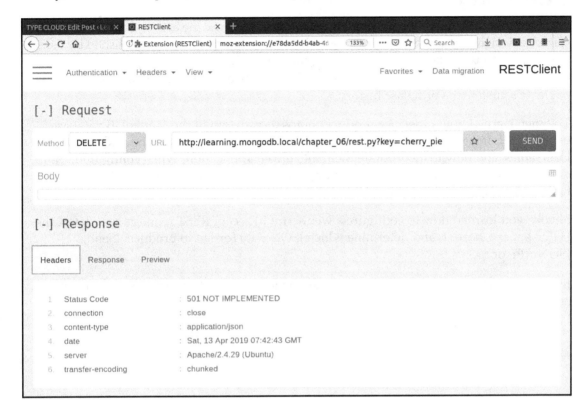

Many developers use Python to listen for REST requests directly. In the REST implementation described in this section, we do not do that. Instead, we assume that a proper web server (for example, Apache or NGINX) is listening to HTTP requests. The implementation we describe in this section leverages the web server's CGI gateway to handle REST requests. The advantage of this approach is that we can use the web server as a high-performance frontend that not only can handle a large volume of requests but also provides much-needed security and protection against malicious web attacks.

To test the REST API code, install a REST client browser plugin on the browser on your local host computer. One such client for Firefox is found at `http://restclient.net/`. If you are working on a customer server, you can also use the `curl` (`https://curl.haxx.se/`) command from the command line to simulate REST requests.

Summary

In this chapter, you were shown implementations that follow the action-domain-responder software design. The first section discussed how pagination can be accomplished by using the `skip` and `limit` parameters, used in conjunction with the `pymongo.collection.find*` methods. The practical application shown was to paginate a customer's purchase history.

You then learned how to service product photos directly out of the MongoDB database using HTML image tags and Base64-encoded data. Next, you learned about jQuery and DataTables, and how to incorporate them into a web application. When configured, the data table makes AJAX requests of your application, which then performs a MongoDB lookup, returning the results as JSON.

Finally, you learned how to configure a web script to accept REST requests that detect the HTTP `Accept` header, and determine which JSON data format to produce: JSend, JSON:API, or `hal+json`.

In the next section, you will learn how to design advanced data structures, which includes embedded objects and arrays.

Section 3: Digging Deeper 3

Section 3 introduces a medium-sized corporation, *Book Someplace, Inc.*, with more complex data needs. This section moves out of the basic and into intermediate to advanced territory. You will learn how to define a dataset for a corporate customer with complex requirements that includes document structures with embedded arrays and objects. You will also learn how to handle tasks involving multiple collections, which require taking advantage of advanced MongoDB features, including aggregation and map-reduce.

This section contains the following chapters:

- Chapter 7, *Advanced MongoDB Database Design*
- Chapter 8, *Using Documents with Embedded Lists and Objects*
- Chapter 9, *Handling Complex Queries in MongoDB*

Advanced MongoDB Database Design

7

Part three of this book, *Digging Deeper*, introduces a large corporation called Book Someplace, Inc. that has complex data needs. This part moves us from the basics and clearly into intermediate to advanced territory. In this part, you'll learn how to define a dataset for a corporate customer with complex requirements, including document structures with embedded arrays and objects. You'll also learn how to handle tasks involving multiple collections, which requires taking advantage of advanced MongoDB features, including aggregation and map-reduce.

In this chapter, we'll expand on the design principles we discussed in `Chapter 4`, *Fundamentals of Database Design*, which also apply to this company. A major difference, however, is that in this chapter, you'll learn how to refine your data structures by using embedded documents and arrays. In addition, since web application requirements are more complicated, this chapter shows you how to use MongoDB domain service classes inside applications based on the popular Django web application framework.

In this chapter, we will cover the following topics:

- Reviewing customer requirements
- Building document structures
- Developing the corresponding Python entity classes
- Defining domain service classes
- Using MongoDB with Django

Let's get started!

Technical requirements

The following are the minimum recommended pieces of hardware for this chapter:

- A desktop PC or laptop
- 2 GB of free disk space
- 4 GB of RAM
- 500 Kbps or faster internet connection

The following are the software requirements for this chapter:

- OS (Linux or Mac): Docker, Docker Compose, Git (optional)
- OS (Windows): Docker for Windows and Git for Windows
- Python 3.x, PyMongo driver, Apache, and Django (already installed in the Docker container used for this book)

Installation of the required software and how to restore the code repository for this book was explained in Chapter 2, *Setting Up MongoDB 4.x*. To run the Python code examples in this chapter from the command line, open a Terminal window (Command Prompt) and enter these commands:

```
cd /path/to/repo
docker-compose build
docker-compose up -d
docker exec -it learn-mongo-server-1 /bin/bash
```

To run the demo website described in this chapter, follow the setup instructions provided in Chapter 2, *Setting Up MongoDB 4.x*. Once the Docker container for chapters 3 through 12 has been configured and started, there are two more things you need to do:

1. Add the following entry to the local hosts file on your computer:

> The local hosts file is located at
> C:\windows\system32\drivers\etc\hosts on Windows
> and /etc/hosts for Linux and Mac.

```
172.16.0.11   booksomeplace.local
172.16.0.11   chap07.booksomeplace.local
```

2. Open a browser on your local computer (outside of the Docker container) to this URL:

```
http://chap07.booksomeplace.local/
```

3. Click on the link for `Chapter 7`.

   When you are finished working with the examples covered in this chapter, return to the Terminal window (Command Prompt) and stop Docker, as follows:

   ```
   cd /path/to/repo
   docker-compose down
   ```

The code used in this chapter can be found in this book's GitHub repository at `https://github.com/PacktPublishing/Learn-MongoDB-4.x/tree/master/chapters/07` as well as `https://github.com/PacktPublishing/Learn-MongoDB-4.x/tree/master/www/chapter_07`.

Reviewing customer requirements

The first step in defining database document structures, as discussed in `Chapter 4`, *Fundamentals of Database Design*, is to review the customer's requirements. To start, here is a brief overview of the new company we'll be working with, Book Someplace, Inc.

Introducing Book Someplace, Inc.

Book Someplace, Inc. is a fictitious web-based company whose core business is to handle hotel bookings for a large international customer base. The company makes its money by taking a small percentage of the payment made for every confirmed booking. In order to differentiate itself from run-of-the-mill online travel agencies, Book Someplace, Inc. offers these unique features:

- A massive database of worldwide properties
- The ability to view a cluster of properties on a map based on geospatial data
- Customer reviews and uploaded photos
- A customer loyalty program that gives progressively bigger discounts to repeat customers
- Being able to book reservations for a range of dates
- Processes payments

The stakeholders in the application to be developed fall into categories very similar to that of *Sweets Complete*:

- Customers:
    - Enroll as members
    - View and book properties
    - Manage their bookings
- Partners:
    - Offer properties to be booked
    - View reservations made
    - Monitor payments
- Administrative assistants:
    - Sign up partners
    - Create entries for new properties offered by partners
    - Handle payment issues and refunds
- Management:
    - Financial reports
    - Information on customer and partner demographics
    - Correlate confirmed bookings with partners

Now that we have a basic idea of the business and what is needed by the various stakeholders, we'll turn our attention to the data flow.

What data is produced by the application?

The following table summarizes the data that is produced by the application:

Stakeholder	Outputs
Customers	The customer's main interest is to view a list of available properties based on location and date.
Partners	Partners are most interested in bookings; that is, which room types are booked on what dates. They are also interested in payments that have been processed.
Admins	Administrative assistants need to review statistics that might cause them to recommend partner revenue splits to be upgraded or a customer being given bigger booking discounts.
Management	Management is interested in financial reports as Book Someplace gets a percentage of every booking with a confirmed payment. In addition, Management needs reports that correlate earnings against customer demographics, and the same for partner demographics. Management can then make decisions about marketing to improve earnings among demographics that might have been overlooked.

Now that we know what data is produced by the application, let's check what data goes into the application.

What data goes into the application?

The following table summarizes the data that is consumed by the application:

Stakeholder	Inputs
Customers	When customers first sign up as members, they enter their name and address, as well as contact information. Subsequently, as the customer uses the website, they enter booking information, including the property, room type, how many guests, as well as arrival and departure dates. After staying at a property, customers are asked to provide reviews and upload photos.
Partners	When a partner signs up, they enter the name, address, and contact information for the primary contact. After being approved, the partner then enters one or more properties they have on offer. Property details include the type of property (for example, hotel, resort, bed and breakfast), location, and photos. For each property, the partner needs to enter a description of the room types, as well as how many of that room type are available.
Admins	The administrative assistants handle any payment disputes and communications between the customer and the partner that the customer has a confirmed booking with. Admins also upgrade or downgrade the status of customers based on how often they use the website and how many reviews they provide. Likewise, Admins upgrade or downgrade the statuses of partners based on reviews and the earnings their properties produce for Book Someplace. Admins are also involved in any cancellations and full or partial refunds.
Management	Management makes decisions on retaining or removing partners.

Now that you have an idea of how data flows in and out of the application, it's time to have a look at what needs to be stored in the database.

What needs to be stored?

Based on the information provided previously, we need to more closely define what demographic information is needed for customers and partners. For both, it would make sense to store location information such as the street address, city, state or province, region, country, postal code, and geospatial coordinates. Other demographic information of interest might include gender, age, marital status, and how many children the customer has. It is extremely important that any demographic information is made confidential and optional in order not to violate customer privacy.

Of course, contact information is also needed: first and last name, email address(es), phone number(s), and social media (Facebook, Twitter, and so on). Demographics irrelevant to the needs of the company would include race, religion, and education level.

Information regarding properties would include, for each property, its location (for example, street address, city, and postcode), description, rating, photos, and rooms. Rooms can be grouped into types. How many of each type is extremely critical information that must be stored. Information on a room type should include which floor, view (for example, city view, poolside view, lake view, and river view), size, price, and bed types. Bed types could include single, double, twin, queen-sized, king-sized, and so on. It is important to store how many of each bed type is available for the given room types.

Financial information to be stored would come into play when a customer makes a booking. The number of each room type booked is recorded and payment is arranged. When payment is confirmed, the partner account needs to be credited, and the revenue split between Book Someplace and the partner is calculated.

 As a general rule, do not store credit card information unless there is no other choice. The primary reason for this is that if a security breach occurs, whether due to inside or outside means, the company would face severe liability issues, as well as a loss of reputation. Accordingly, Book Someplace has decided to effect payment processing via a third-party payment provider. Any financial information other than credit card numbers is stored.

Now that we have an idea of what information needs to be stored, we can start defining collections and MongoDB document structures.

Building document structures

In the database design for Book Someplace, there are many blocks of information that could potentially be redundant. One example is address information: this block of information can be applied to customers, partners, and also to properties! If we were dealing with a legacy two-dimensional RDBMS, we would end up with dozens of small tables with an overwhelming number of relationships and join tables.

Rather than thinking in terms of collections and document structures, let's step back for a moment and see whether we can define JSON *objects* that can be used as interchangeable Lego blocks. Let's start with *Name*.

 To arrive at a good MongoDB database design, DevOps should think in terms of *objects* rather than two-dimensional structures based on rows and columns.

Defining a name structure

Although it might not appear evident that a person's name might be defined using objects, reflect on what exactly goes into a person's name: their first, middle, and last names. In addition, a number of people have suffixes added to their names, such as *Junior* or *III*. Furthermore, traditional companies might include a title such as *Mr., Miss, Mrs.,* or *Ms.*

Bear in mind, however, that titles such as *Miss.* or *Mrs.* might also offend certain customers. Why? You might notice that the common formal title for a man is *Mister*, whereas for a woman it could be *Miss* or *Missus*. This means that the title gives away the marital status of a woman, but not for a man! Accordingly, the female title *Ms.* gained popularity, but this title might offend women who come from a more traditional culture!

Here is how the name structure appears for Book Someplace:

```
Name = {
    "title"     : <formal titles, e.g. Mr, Ms, Dr, etc.>,
    "first"     : <first name, also referred to as "given" name>,
    "middle"    : <middle name, optional>,
    "last"      : <last name, "surname" or "family name">,
    "suffix"    : <information included after the last name>
}
```

As in previous chapters, we'll continue with the convention of defining an additional collection called Common that contains small items to facilitate categorization and search. In this case, we add a key called title to the Common collection to ensure uniformity:

```
Common = [
    { "title" : ['Mr','Ms','Dr',etc.] }
]
```

Next, we will define a Location structure.

Defining a location structure

The information needed in a Location structure includes any information that allows us to send regular postal mail. Thus, we need to include fields such as street address, city, state or province, and so forth. This structure also includes geospatial data such as latitude and longitude, allowing us to perform geospatial inquiries to better gauge property distances.

Some addresses, however, also contain a state, province, and other identifiers. The postcode format varies widely from country to country. Other variations exist as well. In the UK, for example, it is not uncommon for the address to include a building name. If a building has apartments, condos, or flats, the floor number and flat number are needed as well.

With all this in mind, here is our first attempt at a JSON `Location` structure:

```
Location = {
    "streetAddress"     : <number and name of street>,
    "buildingName"      : <name of building>,
    "floor"             : <number of name of the floor>,
    "roomAptCondoFlat"  : <room/apt/condo/flat number>,
    "city"              : <city name>,
    "stateProvince"     : <state or province>,
    "locality"          : <other local identifying information>,
    "country"           : <ISO2 country code>,
    "postalCode"        : <postal code>,
    "latitude"          : <latitude>,
    "longitude"         : <longitude>
}
```

In the preceding structure, we added all the possible fields that might apply to a customer, partner, or property location. Next, we'll look at defining a structure for `Contact`.

Defining a contact structure

The `Contact` structure contains information that's most often used to contact the customer or partner. Examples of such information include a phone number and email address. Most people have multiple forms of contact that include additional phone numbers and email addresses, as well as various social media platforms such as LinkedIn, Facebook, and Twitter. Accordingly, it is of interest to create *two* structures: one for primary contact information and another for additional contact information.

Here is how these structures appear:

```
Contact = {
    "email"    : <primary email address>,
    "phone"    : <primary phone number>,
    "socMedia" : <preferred social media contact>,
}

OtherContact = {
    "emails"        : [email_address2, email_address3, etc.],
    "phoneNumbers"  : [{<phoneType> : phone2}, etc.],
```

```
    "socMedias"      : [{<socMediaType> : URL2}, etc. ]
}

Common = [
    { "phoneType"     : ['home', 'work', 'mobile', 'fax'] },
    { "socMediaType" : ['google','twitter','facebook', etc.] },
]
```

As we mentioned earlier, we break `Contact` information down into smaller object structures. `Contact` is for the primary email address, phone number, and so on. `OtherContact` is used to store secondary email addresses, phone numbers, and social media. `Common`, as mentioned previously, contains a set of static identifiers representing what types of phone numbers and what types of social media we plan to store.

Next, we'll have a look at `User` structures.

Additional structures pertaining to users

Additional structures needed for Book Someplace users include information needed to log in, and other information needed to facilitate demographic reports. Here is an example of two additional structures:

```
OtherInfo = {
    "gender" : <genderType>,
    "age"    : <int>
}

LoginInfo = {
    "username" : <name used to login>,
    "oauth2"   : <OAUTH2 login credentials>,
    "password" : <BCRYPT password>
}

Common = [
    { "genderType"    : [{'M' : 'Male'},{'F' : 'Female'},{'X' : 'Other'}] }
]
```

We provide an `oauth2` option in order to allow users to authenticate using Google, Facebook, Twitter, and so on. The password is stored in a secure manner. Another politically sensitive issue is `gender`, so we add an option; that is, `X`.

Now, let's turn our attention to defining a `Property` structure.

Defining a property information structure

A key part of what Book Someplace needs to store is information regarding properties available for booking. A large part of what goes into the property structure has already been defined as `Location`, which gives us the customer's address and geospatial coordinates. In addition, the property should be identified as a type and needs to define the local currency, description, and photos. Also, the property might pertain to a brand that is part of a larger organization. An example of this is *Accor S.A.*, a French hotel chain that includes well-known subsidiaries such as *Ibis, Red Roof Inn, Motel 6, Novotel, Mercure, Sofitel,* and *Fairmont,* among others. Room information is defined as a separate structure.

Here is our first attempt at a structure pertaining to property information:

```
PropInfo = {
    "type"        : <propertyType>,
    "chain"       : <chain>,
    "photos"      : <stored using GridFS>,
    "facilities"  : [<facilityType>,<facilityType>,etc.],
    "description" : <string>,
    "currency"    : <currencyType>
},
```

Customer review information needs to include key areas such as staff attitude and service, cleanliness, and comfort. These can be formulated using a simple 1 to 5 rating system. In addition, we need a place to store information about the positive and negative aspects of the customer's stay. Finally, we include a reference to the customer who made the review, which gives us flexibility for future ad hoc reports. For example, if a property has a large number of negative reviews, Management might decide to reach out to customers to gather more information before downgrading the property status.

Here is how a `Review` structure might appear:

```
Review = {
    "customerKey" : <string>,
    "staff"       : <int 1 to 5>,
    "cleanliness" : <int 1 to 5>,
    "facilities"  : <int 1 to 5>,
    "comfort"     : <int 1 to 5>,
    "goodStuff"   : <text>,
    "badStuff"    : <text>
}
```

Finally, as with other structures, there are corresponding common aspects that can be added to an eventual `Common` collection. For property information, this could include the following:

```
Common = [
    { "propertyType" : ['hotel','motel','inn','resort',etc.] },
    { "facilityType" : ['pool','parking','WiFi',etc.] },
    { "chain"        : ['Accor','Hyatt','Hilton',etc.] },
    { "currencyType" : ['AUD','CAD','EUR','GBP','INR',etc.] }
]
```

As we mentioned earlier, of primary interest when defining properties is `room`; a property without rooms cannot be rented! Accordingly, we will now turn our attention to defining structures to represent a `room`.

Defining rooms

Defining room information is another key aspect of the Book Someplace application. Properties feature a certain number of rooms, each of a certain type. The partner offering the property is responsible for defining what room types are available, and how many of each. Book Someplace then accepts bookings for the property based on room type. At any given moment, the list of room types for a property are marked as either `available` or `booked`.

To go even further, each `RoomType` has a certain number of bed types. Bed types include `single`, `double`, `queen`, `king`, and so forth. Each room type has one or more bed types. Finally, here is where we define the price. The partner is responsible for providing information about any local government taxes or fees that are imposed. To keep things simple, price information is stored in a single currency. Currency conversion occurs when price information is presented to customers.

Here is our first draft of the `RoomType` structure:

```
RoomType = {
    "roomTypeKey"  : <string>,
    "type"         : <roomType>,
    "view"         : <string>,
    "description"  : <string>,
    "beds"         : [<bedType>,<bedType>,etc.],
    "numAvailable" : <int>,
    "numBooked"    : <int>,
    "roomTax"      : <float>,
    "price"        : <float>
}
```

```
Common = [
    { "bedType"  : ['single','double','queen','king'] },
    { "roomType" : ['premium','standard','poolside',etc.] }
]
```

It's important to note that `price` is in the currency specified in `PropInfo.currency`. As with other the structures shown earlier, data types other than `string`, `int`, `float`, or `boolean` are either objects in and of themselves, or are represented by the `Common` collection. Thus, when a `RoomType` instance is created, the `Common` collection is consulted, and the property owner is presented with a list of room types to choose from: `premium`, `standard`, `poolside`, and more.

Now, it's time to have a look at the structure that's used to represent `Booking`.

Defining a structure for booking

Booking information includes information such as the arrival and departure dates, checkout time, the date after which a refund is not possible, as well as reservation and payment status. For simplicity, in this illustration, payments is in the currency associated with the property (see `PropInfo.currency`). Here is our first draft for `BookingInfo`:

```
BookingInfo = {
    "arrivalDate"        : <yyyy-mm-dd hh:mm>,
    "departureDate"      : <yyyy-mm-dd hh:mm>,
    "checkoutTime"       : <hh:mm>,
    "refundableUntil"    : <yyyy-mm-dd hh:mm>,
    "reservationStatus"  : <rsvStatus>,
    "paymentStatus"      : <payStatus>,
}
```

As with the other structures, certain common elements are defined. Here, we can see common elements associated with booking information:

```
Common = [
    { "rsvStatus" : ['pending','confirmed','cancelled'] },
    { "payStatus" : ['pending','confirmed','refunded'] }
]
```

Due to government regulations and security awareness, many hotels must also collect identification information on guests. This information can be collected upon check-in, however, and does not need to be part of the application we are developing, which is focused on bookings only. Now, we'll have a look at a set of intermediary structures needed to make the system work.

Defining intermediary structures

When a booking is made, there is no need to store the entire Customer document, nor is there a need to store the entire Property document. Instead, we define intermediary structures that leverage the lower level structures we just described. The first is a set of structures that tie a customer to a booking.

 Using this approach, we minimize redundancy by using only subsets of Customer and Property documents when making a booking. At the same time, we avoid having to join two collections together in a relationship, which would introduce excessive overhead and drag down performance.

Customer booking structures

The first in this set of structures defines who is in the *customer party*. If the customer is the only room occupant, this structure is not used. Otherwise, when a booking is made, an array of CustParty documents is assembled. In the structure shown here, we are leveraging the lower-level structure known as OtherInfo:

```
CustParty = {
    "name"  : <Name>,
    "other" : <OtherInfo>
}
```

When the booking is made, we need to store information about the customer making the booking. In the `CustBooking` document structure shown here, you can see that we are leveraging the previously defined `Name`, `Location`, and `Contact` structures. Also, this structure consumes the `CustParty` structure shown in the preceding code block:

```
CustBooking = {
    "customerKey"     : <string>,
    "customerName"    : <Name>,
    "customerAddr"    : <Location>,
    "customerContact" : <Contact>,
    "custParty"       : [<CustParty>,<CustParty>,etc.]
}
```

Now, let's look at the property booking structures.

Property booking structures

When a booking is made, we also need to tie the booking to a property. The `PropBooking` structure includes the property key (for later reference), as well as `Name` and `Location`, both of which were previously defined. Here is the structure:

```
PropBooking = {
    "propertyKey"     : <string>,
    "propertyName"    : <Name>,
    "propertyAddr"    : <Location>
    "propertyContact" : <Contact>
}
```

In addition, when a booking is made, we need to identify the `roomType` that was booked, as well as the quantity of that room type the customer wishes to reserve.

By including the `price` property, we alleviate the need to perform an additional booking when referencing booking information. Note that we include the `roomType` field, drawn from `Common.roomType`, as well as the `roomTypeKey` field, drawn from `RoomType.roomTypeKey`. The reason why we include both is that `roomType` is for display, whereas `roomTypeKey` is used in a link that allows a customer to view more details about that room type. Here is the `RoomBooking` structure:

```
RoomBooking = {
    "roomType"    : <string>,
    "roomTypeKey" : <string>,
    "price"       : <float>,
    "qty"         : <int>
}
```

> To maximize flexibility, we also include the `customerKey` and `propertyKey` fields in a booking, in case there is a future need to extract additional information. The inclusion of these keys allows us to either perform a secondary lookup or join two or three collections together.

Putting it all together

Now, it's time to put all these pieces together. We'll use the basic data structures we defined previously to define document structures for the following collections: `Customer`, `Partner`, `Property`, and `Booking`. We'll start with `Customer`.

Customer collection document structure

Each collection features a unique identifying key that's based on an arbitrary algorithm. In this case, the `customerKey` field includes the first four letters of the first name, the first four letters of the last name, and a phone number. This gives us an alternative way to uniquely identify a customer over and above the autogenerated MongoDB `ObjectId` property.

In addition, we'll leverage the following previously defined subclasses: `Name`, `Location`, `Contact`, `OtherContact`, `OtherInfo`, and `LoginInfo`. In addition, `totalBooked` and `discount` are defined to support the loyalty program. Here is the final collection structure:

```
Customer = {
    "customerKey"   : <string>,
    "name"          : <Name>,
    "address"       : <Location>,
    "contact"       : <Contact>,
    "otherContact"  : <OtherContact>,
    "otherInfo"     : <OtherInfo>,
    "login"         : <LoginInfo>,
    "totalBooked"   : 0,
    "discount"      : 0
}
```

Now, let's look at the `Partner` collection document structure.

Partner collection document structure

The `Partner` structure is identical to `Customer` with the exception of the `businessName` field. It is important to differentiate between `Customer` and `Partner` because customers can make bookings whereas partners cannot. If a partner wishes to also make a booking, the partner must first register as a customer!

Another key difference is that management assigns each `Partner` a `revenueSplit`, representing a percentage of the total paid. The revenue split is based upon the property ratings, as well as its popularity. As customers make payments on bookings, the `acctBalance` of the partner increases. When the partner chooses to withdraw from their account, the request is passed to the Accounts Payables department of Book Someplace. When the withdrawal is processed, the account balance is adjusted.

Here is the `Partner` collection document structure:

```
Partner = {
    "partnerKey"    : <string>,
    "businessName"  : <string>,
    "revenueSplit"  : <float>,
    "acctBalance"   : <float>,
    "name"          : <Name>,
    "address"       : <Location>,
    "contact"       : <Contact>,
    "otherContact"  : <OtherContact>,
    "otherInfo"     : <OtherInfo>,
    "login"         : <LoginInfo>
}
```

Now, let's look at the `Property` collection document structure.

Property collection document structure

The `Property` collection structure actually shares some similarities with `Partner` in that it has its own contact information, which is most likely not the same as the contact information for the `Partner` owning the property. Property contact information is made available on the website, whereas customer and partner contact information is reserved for Book Someplace staff usage only.

In addition, each property has a `Location` and its own unique identifying `propertyKey`. Also, for reference, the owning `partnerKey` is provided. Each property document includes a list of `RoomType` documents, `review` documents, and statistical information, including `rating` and `totalBooked`:

```
Property = {
    "propertyKey"  : <string>,
    "partnerKey"   : <string>,
    "propName"     : <name of this property>,
    "propInfo"     : <PropInfo>,
    "address"      : <Location>,
    "contactName"  : <Name>,
    "contactInfo"  : <Contact>,
    "rooms"        : [<RoomType>,<RoomType>,etc.],
    "reviews"      : [<Review>,<Review>,etc.],
    "rating"       : <int>,
    "totalBooked"  : <int>
}
```

Now, let's look at the `Booking` collection document structure.

Booking collection document structure

The `Booking` collection document structure leverages the intermediary classes discussed earlier in this chapter. It brings together the appropriate information from a customer and a property. The customer then adds information such as arrival and departure dates, people in the party (for example, a spouse and children), as well as which room type and how many of that type. The total price is then calculated based upon room type information, including any local taxes or fees.

Here is the `Booking` document structure:

```
Booking = {
    "bookingKey"   : <string>,
    "customer"     : <CustBooking>,
    "property"     : <PropBooking>,
    "bookingInfo"  : <BookingInfo>,
    "rooms"        : [<RoomBooking>,<RoomBooking>,etc.],
    "totalPrice"   : <float>
}
```

Now, let's look at the `Common` collection document structure.

Common collection document structure

The last collection to be detailed here is `Common`. Each document in this collection has this extremely simple structure:

```
Common = { "key" : [value] }
```

The keys and values that populate this collection were discussed earlier. We are now in a position to translate JSON document structures into Python entity classes.

Developing the corresponding Python entity classes

The advantage of using an object-oriented design approach is that the document structures described earlier in this chapter can be simply converted into Python classes. We can then group related classes into modules. The disadvantage of this approach is that we can no longer rely on the Pymongo database driver to automatically populate our classes for us. The burden of producing instances of the final entity classes – `Customer`, `Partner`, `Property`, and `Booking` – devolves from the domain service classes.

For the purposes of this chapter, we'll focus only on the most complicated entity class: `booking`. The full implementations of the classes associated with the other collections can be found at `/path/to/repo/chapters/07/src/booksomeplace/entity`.

Defining the booking module

The `booksomeplace.entity.booking` module and its main class, `Booking`, is the most complicated of all. The reason is that a booking brings together subclasses from both `Customer` and `Property`. The class definition is located in a file called `/chapters/07/src/booksomeplace/entity/booking.py`. We start by defining a subclass called `PropBooking` to bring together property-related subclasses. In addition, we add `propertyName` for quick reference and `propertyKey` for future reference:

```
# booksomeplace.entity.booking
import os,sys
sys.path.append(os.path.realpath("../../../src"))
from booksomeplace.entity.base import Base, Name, Location, Contact
from booksomeplace.entity.user import OtherInfo

class PropBooking(Base) :
    fields = {
```

```
        'propertyKey'     : '',
        'propertyName'    : '',
        'propertyAddr'    : Location(),
        'propertyContact' : Contact()
    }
```

Next, we define a subclass called CustBooking that allows us to reference customer information. This class, in turn, defines an array of CustParty instances that hold information about how many guests will show up, as well as their ages and genders:

```
class CustParty(Base) :
    fields = {
        'name'  : Name(),
        'other' : OtherInfo()
    }


class CustBooking(Base) :
    fields = {
        'customerKey'     : '',
        'customerName'    : Name(),
        'customerAddr'    : Location(),
        'customerContact' : Contact(),
        'custParty'       : []
    }
```

The BookingInfo subclass includes information pertinent to the booking itself, including dates, times, and the payment status, as well as the overall status of the reservation:

```
class BookingInfo(Base) :
    fields = {
        'arrivalDate'       : '',
        'departureDate'     : '',
        'checkoutTime'      : '',
        'refundableUntil'   : '',
        'reservationStatus' : '',
        'paymentStatus'     : '',
    }
```

Although this has a small number of fields, the RoomBooking subclass brings together room types included in this booking, as well as the quantity of each type:

```
class RoomBooking(Base) :
    fields = {
        'roomType'    : '',
        'roomTypeKey' : '',
        'price'       : 0.00,
        'qty'         : 0
    }
```

The main class is `Booking`, which consumes the subclasses described earlier. Its `getKey()` method returns the value of the `bookingKey` property. The `rooms` property is an array of `RoomBooking` instances. `totalPrice` is calculated from the quantity and price values:

```
class Booking(Base) :
    fields = {
        'bookingKey'   : '',
        'customer'     : CustBooking(),
        'property'     : PropBooking(),
        'bookingInfo'  : BookingInfo(),
        'rooms'        : [],
        'totalPrice'   : 0.00
    }

    def getKey(self) :
        return self['bookingKey']
```

And this concludes the code needed to represent a `Booking`. The other pertinent classes include the following:

- `booksomeplace.entity.base.Base`
- `booksomeplace.entity.base.Name`
- `booksomeplace.entity.base.Location`
- `booksomeplace.entity.base.Contact`
- `booksomeplace.entity.user.Customer`
- `booksomeplace.entity.user.Partner`
- `booksomeplace.entity.user.OtherContact`
- `booksomeplace.entity.user.OtherInfo`
- `booksomeplace.entity.user.LoginInfo`
- `booksomeplace.entity.property.Property`
- `booksomeplace.entity.property.RoomType`
- `booksomeplace.entity.property.Review`
- `booksomeplace.entity.property.PropInfo`

Now, let's define the `Common` module.

Defining the Common module

As we mentioned in earlier chapters, the `Common` module and class are used to hold small bits of information used throughout the application. These primarily take the form of choices in forms, including drop-down select elements. They are also used to reduce duplicate definitions, as well as present a way of aggregating information when queries on properties are performed.

Another advantage of keeping such items in a common collection is to facilitate translation into other languages. Accordingly, here is the actual structure of the `Common` class:

```
# booksomeplace.entity.common
import os,sys
sys.path.append(os.path.realpath("../../../src"))
from booksomeplace.entity.base import Base
class Common(Base) :
    fields = {
        'key'   : '',
        'value' : []
    }
```

The actual keys and values were defined earlier in this chapter when we discussed how to define document structures.

Testing an entity class

Now that all the entity classes and dependent subclasses have been defined, the next question that might arise is, how can we *test* such structures? The answer, of course, lies in developing an appropriate *unit test*. As in earlier chapters, we created a directory called `/path/to/repo/chapters/07/test/booksomeplace/entity` that we could define the test scripts in. It would take far too much space to delineate tests for all the classes discussed in this chapter, so we will only focus on developing a test for the `booksomeplace.entity.user.Customer` class.

As expected, we'll begin the test script with the appropriate import statements. Note that in order to perform imports from the classes we wish to test, we need to append `/repo/chapters/07/src` to the system path:

```
# sweetscomplete.entity.user.test_entity_customer
import os,sys
sys.path.append(os.path.realpath("/repo/chapters/07/src"))
import json
import unittest
from booksomeplace.entity.base import Name, Contact, Location
```

```
from booksomeplace.entity.user import Customer, OtherContact, OtherInfo,
LoginInfo
```

We then define the class and two properties that represent a `Customer` entity derived from a set of subclasses, and another that is derived from the defaults:

```
class TestCustomer(unittest.TestCase) :
    customerFromDict = None
    customerDefaults = None
```

We then start the test data for the various subclasses that are consumed by `Customer`, beginning with `testDict`:

```
testDict = {
    'customerKey'   : '00000000',
    'name'          : Name(),
    'address'       : Location(),
    'contact'       : Contact(),
    'otherContact'  : OtherContact(),
    'otherInfo'     : OtherInfo(),
    'login'         : LoginInfo(),
    'totalBooked'   : 100,
    'discount'      : 0.05
}
```

We then add `Location`:

```
testLocation = {
    'streetAddress'    : '123 Main Street',
    'buildingName'     : '',
    'floor'            : '',
    'roomAptCondoFlat' : '',
    'city'             : 'Bedrock',
    'stateProvince'    : 'ZZ',
    'locality'         : 'Unknown',
    'country'          : 'None',
    'postalCode'       : '00000',
    'latitude'         : '+111.111',
    'longitude'        : '-111.111'
}
```

MongoDB has a built-in capability to handle geospatial queries, including latitude and longitude. Please have a look at Chapter 9, *Handling Complex Queries in MongoDB*, the *Handling geospatial data* subsection, for a detailed discussion.

This is followed by the subclasses that define the name and contact information for a customer:

```
testName = {
    'title'  : 'Mr',
    'first'  : 'Fred',
    'middle' : 'Folsom',
    'last'   : 'Flintstone',
    'suffix' : 'CM'
}

testContact = {
    'email'    : 'fred@slate.gravel.com',
    'phone'    : '+0-000-000-0000',
    'socMedia' : {'google' : 'freddy@gmail.com'}
}
```

We then define the test data for the remaining subclasses that represent other contact, login, and demographic information:

```
testOtherContact = {
    'emails'       : ['betty@flintstone.com', 'freddy@flintstone.com'],
    'phoneNumbers' : [{'home' : '000-000-0000'}, \
                       {'work' : '111-111-1111'}],
    'socMedias'    : [{'google' : 'freddy@gmail.com'}, \
                       {'skype' : 'fflintstone'}]
}

testOtherInfo = {
    'gender'      : 'M',
    'dateOfBirth' : '0000-01-01'
}

testLoginInfo = {
    'username' : 'fred',
    'oauth2'   : 'freddy@gmail.com',
    'password' : 'abcdefghijklmnopqrstuvwxyz'
}
```

This is followed by exactly the same test data but rendered as a JSON string. We're only reproducing part of the structure here to conserve space. The three dots represent information that has been removed:

```
testJson = '''{
    "customerKey" : "00000000",
    "address"     : { ... },
    "name" : { ... },
    "contact" : { ... },
```

```
            "otherContact" : { ... },
            "otherInfo" : { ... },
            "login" : { ... },
            "totalBooked"  : 100,
            "discount"     : 0.05
    }'''
```

The unit test `setUp()` method creates two instances of `Customer` – one populated with test data and another that's blank:

```
def setUp(self) :
    self.testDict['name']         = Name(self.testName)
    self.testDict['address']      = Location(self.testLocation)
    self.testDict['contact']      = Contact(self.testContact)
    self.testDict['otherContact'] = OtherContact(self.testOtherContact)
    self.testDict['otherInfo']    = OtherInfo(self.testOtherInfo)
    self.testDict['login']        = LoginInfo(self.testLoginInfo)
    self.customerFromDict = Customer(self.testDict)
    self.customerDefaults = Customer()
    self.maxDiff = None
```

The first test asserts that the `getKey()` method returns the pre-programmed value:

```
def test_customer_key(self) :
    expected = '00000000'
    actual   = self.customerFromDict.getKey()
    self.assertEqual(expected, actual)
```

We then write a test that extracts the customer's last name. As you may recall, the `name` property is actually an instance of the `Name` subclass. Thus, we need to call the `get()` method twice – once on the `customerFromDict` dictionary and then again on the extracted `Name` instance:

```
def test_customer_from_dict(self) :
    expected = 'Flintstone'
    name     = self.customerFromDict.get('name')
    actual   = name.get('last')
    self.assertEqual(expected, actual)
```

We then add a test that ensures that the empty `Customer` instance behaves as expected:

```
def test_customer_from_blank(self) :
    expected = True
    actual   = isinstance(self.customerDefaults.get('address'),
Location)
    self.assertEqual(expected, actual)
```

We then test the `toJson()` method to make sure it matches what is expected:

```
def test_customer_from_dict_to_json(self) :
    expected = json.loads(self.testJson)
    actual   = json.loads(self.customerFromDict.toJson())
    self.assertEqual(expected, actual)
```

The last few lines of the test script ensure we can run it from the command line:

```
def main() :
    unittest.main()

if __name__ == "__main__":
    main()
```

And here are the results:

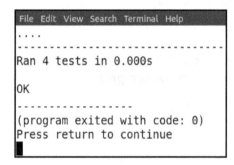

Now, let's look at defining service classes.

Defining domain service classes

Now that the entity classes have been defined, it's time to turn our attention to the domain service classes. As we mentioned earlier in this book, in conjunction with the Sweets Complete web application, domain service classes allow us to connect to the database. They contain methods that are useful for adding, editing, deleting, or querying the various MongoDB collections that have been defined.

Accordingly, as we have determined four primary collections for the Book Someplace web application, we will define four domain service classes. For illustration purposes, we'll discuss how to define a domain service class associated with the `Customers` collection.

Base domain service class

The first class that needs to be defined is one that other domain service classes inherit from. The structure of this class is similar to the one shown earlier in this book. In this case, however, instead of requiring the calling program to define a MongoDB connection and supplying the connection to the constructor, we'll define a constructor that creates its own database and collection from the `config.config.Config` class described earlier in this book.

We'll start by defining the necessary import and class properties:

```
# booksomeplace.domain.base
from db.mongodb.connection import Connection
class Base :
    db          = None
    collection  = None
    dbName      = 'booksomeplace'
    collectName = 'common'
```

As mentioned previously, the `__init__()` method creates an internal `Connection`, from which we can define database and collection properties:

```
def __init__(self, config) :
    if config.getConfig('db') :
        conn = Connection(config.getConfig('db'))
    else :
        conn = Connection()
    self.db = conn.getDatabase(self.dbName)
    self.collection = conn.getCollection(self.collectName)
```

Next, we'll define a domain service that represents properties.

Property domain service class

As with other domain service classes described earlier in this book, we'll create a folder called `/path/to/repo/chapters/07/src/booksomeplace/domain`. The filename is `property.py`, while the class is `PropertyService`, which extends `Base`:

```
# booksomeplace.domain.property
import pymongo
from pymongo.cursor import CursorType
from booksomeplace.domain.base import Base
from booksomeplace.entity.base import Name, Contact, Location
from booksomeplace.entity.property import Property, PropInfo, Review,
RoomType
```

```
class PropertyService(Base) :
    collectName = 'properties'
```

Next, we define a method called `fetchByKey()` that returns a single `Property` instance based on key lookup:

```
def fetchByKey(self, key) :
    query  = {'propertyKey':key}
    result = self.collection.find_one(query)
    if result :
        return self.assemble(result)
    else :
        return None
```

After that, we define a method called `fetchTop10()` that returns a list of the top 10 properties based upon rating and the number of bookings. This method emulates the following *mongo* shell query:

```
db.properties.find({"rating":{"$gt":4}},{}).\
    sort({"totalBooked":-1})\.limit(10);
```

This method is ultimately called by a Django web application and displays the main booking page:

```
def fetchTop10(self) :
    query      = {'rating':{'$gt':4}}
    projection = None
    sortDef    = [('totalBooked', pymongo.DESCENDING)]
    skip       = 0
    limit      = 10
    result     = self.collection.find(query, projection, \
        skip, limit, False, CursorType.NON_TAILABLE, sortDef)
    properties = []
    if result :
        for doc in result :
            properties.append(self.assemble(doc))
    return properties
```

Finally, and perhaps most importantly, we need to define a method called `assemble()`. It accepts a dictionary argument and assembles a `Property` instance using the subclasses discussed earlier in this chapter. As shown in the following code, this method simply checks for the presence of specific keys and creates an instance of the appropriate subclass:

```
def assemble(self, doc) :
    prop = Property(doc)
    if 'propInfo' in doc    : prop['propInfo'] =
PropInfo(doc['propInfo'])
```

```
        if 'address' in doc      : prop['address'] =
    Location(doc['address'])
        if 'contactName' in doc : prop['contactName'] =
    Name(doc['contactName'])
        if 'contactInfo' in doc : prop['contactInfo'] =
    Contact(doc['contactInfo'])
        if 'rooms' in doc :
            prop['rooms'] = []
            for room in doc['rooms'] :
                prop['rooms'].append(RoomType(room))
        if 'reviews' in doc :
            prop['reviews'] = []
            for review in doc['reviews'] :
                prop['reviews'].append(Review(room))
        return prop
```

In the next section, we'll discuss how to use the MongoDB-related classes inside a standard Django web application.

Using MongoDB with Django

Most Python developers, at some point, use a *framework* to facilitate rapid code development. Determining which Python web framework is the first task to be performed. As you can tell from the title of this section, we have settled on *Django*. However, the methodology used to make this determination is what we're interested in here.

Determining which web framework to use

Most code libraries now use GitHub to host a repository for their source code. Searching GitHub for statistics on various popular Python web frameworks reveals the statistics (at the time of writing) summarized in the following table:

Framework	Used	Watch	Star	Fork	Contributors	Last Commit
django	406,000	2,200	49,200	21,300	1,883	-3
flask	443,000	2,300	50,300	13,500	575	-25
pyramid	--	176	3,400	867	266	-3
cherrypy	5,300	66	1,200	289	105	-12

Here is a brief explanation of what these statistics mean:

- **Used**: The number of repositories using this source code.
- **Watch:** This represents developers who have decided they want to be informed of any changes to the source code.
- **Star:** Much like a *Like* on Facebook.
- **Fork:** How many times developers have created duplicates of the source code to form a basis for their own projects.
- **Contributors:** How many developers are involved in the maintenance and development of the framework.
- **Last Commit:** At the time of writing, this specifies how many days ago an update was made to the code. This is an important statistic to watch. If the last commit was more than 1 year ago, chances are the project has been abandoned. A value of 0 means there was a commit on the same day. You need to go back several days in a row, however, to get an accurate picture.

Accordingly, although a case could be made for choosing *flask*, the one chosen for this illustration is *django* as there are more than three times the number of contributors, and it appears the updates are more frequent.

The next logical step would seem to be to choose a MongoDB integration library. This would allow us to use all the native Django database access libraries. As with many frameworks, however, Django is oriented toward legacy, two-dimensional relational databases. For the most part, the MongoDB integration libraries ultimately simply *translate* legacy SQL statements into a MongoDB equivalent. This is not at all desirable if your objective is to leverage the speed and power of MongoDB. Accordingly, for the purposes of what we're trying to illustrate, we will not use the native Django database libraries and instead rely on the custom Python modules discussed here.

Although we will not use the built-in Django database libraries, it's still a good idea to configure Django to use a database. The reason for this is that some internal Django web functionality relies upon an RDBMS. For the purposes of this book, you can install Django and use the default SQLite for its database. Remember that the SQLite database is *only* to be used at a minimal level. The main functionality of this website is to use MongoDB and the modules we will be developing throughout this book!

Documentation references: *django* (`https://github.com/django/django`), *flask* (`https://github.com/pallets/flask`), *pyramid* (`https://github.com/Pylons/pyramid`), and *cherrypy* (`https://github.com/cherrypy/cherrypy`).

Installing the web framework and MongoDB library

In this book, we will not cover how to install Django. For those of you who are new to Django, the best way to install Django is to follow the official installation documentation. For the purposes of this book, you should follow the installation guidelines documented in the Django `pip` installation documentation. In addition, we are using an Apache 2.4 web server.

In the Docker container we'll be using to demonstrate the concepts in this book, Python, PyMongo, Apache, `mod_wsgi`, and Django are already installed. A discussion of how to configure the web server to use Django is included since you need to know how to do this when setting up Django on a customer website.

Django installation documentation: `https://docs.djangoproject.com/en/2.2/topics/install/#how-to-install-django`
Django *pip* installation: `https://docs.djangoproject.com/en/2.2/topics/install/#installing-an-official-release-with-pip`

Apache configuration

For development purposes, you can simply use the Python built-in web server. In fact, after creating a Django project, a script is provided for that purpose. Simply change to the directory containing your Django project and run the following command:

```
python manage.py runserver <port>
```

This is not suitable for a production site, however. Accordingly, for the purposes of this book, we have configured the Apache web server installed in the demonstration Docker container so that it uses `mod_wsgi` to serve files from the Django project. You can find the Django documentation on the installation and configuration of `mod_wsgi` here: `https://docs.djangoproject.com/en/2.2/howto/deployment/wsgi/modwsgi/#how-to-use-django-with-apache-and-mod-wsgi`

In order to contain the required directives in a single location, we created a separate configuration file called `booksomeplace.local.conf` that defines an Apache virtual host that matches our `hosts` file entry, as discussed in the previous subsection. This file needs to be included in the primary Apache configuration file, which is either `apache2.conf` or `httpd.conf`, depending on the operating system and/or Linux distribution.

The first web server configuration directive loads the `wsgi` module. Please note that `XX` reflects the version of Python (for example, 37), while `YY` reflects the installation architecture (for example, x86_64):

```
LoadModule wsgi_module "/path/to/mod_wsgi-pyXX.cpython-XXm-YY-linux-gnu.so"
```

The document root for the virtual host is the main Django project directory, which, in this case, is under `/path/to/repo/www/chapter_07/booksomeplace_dj` in this book's repository. Note that we also need to define a `Directory` block that grants web server permissions to browse this directory:

```
<VirtualHost *:80>
    ServerName booksomeplace.local
    DocumentRoot "/repo/www/chapter_07/booksomeplace_dj"
    <Directory "/repo/www/chapter_07/booksomeplace_dj">
        AllowOverride All
        Require all granted
    </Directory>
```

Next, still within the virtual host definition, we'll include three `mod_wsgi` directives. The first, `WSGIScriptAlias`, maps all requests to this virtual host to the Django `wsgi.py` file for this project. We also include a `WSGIDaemonProcess` directive that tells `mod_wsgi` where Python is located, as well as where the main module for the project is located. In addition, we use this directive to indicate the user and group to use at runtime. Finally, we use `WSGIProcessGroup` to tie all the relevant WSGI directives together. In this case, of course, there is only one other directive that this applies to – `WSGIDaemonProcess`:

```
WSGIScriptAlias / /repo/www/chapter_07/booksomeplace_dj/wsgi.py
WSGIDaemonProcess booksomeplace python-home="/usr" \
    user=www-data group=www-data \
    python-path="/repo/www/chapter_07"
WSGIProcessGroup booksomeplace
```

Finally, we close the `Directory` block with an `Alias` directive, which allows us to serve static files out of a common directory called `assets`. This directory contains the CSS, JavaScript, and images needed for the website. We added another `Directory` directive that gives Apache the rights to browse through the `assets` directory:

```
Alias "/static" "/repo/www/chapter_07/assets"
<Directory  "/repo/www/chapter_07/assets">
    Require all granted
</Directory>
</VirtualHost>
```

Documentation references for Apache:

Directory directive: `http://httpd.apache.org/docs/2.4/mod/core.html#directory`

Alias: `https://httpd.apache.org/docs/current/mod/mod_alias.html`

Refer to the following documentation on serving static files (for example, images) in Django: `https://docs.djangoproject.com/en/2.2/howto/static-files/`

Book Someplace Django project structure

The Django project for Book Someplace is located in the `/path/to/repo/www/chapter_07` directory structure. The main project is `booksomeplace_dj`. This is where `wsgi.py`, the initial point of entry, is located. At this point, we have defined two Django apps: `home` and `booking`. The directory structure for `booksomeplace_dj` is shown on the left in the following diagram. In the middle and to the right of the diagram are the identical directory structures for the `home` and `booking` Django apps:

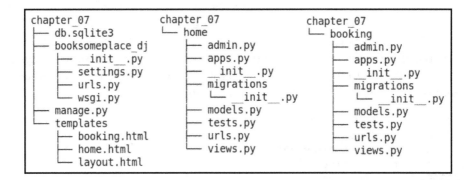

Note that for the purposes of illustrating Book Someplace, we retained the default SQLite database.

Modifications that are needed over and above the defaults will be discussed in the next section.

Making modifications to the main project

In order to provide mappings to the two URLs currently defined, modifications were made to the main project, booksomeplace_dj, and the two apps, home and booking. First of all, a urls.py file was created and placed in the directory for each of the two apps. The contents of the files are the same, as shown here:

```
from django.urls import path
from . import views
urlpatterns = [
    path('', views.index, name='index'),
]
```

Also, in the main project directory, /path/to/repo/www/chapter_07/booksomeplace_dj, the urls.py file was modified to provide mappings to the two apps, as shown here:

```
from django.contrib import admin
from django.urls import path, include
urlpatterns = [
    path('admin/', admin.site.urls),
    path('booking/', include('booking.urls')),
    path('', include('home.urls')),
]
```

Still in the directory for booksomeplace_dj, another file called settings.py modified the ALLOWED_HOSTS constant, adding booksomeplace.local as a host. The BASE_DIR constant ends up pointing to /path/to/repo/www. The SRC_DIR constant is used to indicate the location of the source code, other than that associated with Django:

```
ALLOWED_HOSTS = ['booksomeplace.local']
BASE_DIR = os.path.dirname(os.path.dirname(os.path.abspath(__file__)))
SRC_DIR = os.path.realpath(BASE_DIR + '/../../chapters/07/src')
```

We disable the Django cache by setting the built-in cache to `DummyCache`. This is only for development purposes:

```
CACHES = {
    'default': {
        'BACKEND': 'django.core.cache.backends.dummy.DummyCache',
    }
}
```

In the same file, the `TEMPLATES` constant tells Django where to find HTML templates:

```
TEMPLATE_DIR = os.path.realpath(BASE_DIR + '/templates')
TEMPLATES = [ {
        'BACKEND': 'django.template.backends.django.DjangoTemplates',
        'DIRS': [ TEMPLATE_DIR ],
        'APP_DIRS': True,
        'OPTIONS': {
            'context_processors': [
                'django.template.context_processors.debug',
                'django.template.context_processors.request',
                'django.contrib.auth.context_processors.auth',
                'django.contrib.messages.context_processors.messages',
            ],
        },
    },
]
```

And finally, we set the `STATIC_URL` constant, which then rewrites template requests for images, CSS, and JavaScript by prepending the `/static/` path. The Apache `Alias` directive (discussed earlier) maps such requests to the `assets` directory:

```
STATIC_URL = '/static/'
```

Now that the project modifications have been covered, let's have a look at the output: defining the `view` module.

Presenting the view

Finally, to present the view for bookings, we modify `/path/to/repo/www/chapter_07/booking/view.py`. In this file, we create an instance of the `PropertyService` class described previously. We then call the `fetchTop10()` method and pass the resulting `MongoDB\Cursor` instance to the template associated with this app:

```
from django.template.loader import render_to_string
from django.http import HttpResponse
from config.config import Config
from booksomeplace.domain.property import PropertyService
def index(request):
    config = Config()
    propService = PropertyService(config)
    main = render_to_string('booking.html', { 'top_10' :
propService.fetchTop10() })
    page = render_to_string('layout.html', {'contents' : main})
    return HttpResponse(page)
```

Now that you have an understanding of what goes into the view class, its time to have a look at the actual view template.

Displaying results in the view template

As mentioned in the configuration, all view templates are located in the `/path/to/repo/www/chapter_07/templates` directory. We modify the `booking.html` template, iterating through the results of `fetchTop10()`. Note that we are not showing the entire template here. In the template, we define a `for` loop that displays only the `property` name, and 10 words from the description:

```
<h3>Top Ten</h3>
<ul>
{% for property in top_10 %}
    <li>
        <b>{{ property.propName }}</b>
        <br>{{ property.propInfo.description|truncatewords:"10" }}
    </li>
{% endfor %}
</ul>
```

Here are the results provided by the browser when using the
`http://booksomeplace.local/booking/` URL:

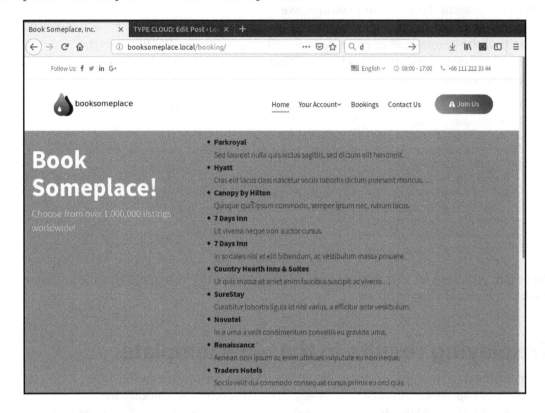

As you can see, the name and abbreviated description of the most popular properties are listed.

Summary

In this chapter, you learned how to review the requirements of an online booking agency with complex needs. You then learned that, due to the object-oriented nature of MongoDB, the document structure design process is completely free of the constraints imposed by legacy two-dimensional RDBMS tables. Accordingly, you can really dig deep into customer requirements and design subclasses from a very low level. These reusable data structures can then be group together like building blocks to form complex data structures. Once defined, you learned how easy it is to translate these structures into Python entity modules and classes.

You then learned how to define domain service classes that consume a MongoDB client and contain key methods needed by the web application. Finally, you learned how to modify a basic Django application so that it uses `mod_wsgi` under Apache. You were then shown how to incorporate the property domain service class to display a list of the top 10 properties.

In the next chapter, you'll learn how to work with the complex document structures defined in this chapter, especially embedded objects and arrays.

8

Using Documents with Embedded Lists and Objects

This chapter focuses on how to work with MongoDB documents that contain embedded lists (arrays) and objects. As you learned in the previous chapter, Chapter 7, *Advanced MongoDB Database Design*, the complex needs of *Book Someplace, Inc.* resulted in a database document structure built around a series of embedded documents. In addition, certain fields are stored as arrays. To further complicate application development, some fields consist of a list of objects! An example of this is the list of reviews that is associated with each property. In this chapter, you will learn how to handle create, read, update, and delete operations that involve embedded lists and objects. You will also learn how to perform queries involving embedded objects and lists.

This chapter covers the following topics:

- Inserting a document containing embedded objects and lists
- Updating arrays embedded within existing database documents
- Updating embedded documents
- Removing elements from embedded arrays
- Creating view methods that return JSON in response to AJAX queries

In the MongoDB documentation, you will see the term *array*. This corresponds to a Python *list*. In this chapter, we use the two terms interchangeably.

Technical requirements

The minimum recommended hardware is as follows:

- Desktop PC or laptop
- 2 GB free disk space
- 4 GB of RAM
- 500 Kbps or faster internet connection

The software requirements are as follows:

- OS (Linux or Mac): Docker, Docker Compose, and Git (optional)
- OS (Windows): Docker for Windows and Git for Windows
- Python 3.x, the PyMongo driver, and Apache (already installed in the Docker container used for the book)

The installation of the required software and how to restore the code repository for the book is explained in Chapter 2, *Setting Up MongoDB 4.x*. To run the code examples in this chapter, open a terminal window (Command Prompt) and enter these commands:

```
cd /path/to/repo
docker-compose build
docker-compose up -d
docker exec -it learn-mongo-server-1 /bin/bash
```

To run the demo website described in this chapter, once the Docker container for Chapters 3 through 12 has been configured and started, there are two more things you need to do:

1. Add the following entry to the local hosts file on your computer:

   172.16.0.11   chap08.booksomeplace.local

   The local hosts file is located at
   C:\windows\system32\drivers\etc\hosts on Windows, and for
   Linux and Mac, at /etc/hosts.

2. Open a browser on your local computer (outside of the Docker container) and go to this URL:

   http://chap08.booksomeplace.local/

When you are finished working with the examples covered in this chapter, return to the terminal window and stop Docker as follows:

```
cd /path/to/repo
docker-compose down
```

The code used in this chapter can be found in the book's GitHub repository at `https://github.com/PacktPublishing/Learn-MongoDB-4.x/tree/master/chapters/08` and `https://github.com/PacktPublishing/Learn-MongoDB-4.x/tree/master/www/chapter_08`.

Overview

Before we get into the technical details of the implementations described in this chapter, it's necessary to provide an overall view of what we plan to achieve. Django provides an object-to-form mapping by way of its `ModelForm` class. However, in order to use this class we would need to create Django `Models`, which correspond to the `entity` classes we have described to this point.

The problem with using Django `Models` and `ModelForm` classes is that we end up being tied to SQL and the RDBMS way of thinking, which imposes incredible constraints on a MongoDB implementation, and in effect defeats the purpose of using MongoDB in the first place. Such an approach introduces a massive amount of overhead, with additional functionality that is never used. Instead, our proposed solution for adding data to MongoDB documents with embedded objects and arrays is to do the following:

- Create a Django form class that corresponds to each `entity` class.
- Move the posted data from each sub-form into its corresponding `entity` object.
- Perform any additional processing on the `entity` object as needed.
- Move the `entity` object into its position as an object embedded in a MongoDB object.

In the case of editing documents, the reverse operation is performed:

- MongoDB database to objects
- Objects to sub-forms
- Form fields are modified
- Form posting to objects
- Objects to the MongoDB database

With this solution, we adhere to a strict object-oriented approach from beginning to end, which ensures flexibility, and the ability to perform adequate testing. The following diagram visually captures the flow:

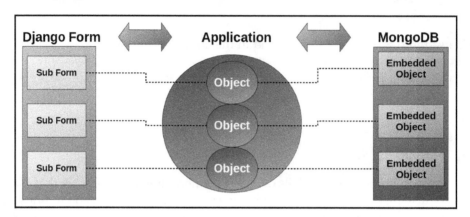

In this chapter, we define four tasks that are critical to the day-to-day operations of *Book Someplace*: adding a new property, adding room types, updating property contact information, and removing a property listing. The corresponding skills these tasks require include inserting a document with embedded objects and lists, adding to an embedded list, updating an embedded object, and deleting a document.

Let's start by examining adding a new property, which involves inserting a document with embedded objects and arrays.

Adding a document with embedded objects and arrays

In this scenario, a partner has logged into the website and wants to add a new property. As you recall from our discussion on the data structure for *Book Someplace*, documents in the properties collection have the following data structure:

```
Property = {
    "propertyKey" : <string>,
    "partnerKey"  : <string>,
    "propName"    : <name of this property>,
    "propInfo"    : <PropInfo>,
    "address"     : <Location>,
    "contactName" : <Name>,
    "contactInfo" : <Contact>,
```

```
    "rooms"       : [<RoomType>,<RoomType>,etc.],
    "reviews"     : [<Review>,<Review>,etc.],
    "rating"      : <int>,
    "totalBooked" : <int>
}
```

You might note that each `Property` documents includes embedded object classes, all under `booksomeplace.entity.*:`, such as `PropInfo`, `Location`, `Name`, and `Contact`. In addition, there are two lists: `rooms`, a list of `RoomType` objects, and `reviews`, a list of `Review` objects. Because the property is new, we do not allow a partner to enter information pertaining to `reviews`, `rating`, nor `totalBooked`. These are ultimately generated by the customers who book the property. Further, the `propertyKey` and `partnerKey` fields are generated by the application, which leaves only basic property information and room types to be defined by the partner.

We do not cover the details of login, authentication, and authorization in Django as that is beyond the scope of this book. If you would like more details on the Django authentication framework, please consult the documentation at `https://docs.djangoproject.com/en/2.2/topics/auth/`.

In `Chapter 11`, *Administering MongoDB Security*, we discuss ways to tie Django authentication into a MongoDB-based application.

There is an excellent book on Django published by Packt entitled *Django 2 Web Development Cookbook - Third Edition*, which includes a good section on authentication (`https://www.packtpub.com/web-development/django-2-web-development-cookbook-third-edition`).

Continuing with our focus on the object-oriented approach, the next logical step is to define Django `Form` classes to match each subclass representing a single property. For this purpose, we create a new Django app called property. In that app we define property-related forms in the `forms.py` module, located in `/path/to/repo/www/chapter_08/property`.

As always, we start with a set of `import` statements:

```
from django import forms
from django.core import validators
from django.core.validators import RegexValidator, EmailValidator
import os,sys
from config.config import Config
from booksomeplace.entity.base import Name, Location, Contact
from booksomeplace.entity.property import Property, PropInfo, RoomType
from booksomeplace.domain.partner import PartnerService
from booksomeplace.domain.property import PropertyService
from booksomeplace.domain.common import CommonService
```

In this illustration, we do not use Django form classes based on `ModelForm`. The reason is because `django.db.models` (https://docs.djangoproject.com/en/2.2/topics/db/models/#module-django.db.models) are inextricably linked to the relational model, and would only cause performance degradation and conversion problems when Django attempts to connect form output to classes used by our MongoDB database. Accordingly, we prefer to use standard Django forms, which are then linked to our own custom `entity` classes using a custom form service (discussed later in this section).

Defining forms for property subclasses

Next, we define a `NameForm` class that works with `booksomeplace.entity.base.Name`. In this class, we first create an instance of `booksomeplace.domain.common.CommonService` in order to retrieve values for titles (Mr, Ms, and so on). You see that we also implement `RegexValidator` for first and last names:

```
class NameForm(forms.Form) :
    config    = Config()
    commonSvc = CommonService(config)
    titles    = [(item, item) for item in commonSvc.fetchByKey('title')]
    alphaOnly = RegexValidator(r'^[a-zA-Z]*$', 'Only letters allowed.')
    name_title = forms.ChoiceField(label='Title',choices=titles)
    name_first = forms.CharField(label='First Name', \
                                 validators=[alphaOnly])
    name_middle= forms.CharField(label='M.I.',required=False)
    name_last  = forms.CharField(label='Last Name',validators=[alphaOnly])
    name_suffix= forms.CharField(label='Suffix',required=False)
```

In this illustration, we do not implement validators for all form fields. In a production environment, however, it is highly recommended that you filter, validate, and sanitize all incoming data!

In a similar manner, we define a `PropInfoForm` form class to capture property information. The drop-down lists needed include property types, facilities, and currencies. These are drawn from the common collection:

```
class PropInfoForm(forms.Form) :
    config    = Config()
    commonSvc = CommonService(config)
    facTypes  = [(item, '... ' + item) for item in \
                   commonSvc.fetchByKey('facilityType')]
    propTypes = [(item, item) for item in \
                   commonSvc.fetchByKey('propertyType')]
    currTypes = [(item, item) for item in commonSvc.fetchByKey('currency')]
    alnumPunc = RegexValidator(r'^[\w\.\-\,\"\' ]+$',
        'Only letters, numbers, spaces and punctuation are allowed.')
```

This class must also define fields associated with the document structure for the `properties` collection:

```
info_type  = forms.ChoiceField(label='Property Type', \
    choices=propTypes)
info_chain = forms.CharField(label='Chain', \
    validators=[validators.validate_slug],required=False)
info_photos = forms.URLField(label='Photo URL',required=False)
info_other_fac = forms.CharField(label='Other Facilities', \
    required=False)
info_currency = forms.ChoiceField(label='Currency',choices=currTypes)
info_taxFee = forms.FloatField(label='Taxes/Fees')
info_description= forms.CharField(
label='Description',validators=[alnumPunc],
widget=forms.Textarea(attrs={'rows' : 4, 'cols' : 80}))
```

Of special interest is how the property facilities are treated. These are displayed as a row of check boxes in the final form. However, we need to account for the fact that more than one can be marked, thus we need the Django `MultipleChoiceField` form class and the `CheckboxSelectMultiple` widget. When the form is submitted, this field, as it is a list, needs additional processing, described later in this section:

```
info_facilities = forms.MultipleChoiceField(
    label='Facilities',choices=facTypes,
    widget=forms.CheckboxSelectMultiple( \
        attrs={'class': 'check_horizontal'}))
```

As you would expect, the form that corresponds to the `Location` subclass is uneventful, as shown here:

```
class LocationForm(forms.Form) :
    iso2 = RegexValidator(r'^[A-Z]{2}$', \
      'Country code must be a valid ISO2 code (2 letters UPPERCASE).')
    location_streetAddress = forms.CharField(label ='Street Address')
    location_buildingName = forms.CharField(label ='Building Name',\
        required=False)
    location_floor = forms.CharField(label ='Floor',required=False)
    location_roomAptCondoFlat = forms.CharField(label ='Room/Condo', \
        required=False)
    location_city          = forms.CharField(label ='City')
    location_stateProvince = forms.CharField(label ='State/Province')
    location_locality      = forms.CharField(label ='Locality', \
        required=False)
    location_country       = forms.CharField(label ='Country',
        validators=[iso2])
    location_postalCode    = forms.CharField(label ='Postal Code')
    location_latitude      = forms.CharField(label ='Latitude')
    location_longitude     = forms.CharField(label ='Longitude')
```

Likewise, the form to capture contact information is very straightforward, as you can see here:

```
class ContactForm(forms.Form) :
    config    = Config()
    commonSvc = CommonService(config)
    socTypes  = [(item, item) for item in \
        commonSvc.fetchByKey('socMediaType')]
    validEmail = EmailValidator(message="Must be a valid email address")
    contact_email    = forms.EmailField(label ='Email', \
        validators=[validEmail])
    contact_phone    = forms.CharField(label='Phone')
    contact_soc_type = forms.ChoiceField(label='Social Media', \
        choices=socTypes,required=False)
```

```
contact_soc_addr = forms.CharField(label='Social Media Identity',
    required=False)
```

The `property` form itself is an anti-climax. It only needs to capture the partner key and property name:

```
class PropertyForm(forms.Form) :
    config      = Config()
    partSvc     = PartnerService(config)
    partKeys    = partSvc.fetchPartnerKeys()
    alnumPunc = RegexValidator(r'^[\w\.\-\,\"\' ]+$', \
        'Only letters, numbers, spaces and punctuation are allowed.')
    prop_partnerKey = forms.ChoiceField(label='Partner',choices=partKeys)
    prop_propName   = forms.CharField(label='Property Name', \
        validators=[alnumPunc])
```

Let's now define templates to match sub-forms in the next section.

Defining templates to match sub-forms

To foster re-usability, each sub-form is associated with its own template. As an example, the `Name` subclass is associated with `NameForm`, which in turn is rendered using the `property_name.html` template, shown here:

```
{% load static %}
<h3>Contact Name</h3>
<div class="row">
    <div class="col-md-4">{{ form.name_title.label_tag }}</div>
    <div class="col-md-4">{{ form.name_title }}</div>
    <div class="col-md-4">{{ form.name_title.errors }}</div>
</div>
<div class="row">
    <div class="col-md-4">{{ form.name_first.label_tag }}</div>
    <div class="col-md-4">{{ form.name_first }}</div>
    <div class="col-md-4">{{ form.name_first.errors }}</div>
</div>
<!-- other fields not shown -->
```

Each sub-form has its own template. When all sub-forms are rendered, the resulting HTML is then injected into a master form template, `property.html`, a portion of which is shown here:

```
<form action="/property/edit/" method="post">
    {% csrf_token %}
    <h1>Add/Edit Property</h1>
    Name of Property: {{ prop_name }}
    <div class="row extra" style="background-color: #E6E6FA;">
        <div class="col-md-6">{{ name_form_html }}</div>
        <div class="col-md-6" style="background-color: #D6D6F0;">
            {{ contact_form_html }}
        </div><!-- end col -->
    </div><!-- end row -->
    <div class="row extra" style="background-color: #E5E5E5;">
        <div class="col-md-8">{{ info_form_html }}</div>
        <div class="col-md-4"> </div>
    </div><!-- end row -->
<input type="submit" value="Submit">
</form>
{{ message }}
```

We will define a `FormService` class in the next section.

Defining a FormService class

As we need a way to manage all the sub-forms and subclasses, the most expedient approach is to develop a `FormService` class. We start by defining a property to contain form instances, and also a list of fields:

```
class FormService() :
    forms = {}
    list_fields = ['info_facilities','room_type_beds']
```

The constructor updates the internal `forms` property with instances of each sub-form:

```
def __init__(self) :
    self.forms.update({ 'prop' : PropertyForm() })
    self.forms.update({ 'name' : NameForm() })
    self.forms.update({ 'info' : PropInfoForm() })
    self.forms.update({ 'contact' : ContactForm() })
    self.forms.update({ 'location' : LocationForm() })
```

It's also necessary to define a method that retrieves a specific sub-form:

```
def getForm(self, key) :
    if key in self.forms :
        return self.forms[key]
    else :
        return None
```

We also define a `hydrateFormsFromPost()` method that populates sub-forms with data coming from `request.POST`:

```
def hydrateFormsFromPost(self, data) :
    self.forms['prop'] = PropertyForm(self.getDataByKey('prop', data))
    self.forms['name'] = NameForm(self.getDataByKey('name', data))
    self.forms['info'] = PropInfoForm(self.getDataByKey('info', data))
    self.forms['contact'] = ContactForm( \
        self.getDataByKey('contact', data))
    self.forms['location'] = LocationForm( \
        self.getDataByKey('location', data))
```

However, since the form posting data, captured in the Django `request` object, is returned as a single list, we need a way to filter out fields specific to each sub-form. This is performed by the `getDataByKey()` method, which simply scans the post data for fields with a prefix that matches the sub-form key:

```
def getDataByKey(self, key, data) :
    prefix = key + '_'
    formData = {}
    for field, value in data.items() :
        if field[0:len(prefix)] == prefix :
            if field in self.list_fields :
                value = data.getlist(field)
            formData.update({field : value})
    return formData
```

We also need a method to validate all sub-forms. This is accomplished by the `is_valid()` method:

```
def validateForms(self) :
    count = 0
    valid = 0
    for key, sub_form in self.forms.items() :
        count += 1
        if sub_form.is_valid() : valid += 1
    return count == valid
```

Post validation, we define a method that extracts clean data into a dictionary and ultimately populates the corresponding subclass:

```
def getCleanDataByKey(self, key) :
    prefix    = key + '_'
    start     = len(prefix)
    cleanData = {}
    for formField, value in self.forms[key].cleaned_data.items() :
        docField = formField[start:]
        cleanData.update({docField : value})
    return cleanData
```

Since the code for sub-form hydration starting with `NameForm` is the same, we leave off the code for hydrating the remaining sub-forms. Please have a look at `/path/to/repo/www/chapter_08/property/forms.py` to see the completed code block.

We also need a method to hydrate sub-forms from a `Property` instance. For that purpose, we define a method called `hydrateFormsFromEntity()`:

```
def hydrateFormsFromEntity(self, prop) :
    # hydrate Property form
    formData = {}
    formData.update({'prop_partnerKey' : prop.get('partnerKey')})
    formData.update({'prop_propName'   : prop.get('propName')})
    self.forms['prop'] = PropertyForm(formData)
    # hydrate NameForm
    formData = {}
    subclass = prop.get('contactName')
    for field, value in subclass :
        formData.update({'name_' + field : value})
    self.forms['name'] = NameForm(formData)
    # other sub-forms are hydrated in a manner identical to NameForm
```

Finally, the getPropertyFromForms() method creates a Property entity instance that can then be used by our application:

```
def getPropertyFromForms(self) :
    prop_form = self.forms['prop']
    data = {
        'partnerKey'  : prop_form.cleaned_data.get('prop_partnerKey'),
        'propName'    : prop_form.cleaned_data.get('prop_propName'),
        'propInfo'    : PropInfo(self.getCleanDataByKey('info')),
        'address'     : Location(self.getCleanDataByKey('location')),
        'contactName' : Name(self.getCleanDataByKey('name')),
        'contactInfo' : Contact(self.getCleanDataByKey('contact')),
    }
    return Property(data)
```

Next, let's define the view logic to handle form rendering and posting.

Defining view logic to handle form rendering and posting

Within the Django property app, let's turn our attention to views.py. At the top of the module, the classes we need are imported:

```
from django.http import HttpResponse, HttpResponseRedirect
from django.template.loader import render_to_string
from .forms import FormService, PropertyForm, NameForm,
    PropInfoForm, LocationForm, ContactForm, RoomTypeForm
import os,sys
from config.config import Config
from booksomeplace.entity.base import Base, Name, Location, Contact
from booksomeplace.entity.property import Property, PropInfo, RoomType
from booksomeplace.domain.property import PropertyService
```

The addEdit() method uses the same set of forms for either adding a new room or updating an existing one. The first thing we need to do in this method is to retrieve an instance of the property and form services:

```
def addEdit(request, prop_key = None) :
    prop_name = 'NEW'
    add_or_edit = 'ADD'
    mainConfig  = Config()
    propService = PropertyService(mainConfig)
    formSvc     = FormService()
    message     = 'Please enter appropriate form data'
```

It's important to determine whether or not a `property` key was passed as a parameter. This parameter is defined in the URL, described in the next section. We make a call to the `fetchByKey()` method of the `booksomeplace.domain.property.PropertyService` class. If we get a hit, we record the property name, and assign `add_or_edit` a value of `EDIT`. In addition, we call the `hydrateFormsFromEntity()` form service method, which populates all sub-forms from a `Property` instance:

```
if prop_key :
    prop = propService.fetchByKey(prop_key)
    if prop :
        prop_name = prop.get('propName')
        add_or_edit = 'EDIT'
        formSvc.hydrateFormsFromEntity(prop)
```

It is also necessary to check and see if the request method was `POST`. If so, we use the form service to populate all sub-forms with the appropriate request data. We then use the form service to validate all sub-forms. If validation is successful, and the operation is to add a property, we call the `save()` method of the `PropertyService` class. If the save operation was successful, we redirect to the page that lets us start adding rooms to the property:

```
if request.method == 'POST':
    formSvc.hydrateFormsFromPost(request.POST)
    if formSvc.validateForms() :
        prop = formSvc.getPropertyFromForms()
        if add_or_edit == 'ADD' :
            if propService.save(prop) :
                message = 'Property added successfully!'
                return HttpResponseRedirect('/property/room/add/' + \
                    prop.getKey())
            else :
                message = 'Sorry! Unable to save form data.'
```

Otherwise, we call the `edit()` method (discussed in the next subsection) to perform an update. If any operation fails, an appropriate message is displayed:

```
        else :
            if propService.edit(prop) :
                message = 'Property updated successfully!'
                return HttpResponseRedirect('/property/list/' + \
                    prop.getKey())
            else :
                message = 'Sorry! Unable to save form data.'
    else :
        message = 'Sorry! Unable to validate this form.'
```

If the request method is not POST, we assume forms just need to be rendered as they stand. Because we are using two levels of form templates, we use the Django render_to_string() method. The results are then rendered into the master template, layout.html:

```
context = { 'message' : message, 'prop_name' : prop_name }
for key, sub_form in formSvc.forms.items() :
    form_var = key + '_form_html'
    form_file = 'property_' + key + '.html'
    context.update({form_var : render_to_string(form_file, \
                    {'form': sub_form})})
prop_master = render_to_string('property.html', context)
page = render_to_string('layout.html', {'contents' : prop_master})
return HttpResponse(page)
```

Next, let's define a route to addEdit().

Defining a route to addEdit()

In order to process requests to add a property, we need to define a route to the appropriate view. This is accomplished in the urls.py file. The main website routes are defined in the project's urls.py file, found in the /path/to/repo/www/chapter_08/booksomeplace_dj directory. Also note the added entry for the property app:

```
from django.contrib import admin
from django.urls import path, include
urlpatterns = [
    path('admin/', admin.site.urls),
    path('booking/', include('booking.urls')),
    path('property/', include('property.urls')),
    path('', include('home.urls')),
]
```

In the `property/urls.py` file, each entry also uses the special `path()` function, which takes three arguments:

- The URL fragment that is appended to `/property`
- The view method to be called if this URL is matched
- A key that is assigned to this route

In order to add a new property, from the client we would enter the URL `/project/edit/`. If, on the other hand, we had a specific property to edit, the URL would be `/project/edit/<PROPKEY>/`, as you can see from the file shown here:

```
from django.urls import path
from . import views
urlpatterns = [
    path('edit/', views.addEdit, name='edit_property_new'),
    path('edit/<slug:prop_key>/', views.addEdit, name='edit_property'),
]
```

Let's now generate the property key.

Generating the property key

Before a new property can be saved, a property key needs to be generated from the property name. The best place to do this is in the `entity` class itself. Accordingly, we add a new method, `generatePropertyKey()`, to the `booksomeplace.entity.property.Property` class, as shown here. If for some reason the property name is less than four characters long, we append an underscore (_) character.

We also remove any spaces and make the first four letters uppercase. Finally, a random four-digit number is appended:

```
def generatePropertyKey(self) :
    import random
    first4 = self['propName']
    first4 = first4.replace(' ', '')
    first4 = first4[0:4]
    first4 = first4.ljust(4, '_')
    self['propertyKey'] = first4.upper() + \
        str(random.randint(1000, 9999))
    return self['propertyKey']
```

Next, we will update the property domain service.

Updating the property domain service

We also need to add a `save()` method to the `booksomeplace.domain.property.PropertyService` class. This method needs to accept a `Property` instance as an argument, retrieved from cleaned form data using the `getPropertyFromForms()` form service method described earlier in this section. As our `PropertyService` class is already geared toward working with `entity` classes, the code is absurdly simple, as you can see here:

```
def save(self, prop) :
    prop.generatePropertyKey()
    return self.collection.insert_one(prop)
```

Next, we will learn to add a new property.

Adding a new property

We can then enter the URL
`http://chap08.booksomeplace.local/property/edit/` into the browser. Here is the
resulting screen, showing the sub-forms for each subclass:

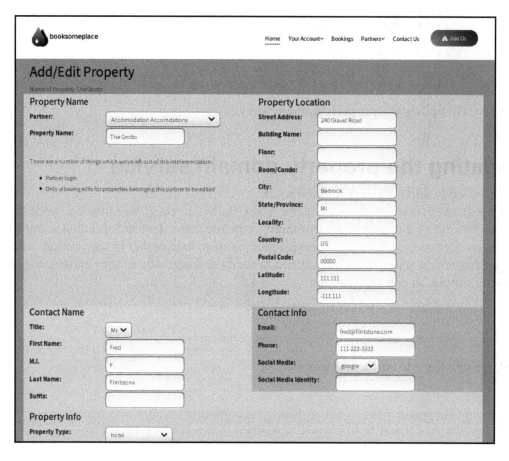

If any of the fields fail validation, the form is re-displayed, with error messages next to the
appropriate form fields. Once the form has been completed successfully, the user is
redirected to a form where room types for this property can be added.

On a production website, you would also need to invoke authentication
and authorization Django middleware. This would ensure that only
partners or admins would be allowed to enter and update property
information. Further, the authentication process could capture the partner
key, which would eliminate the need for the partner key to be a form field.

Adding to arrays embedded within documents

In this sub-section, we address adding rooms to a property. Room definitions take the form of an array (list) of `RoomType` objects. Thus, the task at hand is doubly complicated in that we need to work not only with embedded objects, but embedded lists as well.

Defining an add room form

The `/path/to/repo/www/chapter_08/property/forms.py` file represents a module containing a form to capture room type information that needs a set of multiple check boxes for room and bed types. The actual lists of types are drawn from `common` collection documents under the `roomTypes` and `bedTypes` keys. We use the `CommonService` domain service class to perform lookups for both keys. Each document in the common collection has two fields: key and value. The information contained in the value field is in the form of a list.

We also need a template to match the form. The Django template language is used, and provides safety measures in case of an error. The complete form template is located at `/path/to/repo/www/chapter_08/templates/property_room_type.html`.

Defining the view method and URL route

We need a way to enter room types for a given property. This is seen in the `/path/to/repo/www/chapter_08/property/views.py` file. The `addRoom()` method looks for a URL parameter representing the property key. As you recall from the discussion in the earlier sub-section, after filling in the form for a new property, the *Book Someplace* admin is directed to this view.

The first logical block in this method verifies the incoming property key parameter. If it is present and correct, we use the `PropertyService` to perform a lookup, and retrieve a `Property` entity instance:

```
def addRoom(request, prop_key = None):
    error_flag = True
    prop_name  = 'Unknown'
    message    = 'Please enter room information'
    propService = PropertyService(Config())
    if prop_key :
        prop = propService.fetchByKey(prop_key)
        if prop :
            prop_name = prop.get('propName')
            error_flag = False
        else :
            message = 'ERROR: property unknown: unable to add rooms'
    else :
        message = 'ERROR: property unknown: unable to add rooms'
```

The next several lines of code create an instance of the room form described earlier, and proceed if the request method is POST, and `error_flag` is `False`. `error_flag` is only set to `False` when a valid `Property` instance has been retrieved:

```
room_form = RoomTypeForm()
if request.method == 'POST' and not error_flag :
    if 'done' in request.POST :
        return HttpResponseRedirect('/property/list/' + prop.getKey())
    room_form = RoomTypeForm(request.POST)
```

If the room form is valid, we retrieve clean data from the form:

```
if room_form.is_valid() :
    start = len('room_type_')
    cleanData = {}
    for formField, value in room_form.cleaned_data.items() :
        docField = formField[start:]
        cleanData.update({docField : value})
```

We use the `PropertyService` to save the room, passing the `RoomType` instance and the property key as arguments:

```
    if propService.saveRoom(RoomType(cleanData), prop_key) :
        message = 'Added room successfully'
    else :
        message = 'Sorry! Unable to save form data.'
else :
    message = 'Sorry! Unable to validate this form.'
```

If the request was not POST, or if the form fails validation, we render the room type form and produce a response:

```
room_form_html = render_to_string('property_room_type.html', {
    'form' : room_form,
    'message' : message,
    'prop_key' : prop_key,
    'prop_name' : prop_name,
    'error_flag' : error_flag
})
page = render_to_string('layout.html', {'contents' : room_form_html})
return HttpResponse(page)
```

We also need a path to the new view method. This is defined in /path/to/repo/www/chapter_08/property/urls.py, as shown here. Please note that to conserve space we do not show other paths:

```
urlpatterns = [
    # other paths not shown
    path('room/add/<slug:prop_key>/', views.addRoom,
name='property_add_room'),
]
```

Now we are ready to examine adding an embedded array in MongoDB.

Updating the database

The last piece of the puzzle is to define a method in the property domain service class that accepts form data and updates the database. Because we are adding room types to an embedded list within a document in the properties collection, the database operation we need to perform is not an insert, but rather an *update*.

MongoDB update operators

MongoDB update operators (https://docs.mongodb.com/manual/reference/operator/update/#update-operators) are needed in order to effect changes when performing an update. Update operators are categorized according to what element they affect: *fields*, *arrays*, *modifiers*, and *bitwise*.

The following table summarizes the more important update operators:

Affects ...	Operator	Description
Fields	`$rename`	Allows you to rename a field
	`$set`	Overwrites the current value of the field with the new, updated, value
	`$unset`	Deletes the current value of a field
Arrays	`$`	Placeholder for first element that matches the filter query criteria
	`$[]`	Placeholder for *all* elements that match the filter query criteria
	`$addToSet`	Adds an element to the array only if it doesn't already exist
	`$push`	Adds an item to the array
Modifiers	`$each`	Works with `$push` to append multiple items to the array
	`$sort`	Works with `$push` to re-order the array
Bitwise	`$bit`	Used to perform bitwise operations AND, OR, and XOR on integer values

Next, let's look at the domain service method to add rooms.

Domain service method to add rooms

Before we can define the domain service to add a room, we need to ensure that individual room types can be independently selected. This is accomplished by generating a unique room type key. We thus add a new method, `generateRoomTypeKey()`, to the `booksomeplace.entity.property.RoomType` class, as shown here:

```
def generateRoomTypeKey(self) :
    import random
    first4 = self['type']
    first4 = first4.replace(' ', '')
    first4 = first4[0:4]
    first4 = first4.ljust(4, '_')
    self['roomTypeKey'] = first4.upper() + \
                          str(random.randint(1000, 9999))
    return self['roomTypeKey']
```

We are now in a position to add a `saveRoom()` method to `PropertyService` in the `booksomeplace.domain.property` module. This method adds the new `RoomType` to the embedded list in the `properties` collection. Note the use of the `$push` array operator:

```
def saveRoom(self, roomType, prop_key) :
    success = False
    prop = self.fetchByKey(prop_key)
    if prop :
        updateDoc = {'$push' : { 'rooms' : roomType }}
        query = {'propertyKey' : prop.getKey()}
        success = self.collection.update_one(query, updateDoc)
    return success
```

We will use the form in the next section.

Using the form

In the browser on our local computer, we now test our new feature by entering this URL:

`http://chap08.booksomeplace.local/property/room/add/MAJE2976/`

We are then presented with the `RoomType` form, as shown in the following screenshot:

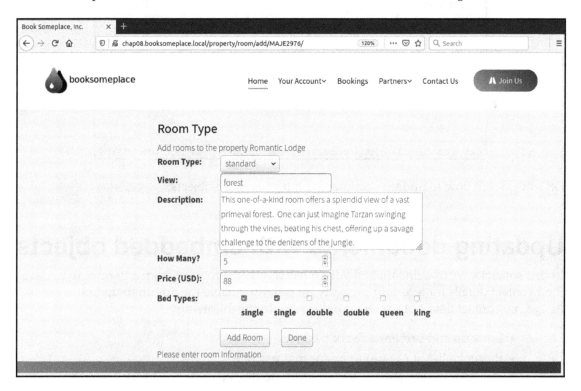

We can confirm the new addition from the *mongo* shell, as shown here:

```
File  Edit  View  Search  Terminal  Help
         {
                         "roomTypeKey" : "STAN9675",
                         "type" : "standard",
                         "view" : "forest",
                         "description" : "This one-of-a-kind room offers a
  splendid view of a vast primeval forest.  One can just imagine Tarzan sw
inging through the vines, beating his chest, offering up a savage challen
ge to the denizens of the jungle.",
                         "beds" : [
                                         "single",
                                         "single"
                         ],
                         "numAvailable" : 5,
                         "numBooked" : 0,
                         "price" : 88
                 }
         ]
}
> db.properties.findOne({"propertyKey":"MAJE2976"},{"_id":0,"rooms":1});
```

Let's now learn how to update documents with embedded objects.

Updating documents with embedded objects

In this scenario, we imagine that an admin has received a request from a partner to update their contact details for a certain property. In order to enable the admin to update `Property` contact details, the web app needs to do the following:

- Look up and present a list of partners.
- Present a list of `properties` for that partner.
- Get current information on the property that needs be updated.
- Present two web forms named `NameForm` and `ContactForm`.
- Collect form data into `Contact` and `Name` instances.
- Update the `properties` collection.

The traditional approach would be to first look up a list of partners and then direct the admin to a screen where they would choose the partner to work on. The form posting would then present the admin with another form where they could choose from a list of properties for that partner. Finally, a third form would appear with the information to be updated. In total, this represents a cumbersome three-stage process, during which the application would need to carry along both the partner and property keys, either through hidden form properties or by using a session mechanism.

By incorporating JavaScript functions, we can accomplish the same thing using just a single form. The JavaScript functions make AJAX requests to backend views that deliver JSON responses. Using the jQuery `change()` method, we configure the form such that when the element containing the list of partners changes, an AJAX request is made to repopulate the list of properties such that the list represents only properties for the selected partner. In a similar manner, once the property has been selected, we populate all pertinent contact detail fields.

We start this discussion by presenting the form field that presents the list of partners.

Choosing the partner

In order to choose a partner to work with, we simply create a form with a single field, which we can populate with values obtained from the database. In order to accomplish this, we need the following:

- A Django form with a `ChoiceField()`
- A domain service method that performs the lookup that populates the field

For this purpose, in the `property` app within our Django website, we define the following form:

```
class ChoosePartnerForm(forms.Form) :
    partSvc     = PartnerService(Config())
    partKeys    = partSvc.fetchPartnerKeys()
    alnumPunc = RegexValidator(r'^[\w\.\-\,\"\' ]+$',
        'Only letters, numbers, spaces and punctuation are allowed.')
    prop_partnerKey = forms.ChoiceField(label='Partner',choices=partKeys)
```

We also need a method in the partner domain service to retrieve partner names and keys for display, which we call `fetchPartnerKeys()`. We use the `projection` argument to the `pymongo.collection.find()` method to limit results to the `partnerKey` and `businessName` fields. We also define sort parameters so that partner business names are presented in alphabetical order. The method returns a dictionary that consists of key/value pairs, where the key is the `partnerKey` field and the value is the `businessName`:

```
def fetchPartnerKeys(self) :
    query       = {}
    projection = {'partnerKey' : 1, 'businessName' : 1}
    sortDef     = [('businessName', pymongo.ASCENDING)]
    skip        = 0
    limit       = 0
    cursor      = self.collection.find(query, projection, skip, \
```

```
                     limit, False, CursorType.NON_TAILABLE, sortDef)
temp        = {}
for doc in cursor :
    temp.update({doc['partnerKey'] : doc['businessName']})
return temp.items()
```

Next, we present the methods that produce JSON results in response to AJAX queries.

Partner and property AJAX lookups

In order to respond to an AJAX query, we define matching domain service methods that look up requested values. Also needed are the corresponding view methods to call the domain service methods and produce JSON responses. Accordingly, in booksomeplace.domain.partner.PartnerService we define a method called fetchByPartnerKey() that retrieves a list of property keys and names:

```
def fetchByPartnerKey(self, key) :
    query  = {'partnerKey' : key}
    proj   = {'propertyKey' : 1, 'propName' : 1, '_id' : 0}
    result = self.collection.find(query, proj)
    data   = []
    if result :
        for doc in result :
            data.append(doc)
    return data
```

The other method needed is fetchContactInfoByKey(), called when a property has been selected. Again we use the projection argument to limit the information returned to the embedded object's *Name*, represented by contactName, and *Contact*, represented by contactInfo.

In this illustration we use the PyMongo find_one() collection method, as the property key is unique:

```
def fetchContactInfoByKey(self, key) :
    query  = {'propertyKey' : key}
    proj   = {'contactName' : 1, 'contactInfo' : 1, '_id' : 0}
    result = self.collection.find_one(query, proj)
    return result
```

In the Django `booking` app, found in the `/path/to/repo/www/chapter_08/` directory, we define two view methods. The first method, `autoPopPropByPartKey()`, automatically populates the list of properties based upon the partner key. In this method, shown here, we return a JSON response:

```
def autoPopPropByPartKey(request, part_key = None) :
    import json
    propSvc = PropertyService(Config())
    result  = []
    if part_key :
        result = propSvc.fetchByPartnerKey(part_key)
    return HttpResponse(json.dumps(result),
content_type='application/json')
```

The second important method is `autoPopContactUpdate()`, which populates the appropriate contact information field once a property is chosen:

```
def autoPopContactUpdate(request, prop_key = None) :
    import json
    propSvc = PropertyService(Config())
    result  = []
    if prop_key :
        result = propSvc.fetchContactInfoByKey(prop_key)
    return HttpResponse(json.dumps(result),
content_type='application/json')
```

After creating these methods, we add a URL that maps to each method described just now. The first path requires a partner key to produce results. The other URL path takes a property key. These paths are placed in the `urls.py` file located in the `/path/to/repo/www/chapter_08/booking` directory, shown here:

```
urlpatterns = [
    path('json/by_part/<slug:part_key>/', views.autoPopPropByPartKey,
        name='property_json_props_by_part_key'),
    path('json/update_contact/<slug:prop_key>/', \
        views.autoPopContactUpdate, \
        name='property_json_update_contact'),
    # other paths not shown
]
```

The next step is to define the contact information form and view, which brings everything together.

You can actually test to see whether the AJAX response method is working properly by simply calling the URL directly from your browser and supplying a valid partner key: `http://chap08.booksomeplace.local/property/json/by_part/<slug:part_key>/`.

Contact information form and view

In order to put all the pieces described in this sub-section together, we need the following:

- A view to render the sub-forms needed to collect contact information.
- View logic to capture and validate form post data.
- A template to incorporate the sub-forms and JavaScript functions.
- A domain service to update the `properties` collection.

A new view method, `updateContact()`, in `/path/to/repo/www/chapter_08/property/views.py`, creates instances of the form service described in the previous sub-section. In addition, it retrieves instances of both the partner and property domain services. The `check` variable represents the number of validation checks to be performed:

```
def updateContact(request) :
    formSvc = FormService()
    message = ''
    config  = Config()
    partSvc = PartnerService(config)
    propSvc = PropertyService(config)
    check   = 4
```

We then define a sequence of validation checks to be performed if the request method is POST. Before validation we use the form service to *hydrate* `NameForm` and `ContactForm` with post data:

```
if request.method == 'POST' :
    formSvc.hydrateFormsFromPost(request.POST)
    name_form    = formSvc.getForm('name')
    contact_form = formSvc.getForm('contact')
```

The first validation block confirms partner key validity by calling upon the partner domain service `fetchByKey()` method:

```
if 'prop_partnerKey' in request.POST :
    partKey = request.POST['prop_partnerKey']
```

```
        part = partSvc.fetchByKey(partKey)
        if part :
            check -= 1
        else :
            message += '<br>Unable to confirm this partner.'
    else :
        message += '<br>You need to choose a partner.'
```

The second validation block confirms that the property key is valid:

```
    if 'prop_propertyKey' in request.POST :
        propKey = request.POST['prop_propertyKey']
        prop = propSvc.fetchByKey(propKey)
        if prop :
            check -= 1
        else :
            message += '<br>Unable to confirm this property.'
    else :
        message += '<br>You need to choose a property.'
```

The last two validation blocks use pre-defined Django form validators against the name and contact sub-forms:

```
    if name_form.is_valid() : check -= 1
    if contact_form.is_valid() : check -= 1
```

If all validations succeed, we create `Name` and `Contact` instances from the `booksomeplace.entity.base` module. We set these instances into the `Property` entity instance retrieved during validation and call upon the `PropertyService` to perform the update using a method called `edit()` (described later in this chapter). If the operation succeeds, we list the property:

```
    if check == 0 :
        prop.set('contactName', \
            Name(formSvc.getCleanDataByKey('name')))
        prop.set('contactInfo', \
            Contact(formSvc.getCleanDataByKey('contact')))
        if propSvc.edit(prop) :
            message += '<br>Property contact information updated!'
            return HttpResponseRedirect('/property/list/' + \
                prop.getKey())
        else :
            message += '<br>Unable to update form data.'
    else :
        message += '<br>Unable to validate this form.'
```

If the request is not POST, we render the sub-forms and inject the resulting HTML into the primary website layout:

```
context = {
    'message' : message,
    'choose_part_form' : ChoosePartnerForm(),
    'name_form_html' : render_to_string('property_name.html',
        {'form': formSvc.getForm('name')}),
    'contact_form_html' : render_to_string('property_contact.html',
        {'form': formSvc.getForm('contact')})
}
prop_master = render_to_string('property_update_contact.html', context)
page = render_to_string('layout.html', {'contents' : prop_master})
return HttpResponse(page)
```

In order to have Django return the property contact update form, we need to add an entry to urls.py for the property app:

```
urlpatterns = [
    path('contact/update/', views.updateContact,
name='property_update_contact'),
    # other paths not shown
]
```

Let's now define JavaScript functions to generate AJAX requests.

Defining JavaScript functions to generate AJAX requests

The property contact update form is located in /path/to/repos/www/chapter_08/templates/property_update_contact.html. It starts with a standard HTML <form> tag, along with an identifying header:

```
<form action="/property/contact/update/" method="post">
    {% csrf_token %}
    <h1>Update Property Contact</h1>
    <div class="row extra" style="background-color: #C8C8EE;">
        <div class="col-md-6">
            {{ choose_part_form }}
        </div><!-- end col -->
```

In the next several lines, you can see that the option to choose a property is populated from an AJAX request, and is thus not represented as a Python form:

```html
<div class="col-md-6">
    <label>Property:</label>
    <select name="prop_propertyKey" id="ddlProperties" >
        <option value="0">Choose Partner First</option>
    </select>
</div><!-- end col -->
</div><!-- end row -->
```

The remaining HTML is very similar to the templates discussed earlier in this chapter:

```html
<div class="row extra" style="background-color: #E6E6FA;">
    <div class="col-md-6">
        {{ name_form_html }}
    </div><!-- end col -->
    <div class="col-md-6" style="background-color: #D6D6F0;">
        {{ contact_form_html }}
    </div><!-- end col -->
</div><!-- end row -->
<input type="submit" value="Submit">
</form>
```

At the end of the contact form are two JavaScript functions using jQuery. The first monitors any changes to the HTML `select` element used to choose the partner. Should this change, an AJAX request is made that re-populates the HTML `select` element representing properties for that partner:

```javascript
<script src="https://code.jquery.com/jquery-3.4.1.min.js"></script>
<script>
$(document).ready(function() {
    $("#id_prop_partnerKey").change(function() {
        var partKey = $(this).val();
        console.log( "partner key: " + partKey );
        $("#ddlProperties").empty();
        $.ajax({
            type: "GET",
            url: "/property/json/by_part/" + partKey + "/",
            dataType: 'json',
            success: function(data){
                $.each(data, function(i, d) {
                    $("#ddlProperties").append(
                        '<option value="' + d.propertyKey + '">'
                        + d.propName
                        + '</option>');
                });
            }
```

```
        });
    });
```

The second JavaScript function monitors any changes to the auto-generated properties of the HTML `select` element. An AJAX request is made to provide the property key:

```
$("#ddlProperties").change(function() {
    var propKey = $(this).val();
    console.log( "property key: " + propKey );
    $.ajax({
        type: "GET",
        url: "/property/json/update_contact/" + propKey + "/",
        dataType: 'json',
```

If the call was successful, the contact information fields are updated accordingly:

```
        success: function(data){
            $("#id_name_title").val(data['contactName']['title']);
            $("#id_name_first").val(data['contactName']['first']);
            $("#id_name_last").val(data['contactName']['last']);
            $("#id_name_suffix").val(data['contactName']['suffix']);
            $("#id_contact_email").val(data['contactInfo']['email']);
            $("#id_contact_phone").val(data['contactInfo']['phone']);
            var socMedia = data['contactInfo']['socMedia'];
            for (var k in socMedia) {
                $("#id_contact_soc_type").val(k);
                $("#id_contact_soc_addr").val(socMedia[k]);
            }
        }
    }
}); }); });
</script>
```

Next, let's use operators to update contact information.

Using operators to update contact information

In order to apply a contact information update, we use the `$set` update operator. As you recall, updates are performed using `entity` classes as an argument. Accordingly, it makes sense to use the `entity` class to create its own update document. Therefore, we add a method called `getUpdateDoc()` to the `booksomeplace.entity.base.Base` class, as shown here:

```
class Base(dict) :
    def getUpdateDoc(self) :
        doc = {}
        for key, value in self.fields.items() :
```

```
            doc.update({'$set' : {key : self[key]}})
        return doc
    # other methods not shown
```

We then add a method called `edit()` to
`booksomeplace.domain.property.PropertyService` that accepts a `property` entity
as an argument. In order to perform the update, it calls the inherited `getUpdateDoc()`
method:

```
def edit(self, prop) :
    query = {'propertyKey' : prop.getKey()}
    return self.collection.update_one(query, prop.getUpdateDoc())
```

Next, we use the contact update form.

Using the contact update form

To use the contact update form, we enter the URL
`http://chap08.booksomeplace.local/property/contact/update`. The initial form
is as shown here:

When we select a partner, the list of properties changes, as shown here:

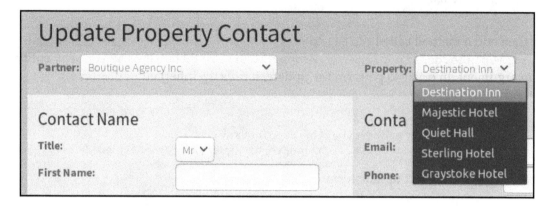

And finally, after selecting a property, contact information is updated, as shown here:

We can then make the needed changes and hit **Submit** to save changes.

In order to test the AJAX queries, open your browser tools and select *Network*. You can then see the requests being made while the web page is open and troubleshoot any JavaScript problems noted by the tools. If you are using the Docker environment described in Chapter 2, *Setting up MongoDB 4.x*, be sure to restart the Apache web server in the running Docker container.

Querying properties of embedded objects

Listing a property does not require a form, and only involves the
`pymongo.collection.findOne()` method called from a property domain service lookup
method. Things get a bit tricky, however, when dealing with room type listings. This
involves parsing an embedded array of `RoomType` objects. What is even trickier is that
within this array of objects is another property, `beds`, itself another array! So, here is our
proposed approach:

- Define a view using the property domain service to perform a straightforward
  lookup on the property by key.
- Define view methods that return a JSON list of room types.
- Add two routes: one for the listing, the other for the method called by jQuery.
- Define a template that displays basic property information.
- Within the template, add a jQuery AJAX request for room type information.

The view method that returns JSON also breaks down the array of bed types embedded
within each `RoomType` object.

Defining a property listing view method

In the Django `property` app, we add to
`/path/to/repo/www/chapter_08/property/views.py` a view method, `listProp()`,
that accepts a property key as an argument. If the property key is present, we use the
property domain service to retrieve a `Property` entity instance:

```
def listProp(request, prop_key = None) :
    propSvc = PropertyService(Config())
    prop    = Property()
    rooms   = []
    message = ''
    if prop_key :
        prop = propSvc.fetchByKey(prop_key)
    # build master form
```

We are now in a position to render the form:

```
info = prop.get('propInfo')
context = {
    'message'   : message,
    'prop_name' : prop.get('propName'),
    'prop_type' : info.get('type'),
```

```
        'prop_des'  : info.get('description'),
        'prop_key'  : prop_key,
    }
    prop_master = render_to_string('property_list.html', context)
    page = render_to_string('layout.html', {'contents' : prop_master})
    return HttpResponse(page)
```

Let's now look at the room types view method.

Room types view method

In the same file in the Django `property` app, we define a method that returns a JSON response to a jQuery AJAX request for room information. Again, as before, if the property key is present in the request, we use the property service to retrieve a `Property` instance:

```
def autoPopRoomsByPropKey(request, prop_key = None) :
    import json
    config  = Config()
    propSvc = PropertyService(config)
    result  = []
    if prop_key :
        temp = propSvc.fetchRoomsByPropertyKey(prop_key)
```

If we are able to get an instance, we flatten the `RoomType.beds` property list into a string:

```
        if temp :
            for rooms in temp :
                beds = ''
                for val in rooms['beds'] :
                    beds += val + ', '
                beds = beds[0:-2]
```

We then formulate a return value as a JSON encoded list of dictionaries:

```
                room_config = config.getConfig('rooms')
                link = '<a target="_blank" href="'
                link += room_config['details_url']
                link += rooms['roomTypeKey'] + '/">Details</a>'
                result.append({'type':rooms['type'], \
                    'view':rooms['view'], 'beds':beds, \
                    'available':rooms['numAvailable'],'link':link})
    return HttpResponse(json.dumps(result), \
        content_type='application/json')
```

You also can see that we modified the `Config` class found in
`/path/to/repo/chapters/08/src/config/config.py` by adding extra configuration
to represent the URL used in the link, giving more details on any given room:

```
# config.config
import os
class Config :
    config = dict({
        'rooms' : {
            'details_url' :
'http://chap08.booksomeplace.local/property/room/details/',
        },
    })
    # other configuration and methods not shown
}
```

Next, we add the appropriate URL routes.

Adding the appropriate URL routes

In the Django `property` app, in the module file
at `/path/to/repo/www/chapter_08/property/urls.py`, we add two routes, as shown
here. The first route gives us access to the property listing. The second route is used by the
jQuery AJAX request. They both define a property key parameter:

```
urlpatterns = [
    path('list/<slug:prop_key>/', views.listProp, name='property_list'),
    path('json/rooms/<slug:prop_key>/', \
        views.autoPopRoomsByPropKey, \
        name='property_json_rooms_by_prop_key'),
    # other URL routes not shown
]
```

Note that we use the convention of a backslash (\) to indicate that two
lines should be on a single line.

Property listing template

The property listing template, `property_list.html`, found in `/path/to/repo/www/chapter_08/templates`, is much like the other Django templates shown earlier in this chapter. We provide the values to the render process, which then produces the actual results:

```html
<div class="container" style="margin-top: -100px;">
    <div class="row">
        <div class="col-md-6">
            <h1>Property Listing</h1>
            <div class="row extra">
                <div class="col-md-2"><b>Name</b></div>
                <div class="col-md-4">{{ prop_name }}</div>
            </div><!-- end row -->
            <div class="row extra">
                <div class="col-md-2"><b>Type</b></div>
                <div class="col-md-4">{{ prop_type }}</div>
            </div><!-- end row -->
            <div class="row extra">
                <div class="col-md-2"><b>Description</b></div>
                <div class="col-md-4">{{ prop_des }}</div>
            </div><!-- end row -->
        </div>
        <div class="col-md-6">
            <img src="{% static 'images/hotel.jpg' %}"
 style="width:100%;"/>
        </div>
    </div>
```

The table contains headers representing room type information, but the actual data is left blank. The reason is that this information is supplied by a jQuery AJAX request (described in the next sub-section):

```html
<div class="row">
    <div class="col-md-12">
    <h3>Room Types</h3>
    <table id="data_table" class="display" width="100%">
        <thead>
            <tr>
                <th>Type</th>
                <th>View</th>
                <th>Beds</th>
                <th>Available</th>
                <th>Info</th>
            </tr>
        </thead>
```

```
      </table>
      {{ message }}
      </div>
   </div><!-- end row -->
</div><!-- end container -->
```

Next, we will cover the jQuery AJAX function requesting room types.

jQuery AJAX function requesting room types

At the end of the same template file, `property_list.html`, we add a header block that ensures the desired version of jQuery is loaded:

```
{% block extra_head %}
<script type="text/javascript" charset="utf8"
src="https://code.jquery.com/jquery-3.3.1.min.js"></script>
{% endblock %}
```

After that, we add a jQuery `$(document).ready()` function that makes an AJAX request to the URL described in the preceding code block. The return value from the request is a list of dictionaries with the properties `type`, `view`, `beds`, `available`, and `link`:

```
<script type="text/javascript" charset="utf8">
$(document).ready( function () {
    $.ajax({
        'type'     : 'GET',
        'url'      : '/property/json/rooms/{{ prop_key }}/',
        'dataType' : 'json',
        'success'  : function(data){
            $.each(data, function(i,row) {
                item = '<tr><td>' + row.type + '</td>'
                item += '<td>' + row.view + '</td>'
                item += '<td>' + row.beds + '</td>'
                item += '<td>' + row.available + '</td>'
                item += '<td>' + row.link + '</td></tr>'
                $('#data_table').append(item);
            });
        }
    });
});
</script>
```

We can now list the results in the next section.

Listing results

Putting it all together, in a browser, enter a URL that contains a valid property key. Here is an example of an appropriate URL:

```
http://chap08.booksomeplace.local/property/list/MAJE2976/
```

The result is shown here:

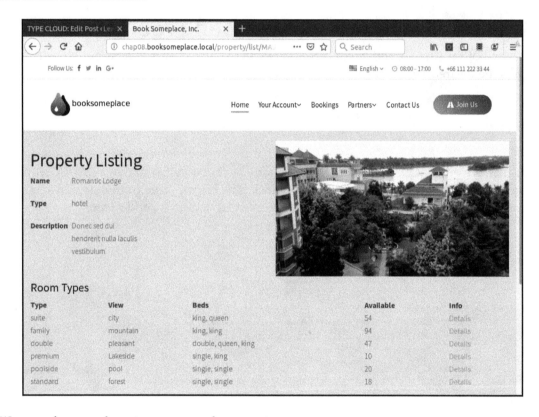

We now focus on how to remove a document.

Removing a document

Removing a property shares many similarities with the process of editing and listing a property. Where the operations diverge, of course, is when removing a property, we rely upon the `pymongo.collection.delete_one()` method. This method is called from the property domain service class.

Defining a property domain service deletion method

The first step we take is to define a method in the `PropertyService` class, found in the `/path/to/repo/chapters/08/src/booksomeplace/domain/property.py` module:

```
def deleteByKey(self, key) :
    query  = {'propertyKey':key}
    return self.collection.delete_one(query)
```

Next, we turn our attention to the process of choosing the partner and the property to be deleted.

Choosing the partner and property

As we mentioned, this process is similar to the process needed to update property contact information. Accordingly, we can simply copy the already existing view and template code from the update contact logic and merge it with the logic needed to list a property. In the Django `property` app, we add the `delProperty()` method to `views.py`. We start by initializing variables:

```
def delProperty(request) :
    propSvc   = PropertyService(Config())
    prop      = Property()
    rooms     = []
    message   = ''
    confirm   = False
    prop_key  = ''
```

Next, we check to see if a property has been chosen:

```
if request.method == 'POST' :
    if 'prop_propertyKey' in request.POST :
        prop_key = request.POST['prop_propertyKey']
        prop = propSvc.fetchByKey(prop_key)
        if prop :
            confirm = True
            message = 'Confirm property to delete'
```

If the `prop_propertyKey` field from the `ChoosePartnerForm` is not present, we next check to see if we are in phase 2, which is where a property has been chosen and we wish to confirm deletion:

```
if 'del_confirm_yes' in request.POST \
    and 'prop_key' in request.POST :
    prop_key = request.POST['prop_key']
    prop = propSvc.fetchByKey(prop_key)
```

If the delete operation is confirmed, we call upon the property domain service to perform the delete and report whether it is a success or failure. We also set the appropriate return message:

```
if prop :
    if propSvc.deleteByKey(prop_key) :
        message = 'Property ' + prop.get('propName')
        message += ' was successfully deleted'
    else :
        message = 'Unable to delete property ' + \
                    prop.get('propName')
else :
    message = 'Unable to delete property ' + \
                prop.get('propName')
```

The last part of the view method pushes forms and other information to the template, rendered as shown:

```
info = prop.get('propInfo')
context = {
    'choose_part_form' : ChoosePartnerForm(),
    'message'    : message,
    'prop_name'  : prop.get('propName'),
    'prop_type'  : info.get('type'),
    'prop_des'   : info.get('description'),
    'prop_key'   : prop_key,
    'confirm'    : confirm,
}
prop_master = render_to_string('property_list_del.html', context)
page = render_to_string('layout.html', {'contents' : prop_master})
return HttpResponse(page)
```

We can now define and choose the confirm delete template.

Defining the choose and confirm delete template

The template for this operation needs to have two parts: one part where the admin chooses the partner and property. The second part displays chosen property information and asks for confirmation. This is accomplished using Django template language. We add an `if` statement that tests the value of a `confirm` property. If it is `False`, we display the sub-form to choose the partner and property. The second sub-form is only shown if we are in the confirmation phase. You can view the full form at `/path/to/repo/www/chapter_08/templates/property_list_del.html`.

Defining a URL route

In the Django property app, we add a route to `urls.py` to access the new view method:

```
urlpatterns = [
    path('delete/', views.delProperty, name='property_delete'),
    # other paths not shown
]
```

We can then enter this in the browser:

`http://chap08.booksomeplace.local/property/delete/`

We are then presented with the sub-form, which allows us to choose a partner and property to delete:

After making the desired choices, we are then shown property information and asked to confirm:

We are then returned to the first form, with a success message shown at the bottom.

Summary

The focus of this chapter was working with embedded objects and arrays (lists). Projects presented in this chapter, for the most part, were integrated into Django. In this chapter, you learned how to add a complex object containing embedded objects and arrays to the database. In order to accomplish this, you learned how to create `entity` classes from form data by defining a form service. The form service performed a process called hydration and moved data between `form` and `entity`, and back again. You were also shown how to associate embedded objects with their own sub-forms, and how to insert sub-forms into a primary form.

You then learned about adding rooms to a property, which involved creating embedded objects, that were in turn added to an embedded array. When learning how to use the `pymongo.collection.update_one()` method to update property contact information, you were also shown how to use AJAX requests to pre-populate fields when the value of other fields changed.

In the next chapter, you will learn to handle complex tasks using the MongoDB Aggregation Framework, as well as how to manage queries involving geo-spatial data and generating financial reports.

Handling Complex Queries in MongoDB

9

This chapter introduces you to the aggregation framework, a feature unique to MongoDB. This feature represents a departure point where MongoDB forges ahead of its legacy **relational database management system (RDBMS)** ancestors. In this chapter, you will also learn about the legacy Map-Reduce capability, which has been available in MongoDB since the beginning. Two other extremely useful techniques covered in this chapter are how to model complex queries using the MongoDB Compass tool and how to conduct geospatial queries.

The focus of this chapter is on financial reports, including revenue reports aggregated by user-selectable criteria, aging reports, and how to perform financial analysis on booking trends in order to create a financial sales projection report for Book Someplace. In this chapter, you will learn how all these revenue reports can assist in performing **risk analysis**. Finally, you will learn how to use geospatial data to generate a revenue report based on geographic area.

The topics covered in this chapter include the following:

- Modeling complex queries using MongoDB Compass
- Using the aggregation framework
- Working with Map-Reduce functions
- Working with geospatial data
- Using aggregation to produce financial reports

Technical requirements

The minimum recommended hardware for this chapter is as follows:

- A desktop PC or laptop
- 2 GB free disk space
- 4 GB of RAM
- 500 Kbps or faster internet connection.

The software requirements are as follows:

- **OS (Linux or macOS)**: Docker, Docker Compose, and Git (optional)
- **OS (Windows)**: Docker for Windows and Git for Windows
- Python 3.x, a PyMongo driver, and Apache (already installed in the Docker container used for this book)
- MongoDB Compass (needs to be installed on your host computer; instructions are in the chapter)

Installation of the required software and how to restore the code repository for the book is explained in Chapter 2, *Setting Up MongoDB 4.x*. To run the code examples in this chapter, open a Terminal window (Command Prompt) and enter these commands:

```
cd /path/to/repo
docker-compose build
docker-compose up -d
docker exec -it learn-mongo-server-1 /bin/bash
```

To run the demo website described in this chapter, once the Docker container for `Chapter 3`, *Essential MongoDB Administration Techniques,* to `Chapter 12`, *Developing in a Secured Environment,* has been configured and started, there are two more things you need to do:

1. Add the following entry to the local `hosts` file on your computer:

```
172.16.0.11    booksomeplace.local
172.16.0.11    chap09.booksomeplace.local
```

The local `hosts` file is located at `C:\windows\system32\drivers\etc\hosts` on Windows, and for Linux and macOS, at `/etc/hosts`.

2. Open a browser on your local computer (outside of the Docker container) to the following URL:

`http://chap09.booksomeplace.local/`

When you are finished working with the examples covered in this chapter, return to the Terminal window (Command Prompt) and stop Docker, as follows:

```
cd /path/to/repo
docker-compose down
```

The code used in this chapter can be found in this book's GitHub repository at `https://github.com/PacktPublishing/Learn-MongoDB-4.x/tree/master/chapters/09`, as well as `https://github.com/PacktPublishing/Learn-MongoDB-4.x/tree/master/www/chapter_09`.

Modeling complex queries using MongoDB Compass

Queries can become quite complex when working with documents containing embedded objects and arrays. In this situation, it might be helpful to use an external tool to model potential queries before committing them to code. One such tool, included in default MongoDB installations, is the **MongoDB Compass** tool. As a first step, we will demonstrate how to conduct queries on fields within embedded objects. After that, we will move on to data contained in embedded arrays.

Understanding MongoDB Compass

There are three versions of MongoDB Compass available, all open source and free to use:

- **Compass**: In the past year, the company producing MongoDB decided to make available a full-featured version for all versions of MongoDB, including Community Edition.
- **Compass Readonly**: Designed to provide the ability to generate ad hoc reports by querying the database. This version has **write** functionality disabled, encouraging companies to allow administrators to perform direct queries without harming the database.
- **Compass Isolated**: Designed for high-security environments.

The discussion in this section focuses on the full version, simply called **Compass** or **MongoDB Compass**.

 For more information, refer to `https://docs.mongodb.com/compass/master/#mongodb-compass`.

Installing MongoDB Compass

To install MongoDB Compass on Windows, proceed as follows:

1. Open a browser to the **Downloads** page for MongoDB Compass.
2. It is recommended that in the **Platform** selection box, you choose the **Microsoft Installer** (**MSI**) option.
3. After the download has completed, proceed with the installation as you would any other Windows application.

To install MongoDB Compass on a Mac, proceed as follows:

1. Open a browser to the **Downloads** page for MongoDB Compass.
2. In the **Platform** selection box, choose the **OS X** option.
3. Drag and drop the downloaded `DMG` file into the `Applications` folder.
4. Eject the disk image.
5. Double-click on the MongoDB Compass icon and proceed with the installation as you would any other Mac OS X application.

To install MongoDB Compass on RedHat/CentOS/Fedora, proceed as follows:

1. Open a browser to the **Downloads** page for MongoDB Compass.
2. In the **Platform** selection box, choose the **RedHat** option.
3. Apply your local package management utility to the downloaded RPM file.

To install MongoDB Compass on Ubuntu/Debian, proceed as follows:

1. Open a browser to the **Downloads** page for MongoDB Compass.
2. In the **Platform** selection box, choose the **Ubuntu** option.
3. Apply your local package management utility to the downloaded DEB file.

 The **Downloads** page for MongoDB Compass can be found at `https://www.mongodb.com/try/download/compass`.

Now, let's have a look at establishing the initial connection.

MongoDB Compass initial connection

The initial connection screen is very straightforward. You need to supply information pertaining to the host upon which your database resides, and any authentication information. In our example, ensure the Docker container you defined for Chapter 3, *Essential MongoDB Administration Techniques*, to Chapter 12, *Developing in a Secured Environment*, is up and running. Here is the initial screen after first starting MongoDB Compass on the host computer:

You can also select **Fill in connection fields individually,** in which case these are the two tab screens you can use:

You can see right away that the fields closely match the `mongo Shell` connection options. Here is a summary of the options available on the MongoDB Compass connection screens:

Option	Notes
Hostname	Enter the DNS or IP address of the server to which you wish to connect.
Port	Enter the appropriate port number. The default is `27017`.
SRV Record	Toggle this switch if the server to which you want to connect has a DNS SRV record. Note that when you toggle this switch, the **Port** option disappears as Compass queries the DNS for more information.
Authentication	Options include a standard username/password or SCRAM-SHA-256. The paid versions also include Kerberos, LDAP, and X.509. Authentication is discussed in more detail in `Chapter 11`, *Administering MongoDB Security.*
Replica Set Name	In this field, you must enter the name of the replica set if the server to which you plan to connect is a member of a replica set. The default is **None**.
Read Preference	This field allows you to influence how Compass performs reads. The default is **Primary**. If you choose **Secondary**, all reads are performed from a secondary server in the replica set. Choose **Nearest** if replica set servers are dispersed geographically and you wish to perform reads from the one that is fewer route hops away. Finally, there are options for **Primary Preferred** and **Secondary Preferred**. The **Preferred** option tells Compass to perform the read from the preferred choice, but allows reads from the other type if the preference is not available.

You might also note a star icon to create a list of **favorites**. Next, we will discuss the SSL connection options, which are also part of the initial connection screen.

SSL connection options

Use the **SSL** option if you need a secure connection. The default is **None**, which corresponds to the **Unvalidated (insecure)** option. If you are making a connection to MongoDB Atlas, use the **System CA/Atlas Deployment** option. If you select **Server Validation**, you will see a browse button appear that lets you choose the certificate authority file for the server running the `mongod` instance.

If you choose **Server and Client Validation**, you are presented with a set of drop-down boxes that let you select the server's certificate authority file, the client certificate file, and the client's private key file. You must also enter the password. Here are the additional fields that appear when you choose **Server and Client Validation**:

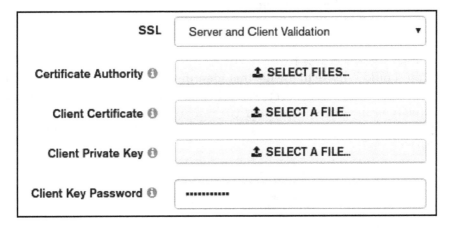

In order to connect via an SSH tunnel, click **SSH Tunnel** and select either **Use Password** or **Use Identity File**. If you choose the former, you must enter the SSH server's hostname, tunnel port, username, and password. Here is what you'll see:

If for **SSH Tunnel** you select **Use Identity File**, in addition to the information shown previously, you should browse to the file that identifies the client. On Linux and macOS, this file is typically located in the user's home directory in a hidden folder named `.ssh`. The name of this file typically starts with either `id_rsa*` or `id_dsa*`, although there are many variations. Here is the dialog that appears after selecting the **Use Identity File** option for **SSH Tunnel**:

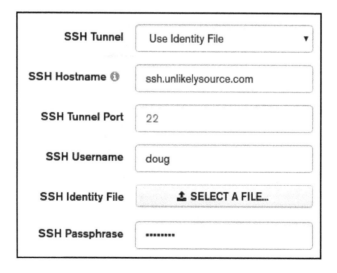

Now that we have covered the SSL connection options, let's learn how to make the connection.

Making the connection

After selecting the appropriate options and entering the required information, click on the **Connect** button. You will see an initial screen that shows you the various databases present on your server. If you have been following the labs and examples in this book, you will see the presence of the `sweetscomplete` and `booksomeplace` databases. Here is the screen after connecting:

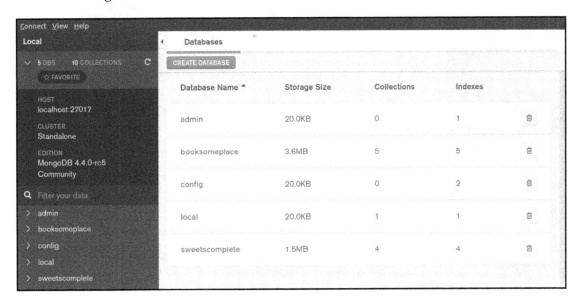

You can see that MongoDB Compass Community Edition can be used for basic database administration, which includes the following:

- Creating or deleting a database
- Creating or deleting a collection
- Adding, editing, and deleting documents in a collection

For the purpose of this section, go ahead and select **booksomeplace**. You will see documents in the collection in either tabular or list format. Here is the screen that shows details of the `bookings` collection:

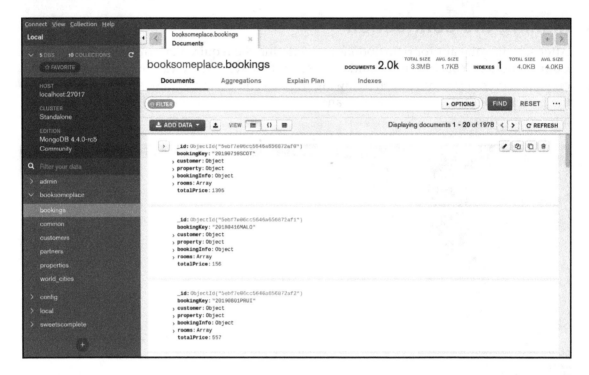

 In this section, our focus here is on modeling complex queries. For that purpose, be sure to have executed the scripts to populate the database with sample data. These scripts can be executed using the `mongo` shell and are located in the `/path/to/repo/sample_data` directory.

You can actually perform a direct import of JSON documents from MongoDB Compass by selecting **Add Data**. However, the documents have to be pure JSON: not JavaScript variables, functions, or object code. For that reason, the `mongo` shell JavaScript files present in the `/path/to/repo/sample_data` directory do not work. Here is the import option dialog:

When you are importing a script from your local computer to MongoDB running in the Docker container, the file you are importing must already exist on your local computer.

Now it's time to have a look at how we can model queries using MongoDB Compass.

Formulating queries involving embedded objects

Management has asked you to create a financial report that provides a summary of revenue generated by users with **Facebook** accounts. Looking at the structure of documents in the `booksomeplace.bookings` collection, you note that the Facebook page for customers is contained within a `socialMedia` array, which itself is contained in an embedded document, `customerContact`. You further note that information in the `socialMedia` array is keyed according to the social media type.

The MongoDB internal document representation closely follows JSON syntax. Accordingly, there are no object class types assigned when the document is stored in the database. You can access object properties by simply appending a dot (`.`) after the field that references the object. So, in our example, the `customerContact` field represents a `Contact` object. It, in turn, is nested in the field referenced by `customer`. In order to reference information in the `socialMedia` property of the embedded `Contact` object, the completed field reference would be as follows:

```
customer.customerContact.socMedia
```

In a similar way, array keys are viewed internally as properties of their owning field. In this case, we wish to locate documents where the array key in the `socMedia` field equals `facebook`. In order to use MongoDB Compass to model a query, connect to the appropriate database, select the target collection, and under the **Documents** tab, click on **Options**. We first enter the appropriate filter. Note that as you begin typing, a list of fields automatically appears, as shown here:

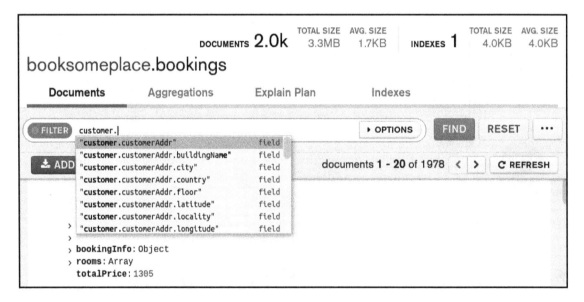

When the **FILTER** button appears red, the query string is in error. Be sure to formulate all query strings in proper JSON syntax, starting and ending with curly braces: `"{ <expression> }"`.

After choosing the field, we enter a colon (:). After typing in the dollar sign ($), a list of operators appears next, as shown:

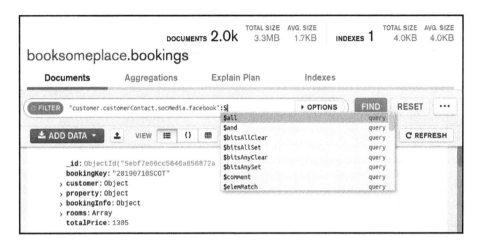

You can then complete the query by filling in the projection (**PROJECT**) and any other appropriate parameters. The **FIND** button does not become active until you ensure that each option is itself a proper JSON document. Once completed, click **FIND**. Here are the results using the sample data:

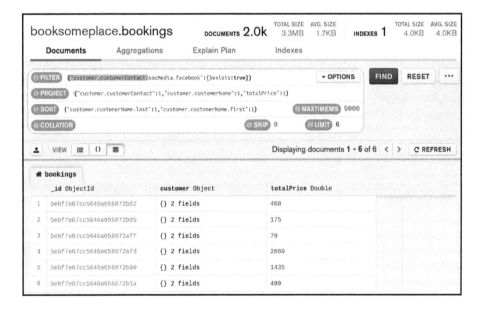

There is limited capability to export the query to a programming language (including Python); however, the current version only exports the JSON document representing the query filter, so it is of marginal use. To perform the export, click on the three dots at the top right (...).

To get a good overview of the completed query document, click on the three dots (...) at the top right, and select **Toggle Query History**. You can view your query history in MongoDB Compass, mark a query as **favorite**, and also copy the entire query document to the clipboard. In addition, if you click on the query itself, it places all the query parameters back into the main query options dialog. Here is the **Past Queries** window (on the right side of the following screenshot):

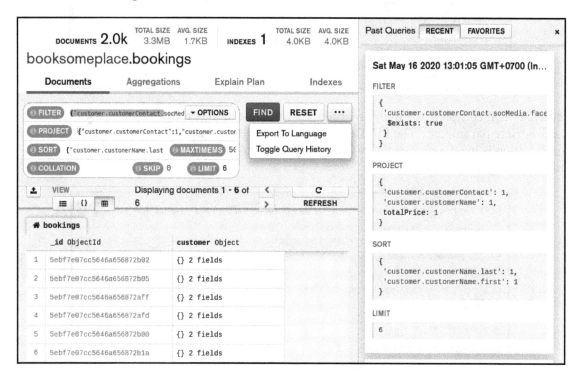

In order to move the query into your preferred programming language, from the **Past Queries** window, select the target query, and then select the icon to copy to the clipboard. We can now use this in a Python script, as in the following (found in `/path/to/repo/chapters/09/compass_query_involving_embedded_objects.py`).

Here is what was returned by copying the query from the MongoDB Compass **Past Queries** window:

```
{
  filter: {
    'customer.customerContact.socMedia.facebook': {
      $exists: true
    }
  },
  project: {
    'customer.customerContact': 1,
    'customer.customerName': 1,
    totalPrice: 1
  },
  sort: {
    'customer.customerName.last': 1,
    'customer.customerName.first': 1
  },
  limit: 6
}
```

We can use this to create a script that produces the total revenue for customers who are on Facebook. As with the scripts you have reviewed in earlier chapters, we start with a set of `import` statements. We also create an instance of `BookingService`:

```
import os,sys
sys.path.append(os.path.realpath("src"))
import pymongo
from config.config import Config
from booksomeplace.domain.booking import BookingService
service = BookingService(Config())
```

We are then able to use the code captured by MongoDB Compass to formulate parameters for the query:

```
total  = 0
query  = {'customer.customerContact.socMedia.facebook': {'$exists': True}}
proj   = {'customer.customerContact': 1, \
          'customer.customerName': 1,'totalPrice': 1}
sort   = [('customer.customerName.last',pymongo.ASCENDING), \
          ('customer.customerName.first',pymongo.ASCENDING)]
result = service.fetch(query, proj, sort)
```

After that, it's a simple matter of displaying the results and calculating the total revenue:

```
pattern = "{:12}\t{:40}\t{:8.2f}"
print('{:12}\t{:40}\t{:8}'.format('Name','Facebook Email','Amount'))
for doc in result :
    name = doc['customer']['customerName']['first'][0] + ' '
    name += doc['customer']['customerName']['last']
    email = doc['customer']['customerContact']['socMedia']['facebook']
    price = doc['totalPrice']
    total += price
    print(pattern.format(name, email, price))
print('{:12}\t{:40}\t{:8.2f}'.format('TOTAL',' ',total))
```

Here is the last part of the results produced:

```
File  Edit  View  Search  Terminal  Help
L Tang         ltang110@telecom.com@facebook.com                   1110.00
L Tang         ltang110@telecom.com@facebook.com                     57.00
L Tang         ltang110@telecom.com@facebook.com                    374.00
L Tang         ltang110@telecom.com@facebook.com                    480.00
L Tang         ltang110@telecom.com@facebook.com                    300.00
L Tang         ltang110@telecom.com@facebook.com                    930.00
A Turner       aturner136@singtel.com@facebook.com                   54.00
A Turner       aturner136@singtel.com@facebook.com                  278.00
A Turner       aturner136@singtel.com@facebook.com                  159.00
A Turner       aturner136@singtel.com@facebook.com                 1557.00
A Turner       aturner136@singtel.com@facebook.com                  111.00
A Turner       aturner136@singtel.com@facebook.com                 2260.00
A Turner       aturner136@singtel.com@facebook.com                  201.00
A Turner       aturner136@singtel.com@facebook.com                 2000.00
W Yang         wyang329@vodafone.com@facebook.com                  2660.00
W Yang         wyang329@vodafone.com@facebook.com                   279.00
W Yang         wyang329@vodafone.com@facebook.com                   272.00
W Yang         wyang329@vodafone.com@facebook.com                   712.00
W Yang         wyang329@vodafone.com@facebook.com                  2672.00
W Yang         wyang329@vodafone.com@facebook.com                  2673.00
W Yang         wyang329@vodafone.com@facebook.com                   795.00
W Yang         wyang329@vodafone.com@facebook.com                  1755.00
----------    ------------------------------------               --------
TOTAL                                                           219742.00
root@server1:/repo/chapters/09# python compass_query_involving_embedded_objects.py
```

Please note that the result shown in the preceding screenshot could potentially be consolidated into a single query using the MongoDB **aggregation framework** or aggregation pipeline (discussed in the next section). This would allow the database to perform the bulk of the processing, which lightens the load on the Python script.

In the next section, we will learn how to build queries involving embedded arrays.

Building queries involving embedded arrays

Management is concerned about customer complaints regarding cleanliness. You have been instructed to produce a report on properties where more than 50% of the reviews for that property report a low cleanliness score (for example, 3 or less).

When conducting queries of any type, it's important to be aware of the database document structure. In this case, we need to examine the `properties` collection. Using MongoDB Compass, all you need to do is to connect to the database and select the **booksomeplace** database, as well as the `properties` collection. In the list view, you can click on the `reviews` field to get an idea of the structure, as shown here:

As you can see, reviews are represented as embedded arrays. Following the logic presented in the earlier sub-section in this chapter, your first thought might be to construct a query document such as this: `{"reviews.cleanliness":{$lt:3}}`:

This is not exactly the information we are looking for, of course, so we need to move the final query into a Python script that produces the final report. After selecting **Toggle Query History**, we copy the query document, as shown here:

```
{
  filter: {'reviews.cleanliness': {$lt: 3}},
  project: {propName: 1,address: 1,reviews: 1}
}
```

We can move the query into Python script. As before, we start with a series of import statements and create an instance of the PropertyService class from the booksomeplace.domain.service.property module:

```
import os,sys
sys.path.append(os.path.realpath("src"))
import pymongo
from config.config import Config
from booksomeplace.domain.property import PropertyService
service = PropertyService(Config())
```

Next, we formulate the query parameters and run the query off the fetch() method in the service. The cutoff variable represents the maximum cleanliness rating score specified by management:

```
cutoff = 3
cat     = 'cleanliness'
key     = 'reviews.' + cat
query   = {key : {'$lt': 3}}
proj    = {'propName': 1,'address':1,'reviews': 1}
sort    = [('propName',pymongo.ASCENDING)]
result  = service.fetch(query, proj, sort)
```

We can now loop through the result set and product a report with the property name, city, country, and average cleanliness score:

```
pattern = "{:20}\t{:40}\t{:5.2f}"
print('{:20}\t{:40}\t{:5}'.format('Property Name','Address','Score'))
print('{:20}\t{:40}\t{:5}'.format('-------------','-------','-----'))
for doc in result :
    name = doc['propName']
    addr = doc['address']['city'] + ' ' + doc['address']['country']
    # total cleanliness scores
    total = 0
    count = 0
    for revDoc in doc['reviews'] :
        total += revDoc[cat]
        count += 1
```

```
avg = 5 / (total/count)
if avg < cutoff :
    print(pattern.format(name,addr[0:40], avg))
```

The result, in a Python
`/path/to/repo/chapters/09/compass_query_involving_arrays.py` script, might appear as shown here. Note that we have only shown the last several lines to preserve space:

```
File Edit View Search Terminal Help
Valley Hall             Cutella AU                                  1.33
Valley Hall             Gadra Road IN                               1.67
Valley Hall             Munroe Falls US                             1.56
Valley Hotel            Hyde Park AU                                1.82
Valley Hotel            Mekinock US                                 1.79
Valley Inn              Стольт / Stolut BG                          1.67
Valley Inn              St. George's BM                             1.52
Valley Keep             West Kootenays (Rossland) CA                1.85
Valley Keep             Kings County (Kingston) CA                  1.43
Valley Lodge            Port Macdonnell AU                          1.79
Valley Stay             Saint-Pierre PM                             1.67
Valley Stay             Broadwood NZ                                1.70
Valley Stay             Соколове UA                                 2.14
Voyage Destination      Hampstead CA                                1.25
Voyage Destination      Edenhope AU                                 1.72
Voyage Hotel            Peterd HU                                   1.59
Voyage Inn              Parrott US                                  1.86
Voyage Inn              Ammandivilai IN                             1.54
Voyage Inn              East York (Leaside) CA                      1.84
Voyage Keep             Mitana IN                                   1.67
Voyage Lodge            Downsview East (CFB Toronto) CA             2.35
Voyage Lodge            Wallace US                                  1.67
Voyage Lodge            Gloucester (Blackburn Hamlet / Pine View    1.44
Voyage Resort           Manchester GB                               1.74
root@server1:/repo/chapters/09# python compass_query_involving_arrays.py
```

MongoDB also provides a number of **projection operators** (`https://docs.mongodb.com/manual/reference/operator/projection/#projection-operators`) that allow you to perform more granular operations on the projection. These operators include the following. `$` returns the first array element, which matches the `query` argument. `$elemMatch` returns the first array element, which matches the argument included in the projection. `$slice` produces a subset of the array. `$meta` returns the `textScore` associated with text searches. The score represents how close the match is.

In the next section, we will learn how to use the aggregation framework.

Using the aggregation framework

The MongoDB aggregation framework allows you to group results into datasets upon which a series of operations can be performed. MongoDB aggregation uses a **pipeline** architecture whereby the results of one operation becomes the input to the next operation in the pipeline. The core of MongoDB aggregation is the collection-level `aggregate()` method.

Working with the MongoDB aggregate() method

The arguments to the `db.<collection_name>.aggregate()` method consist of a series of expressions, each beginning with a **pipeline stage operator** (see the next section). The output from one pipeline stage feeds into the next, in a chain, until the final results are returned. The following diagram shows the generic syntax:

```
db.<collection_name>.aggregate([

    { <pipeline_operator> : { "<field>":<expression> } },      PIPELINE STAGE

    { <pipeline_operator> :                                    PIPELINE STAGE
        { "_id" : "<field>",
          "<field>" : { <expression_operator>:"$<field>"} }
    }

]);                                                            RESULTS
```

When you place a dollar sign ($) in front of a field, the *value* of the field is returned.

Arguably, the most popular pipeline stage operators are `$match`, which performs a similar function to a query filter, and `$group`, which groups results according to the `_id` field specified. As a quick example, before getting into more details, note the following command, which produces a revenue report by country.

The `$match` stage allows documents where `paymentStatus` in the `bookingInfo` embedded object is confirmed. The `$group` stage performs grouping by country. In addition, the `$group` stage uses the `$sum` pipeline expression operator to produce a sum of the values in the matched `totalPrice` fields.

Finally, the `$sort` stage reorders the final results by the `_id` match field (that is, the country code). Here is the query:

```
db.bookings.aggregate([
    { $match : { "bookingInfo.paymentStatus" : "confirmed" } },
    { $group:  { "_id"    : "$customer.customerAddr.country",
                 "total" : { $sum : "$totalPrice" } } },
    { $sort : { "_id" : 1 } }
]);
```

Here is the output from the `mongo` shell:

```
File Edit View Search Terminal Help
MongoDB jed@test[up:72566 secs]
2>use booksomeplace;
switched to db booksomeplace

MongoDB jed@booksomeplace[up:72578 secs]
3>db.bookings.aggregate([
...      { $match : { "bookingInfo.paymentStatus" : "confirmed" } },
...      { $group:  { "_id"    : "$customer.customerAddr.country",
...                   "total" : { $sum : "$totalPrice" } } },
...      { $sort : { "_id" : 1 } }
... ]);
{ "_id" : "AD", "total" : 4770 }
{ "_id" : "AR", "total" : 4500 }
{ "_id" : "AT", "total" : 12329 }
{ "_id" : "AU", "total" : 122733 }
{ "_id" : "AX", "total" : 3351 }
{ "_id" : "BD", "total" : 12319 }
{ "_id" : "BM", "total" : 6687 }
{ "_id" : "BR", "total" : 7679 }
{ "_id" : "BY", "total" : 7870 }
{ "_id" : "CA", "total" : 87550 }
{ "_id" : "CH", "total" : 10729 }
{ "_id" : "CL", "total" : 2804 }
{ "_id" : "CR", "total" : 3474 }
{ "_id" : "CZ", "total" : 19195 }
{ "_id" : "DE", "total" : 3984 }
{ "_id" : "DK", "total" : 9554 }
{ "_id" : "DO", "total" : 14265 }
{ "_id" : "ES", "total" : 3190 }
{ "_id" : "FI", "total" : 14006 }
{ "_id" : "FM", "total" : 4570 }
Type "it" for more

MongoDB jed@booksomeplace[up:72581 secs]
4>
```

Let's now look at the pipeline stage operators.

Pipeline stage operators

You can create multiple stages within the aggregation by simply adding a new document to the list of arguments presented to the `aggregate()` method. Each document in the list must start with a pipeline stage operator. Here is a summary of the more important pipeline stage operators:

Operator	Notes
`$bucket`	Creates sub-groups based on **boundaries**, which allow further, independent processing. As an example, you could create a revenue report with **buckets** consisting of $0.00 to $1.000, $1,001 to $10,000, $10,001 to $100,000, and more.
`$count`	Returns the number of documents at this stage in the pipeline.
`$geoNear`	Used for geospatial aggregation. This single operator internally leverages `$match`, `$sort`, and `$limit`.
`$group`	Produces a single document that has an accumulation of all the documents with the same value for the field identified as `_id`.
`$lookup`	Incorporates documents from other collections in a similar manner to an SQL **LEFT OUTER JOIN**.
`$match`	Serves exactly like a query filter: eliminates documents that do not match the criteria.
`$merge`	Allows you to output to a collection in the same database or in another database. You can also merge results into the same collection that is being aggregated. This also gives you the ability to create **on-demand materialized views**.
`$out`	This operator is similar to `$merge` except that it's been available since MongoDB 2.6. Starting with MongoDB 4.4, this operator can now output to a different database, whereas with MongoDB versions 4.2 and below, it had to be the same database. Unlike `$merge`, this operator cannot output to a sharded collection.
`$project`	Much like the **projection** in a standard MongoDB query, this pipeline stage operator allows you to add or remove fields from the output stream at this stage in the pipeline. It allows the creation of new fields based on information from existing fields.
`$sort`	Reorders the documents in the output stream at this stage in the pipeline. Has the same syntax as the sort used in a standard MongoDB query.
`$unionWith`	With this operator, the pipeline stage combines the output of two collections into a single output stream. It allows you to create the equivalent of the following SQL statement: `SELECT * FROM xxx WHERE UNION ALL SELECT * FROM yyy`
`$unwind`	Deconstructs embedded arrays, sending a separate document for each array element to the output stream at this stage in the pipeline.

Please refer to the following links for more information:

- **Pipeline stage operators**: `https://docs.mongodb.com/manual/reference/operator/aggregation-pipeline/#aggregation-pipeline-stages`
- **SQL** `LEFT OUTER JOIN`: `https://dev.mysql.com/doc/refman/5.7/en/join.html`
- **MongoDB on-demand materialized views**: `https://docs.mongodb.com/master/core/materialized-views/#on-demand-materialized-views`.

Examples of these pipeline stages are used throughout the remainder of this chapter. Before we get to the fun stuff, it's also important to have a quick look at the pipeline expression operators, which can be used within the various pipeline stages.

Pipeline expression operators

Pipeline *expression operators* perform the various operations needed at any given pipeline stage you have defined. The two most widely used operators fall into the **date** and **string** categories. Let's start with the string expression operators.

String expression operators

Although just one set of operators among many, the string operators probably prove to be the ones most frequently used when expression manipulation needs to occur. Here is a summary of the more important of these operators:

Operator	Notes
`$concat`	Concatenates two or more strings together.
`$trim`	Removes whitespace, or other specified characters (for example, CR\LF) from both the beginning and the end of a string. This family also includes `$ltrim` and `$rtrim`, which trims only to the left or right, respectively.
`$regexMatch`	New in MongoDB 4.2, returns TRUE or FALSE if the string matches the supplied regular expression. Also in this family are `$regexFind`, which returns the first string that matches a regular expression, and `$regexFindAll`, which returns all the strings that match the regex.
`$split`	Splits a string based on a delimiter into an array of substrings.

`$strLenBytes`	Returns the number of UTF-8-encoded bytes. Note that UTF-8-encoded characters may be encoded using up to 4 bytes per character. Thus, a word of three characters in Chinese might return a value of 9 or 12, whereas a word of three characters using a Latin-based alphabet might only return a value of 3.
`$substrCP`	Returns a substring that starts at the specified UTF-8 **code point index** (zero-based) for the number of UTF-8-encoded characters indicated by **code point count**.
`$toUpper,` `$toLower`	These operators produce a string of all-**UPPER** or all-**lower** case characters.

As an example, using aggregation pipeline string operators, let's say that as part of a report, you need to produce an alphabetical listing of a customer's name in this format:

```
title + first name + middle initial + last name
```

The first consideration is that we need the `$concat` string operator in order to combine the four name attributes of the title and the first, middle, and last names. We also need `$substrCP` in order to extract the first letter of the middle name, as well as `$toUpper`. Finally, `$cond` (discussed later) is needed in case either the title or middle name is missing.

Here is how the query might be formulated. We start with the `$project` pipeline stage operator, and postulate two new fields. The `last` field contains the value of the last name, and is suppressed in the last state. The `full_name` field is our concatenation:

```
db.customers.aggregate([
    { $project: {
        last : "$name.last",
        full_name: { $concat: [
```

We then describe four expressions. The first and third include a condition, and only return a value if that field contains a value. To obtain the middle initial from the `middle` name field (that is, the first letter of a person's middle name, if there is one), we also apply `$substrCP` to extract the first letter, and make it uppercase. You might also note that in the case of the title and middle name, we do another concatenation to add a space:

```
                    { $cond : {
                        if: "$name.title",
                        then: { $concat : ["$name.title", " "] },
                        else: "" } },
                    "$name.first", " ",
                    { $cond : {
                        if: "$name.middle",
                        then: { $concat:[{ \
                            $toUpper : {$substrCP:[ \
                                "$name.middle",0,1]}},"."]},
```

```
          else: "" } },
       "$name.last" ]
} } },
```

In the next stage, we use the `$sort` pipeline operator to alphabetize the list. In the last stage, we use another `$project` operator to suppress the last name field, which is needed by the previous stage:

```
{ $sort : { "last" : 1 } },
{ $project : { cust_name : "$full_name" } } ]);
```

Here are the first 20 results of the operation:

```
File  Edit  View  Search  Terminal  Help
{ "full_name" : "Ms Leann M. Aguilar" }
{ "full_name" : "Ms Renato H. Allen" }
{ "full_name" : "Latarsha Alvarez" }
{ "full_name" : "Ms Jolyn G. Anderson" }
{ "full_name" : "Mr Junior G. Archer" }
{ "full_name" : "Ms Margret Z. Arroyo" }
{ "full_name" : "Mr Darrel X. Atkinson" }
{ "full_name" : "Tiffani Austin" }
{ "full_name" : "Ms Lan W. Avila" }
{ "full_name" : "Mr Scotty Z. Ayala" }
{ "full_name" : "Mr Shannon C. Ballard" }
{ "full_name" : "Mr Boyd J. Ballard" }
{ "full_name" : "Mr Gilbert G. Barber" }
{ "full_name" : "Ms Danyelle M. Barber" }
{ "full_name" : "Ms Jerica G. Barber" }
{ "full_name" : "Ms Belen S. Barker" }
{ "full_name" : "Rudolph Barnes" }
{ "full_name" : "Ms Jesica D. Barrera" }
{ "full_name" : "Mr Xavier D. Beasley" }
{ "full_name" : "Mr Allen Z. Beasley" }
Type "it" for more

MongoDB jed@booksomeplace[up:72912 secs]
5>
```

You can view the entire query in the `aggregation_examples.js` file in
`/path/to/repo/chapters/09`.

 MongoDB automatically converts string data into a BSON string (`https:/`
`/docs.mongodb.com/master/reference/bson-types/#string`) internally,
which is UTF-8-encoded (`https://tools.ietf.org/html/rfc3629`). For
more information on UTF-8 code points, have a look at `http://www.`
`unicode.org/glossary/#code_point`.

Please refer to the following link for more information on expression
operators: `https://docs.mongodb.com/master/reference/operator/`
`aggregation/#expression-operators`.

Date expression operators

As you have seen in other examples in this chapter and earlier chapters, it is always
possible to work with date strings. The problem with this approach, however, is that the
data you have to work with might not be *clean* in that different date formats may have been
used. This is especially true in situations where the data has accumulated over a period of
time, perhaps from different versions of the web application.

Another problem when trying to perform operations on dates as strings is that performing
date arithmetic becomes problematic. You will learn more about this later in this chapter
when it comes time to present a solution for generating a 30-60-90 day **aging report**, which
is a typical requirement for accounting purposes.

For the purposes of this sub-section, we will examine the more important date expression
operators. There are two operators that create BSON date objects:

- `$dateFromParts`: Creates a BSON date object from parameters that include the
  year, month, day, hour, minute, second, millisecond, and time zone. This
  operator also accepts ISO/week-date parameters, including `isoWeekYear`,
  `isoWeek`, and `isoDayOfWeek`.
- `$dateFromString`: Creates a BSON date object from a string. The two main
  parameters needed for this operator to succeed include `dateString` and
  `format`. You can also specify the time zone. Two additional parameters,
  `onError` and `onNull`, allow you to specify a string or value that appears in place
  of a BSON date object if an error condition or `NULL` value arises. The beauty of
  this arrangement is that processing subsequent date conversions continues,
  instead of an error being thrown.

The format string follows ISO standards, and defaults to the following if not specified:

```
"%Y-%m-%dT%H:%M:%S.%LZ"
```

Thus, using this format, 10:02 the morning of January 1, 2020, would appear as follows, where `Z` indicates **Zulu time** (that is, UTC time):

```
"2020-01-01T10:02:00.000Z"
```

You can also reverse the process using either `$dateToString` or the `$dateToParts` operator. Once you have a BSON date object, use these operators to produce its component parts: `$year`, `$month`, `$dayOfMonth`, `$hour`, `$minute`, `$second`, and `$millisecond`.

There is also a general-purpose conversion operator, `$convert`, which was first introduced in MongoDB 4.0. It takes the following parameters as arguments: `input`, `to`, `onError`, and `onNull`. The value supplied with `input` can be any valid MongoDB expression. The value supplied with `to` can include `double`, `string`, `objectId`, `bool`, `int`, `long`, `decimal`, and finally, `date`.

For more information, please refer to the following links:

- **Date expression operators**: `https://docs.mongodb.com/manual/reference/operator/aggregation/#date-expression-operators`
- **Date format string**: `https://docs.mongodb.com/manual/reference/operator/aggregation/dateFromString/#format-specifiers`

Other important expression operators

Here is a brief summary of other important pipeline expression operators:

Category	Operator	Notes
Arithmetic	`$add`, `$subtract`, `$multiply`, `$divide`, and `$round`	Performs the associated arithmetic operation to returns sums of numbers, and so on. Note that `$add` and `$subtract` also work on dates. It should be noted that `$round` is only available as of MongoDB 4.2.
Array	`$arrayElemAt`, `$first`, and `$last`	Returns the array element at a specified index or the first/last elements. `$first` and `$last` were introduced in MongoDB 4.4.
	`$arrayToObject` and `$objectToArray`	Converts between an array and object.
	`$in`	Returns a Boolean value: TRUE if the item is in the array and FALSE otherwise.
	`$map`	Applies an expression to every array element.
	`$filter`	Returns a subset of array items matching the filter.

Boolean	$and, $or, and $not	Lets you create complex expressions.
Comparison	$eq, $gt, $gte, $lt, $lte, $ne, and $cmp	Provides conditions for filtering results, corresponding to equals, greater than, greater than or equals, and so on. The last operator, $cmp, produces −1 if op1 < op2, 0 if equal, and +1 if op1 > op2.
Conditional	$cond	Provides *ternary* conditional expressions in the following form: { $cond : [<expression>, <value if TRUE>, <value if FALSE>] }
	$ifNull	Provides the first non-null value for the specified field; otherwise, a default you provide.
	$switch	Implements a C language-style **switch** block.
Literal	$literal	Allows you to return a value directly as is, with no modifications.
Object	$mergeObjects	Combines objects into a single document.
Set	$setDifference, $setUnion, and $setIntersection	These operators treat arrays as **sets**, allowing manipulations on the entire array.
Text	$meta	Gains access to text search metadata.
Trigonometry	$sin, $cos, and $tan	First introduced in MongoDB 4.2, this group of operators provides trigonometric functionality, such as sine, cosine, tangent, and so on.
Type	$convert, $toDate, $toInt, and $toString	First introduced in MongoDB 4.0, these operators perform data type conversions.

Accumulator expression operators

Another group of operators, referred to as **accumulators**, are used to collect results such as producing totals. These operators are generally tied to specific aggregation pipeline stages, including $addFields, $group, and $project. Here is a brief summary of the more important accumulators associated with the $group, $addFields, and $project pipeline stages:

Operator	Notes
$sum	Produces a sum of one of the fields in the group. The sum can also be produced from an **expression**, which might involve additional fields and/or operators. Any non-numeric values are ignored.
$function	Allows you to define a JavaScript function, the results of which are made available to this pipeline stage. Only available as of MongoDB 4.4, this expression operation completely eliminates the need for Map-Reduce.
$accumulator	Introduced in MongoDB 4.4, this operator lets you include a JavaScript function to produce results, making this an extremely valuable general-purpose accumulating operator.
$avg	Produces an average of a field or expression from within the group. Any non-numeric values are ignored.
$min	Returns the minimum value of a field or expression from within the group.
$max	Returns the maximum value of a field or expression from within the group.
$push	Returns the value of a field or expression of all the documents within the group in the form of an array.

$stdDevPop	Used in statistical operations, this operator returns the *population* standard deviation of a field or expression from within the group.
$stdDevSamp	Used in statistical operations, this operator returns the *sample* standard deviation of a field or expression from within the group.

In addition, these operators are available only in the $group pipeline stage: $addToSet, $first, $last, and $mergeObjects.

> An example involving both $push and $sum is presented later in this chapter when we discuss generating a revenue report grouped by quarters.

In the next section, we will examine a more *traditional* way of handling aggregation; namely, using **Map-Reduce** functions.

Working with Map-Reduce functions

Prior to the introduction of the aggregation framework, in order to perform grouping and other aggregation operations, database users used a less efficient method: db.collection.mapReduce(). This command has been a part of the MongoDB command set since the beginning, but is more difficult to implement and maintain and suffers from relatively poor performance compared with similar operations using the aggregation framework. Let's now have a look at the mapReduce() method.

Understanding the mapReduce() method syntax

The primary method that allows you to perform aggregation operations using a JavaScript function is mapReduce(). The generic syntax for the collection method is this:

```
db.<collection>.mapReduce(<map func>,<reduce func>, \
                    {query:{},out:"<new collection>"})
```

The first argument is a JavaScript function representing the map phase. In this phase, you define a JavaScript function that calls emit(), which defines the fields included in the operation. The second argument is a JavaScript function that represents the reduce phase. A typical use for this is to produce some sort of accumulation (for example, sum) on the mapped fields.

The third argument is a JSON object with two properties: query and out. query, as you might have guessed, is a filter that limits documents included in the operation. out is the name of a collection into which the results of the operation are written.

An even more flexible set of options is available when using db.runCommand(). Here is the alternative syntax:

```
db.runCommand( {
    mapReduce: <collection>, map: <function>,
    reduce: <function>, finalize: <function>,
    out: <output>, query: <document>, sort: <document>, limit: <number>,
    scope: <document>, jsMode: <boolean>, verbose: <boolean>,
    bypassDocumentValidation: <boolean>, collation: <document>,
    writeConcern: <document>
});
```

You can see that when using db.runCommand(), an additional function, finalize, is available, which is executed after the map and reduce functions.

> The functionality of a Map-Reduce operation can easily be developed using the aggregation framework pipeline stages and operators, including $match, $group, $project, and so forth. MongoDB recommends using the aggregation framework as its function is completely integrated into the core engine. However, one reason to use Map-Reduce is that you have maximum flexibility in how you define the JavaScript functions, which opens complex operations up to a programmatic solution.

Reviewing a Map-Reduce example

A very simple example is enough to give you an idea of how this command works. Let's say that management wants a revenue report for 2019 with a sum for each country. First, we need to define the JavaScript map function. This function *emits* the country code and total price for each document in the `bookings` collection:

```
map_func = function () { \
     emit(this.property.propertyAddr.country, this.totalPrice); }
```

Next, we define the `reduce` function. This function accepts the two values that are emitted, and produces a sum of the total prices paid aggregated by country:

```
reduce_func = function (key, prices) { return Array.sum(prices); }
```

Next, we define the document used as a query filter, as well as the target collection into which the resulting information is written:

```
query_doc = {"bookingInfo.paymentStatus":"confirmed",
"bookingInfo.arrivalDate":/^2019/}
output_to = "country_totals"
```

We are now ready to execute the command using `mapReduce()` as a collection method:

```
db.bookings.mapReduce(map_func, reduce_func, {query:query_doc,
out:output_to });
```

Alternatively, we can run this command using `db.runCommand()`, as follows:

```
sort_doc = { "property.propertyAddr.country" : 1 }
db.runCommand( {
    mapReduce: "bookings",
    map:       map_func,
    reduce:    reduce_func,
    out:       output_to,
    query:     query_doc,
    sort:      sort_doc
});
```

As mentioned earlier, the main difference when using `runCommand()` is that other options, such as `sort`, are available. Once either command is executed, the result is written to a new collection, `country_totals`. Here are the first 20 results:

```
File Edit View Search Terminal Help
MongoDB jed@booksomeplace[up:73383 secs]
9>db.bookings.mapReduce(map_func, reduce_func, { query: query_doc, out: output_to });
{ "result" : "country_totals", "ok" : 1 }

MongoDB jed@booksomeplace[up:73396 secs]
10>db.country_totals.find();
{ "_id" : "FR", "value" : 158 }
{ "_id" : "DE", "value" : 2486 }
{ "_id" : "PK", "value" : 1033 }
{ "_id" : "FM", "value" : 362 }
{ "_id" : "BR", "value" : 3125 }
{ "_id" : "DK", "value" : 3650 }
{ "_id" : "NL", "value" : 5808 }
{ "_id" : "CO", "value" : 1230 }
{ "_id" : "LI", "value" : 3523 }
{ "_id" : "TH", "value" : 355 }
{ "_id" : "SI", "value" : 2468 }
{ "_id" : "UA", "value" : 1078 }
{ "_id" : "SJ", "value" : 2332 }
{ "_id" : "YT", "value" : 2457 }
{ "_id" : "LK", "value" : 3499 }
{ "_id" : "LV", "value" : 197 }
{ "_id" : "FI", "value" : 6354 }
{ "_id" : "AT", "value" : 5580 }
{ "_id" : "MT", "value" : 2768 }
{ "_id" : "JP", "value" : 2203 }
Type "it" for more

MongoDB jed@booksomeplace[up:73442 secs]
11>
```

The complete Map-Reduce query is in the `/path/to/repo/chapters/09/map_reduce_example.js` file.

> MongoDB 4.4 introduces the `$function` aggregation pipeline operator (`https://docs.mongodb.com/master/reference/operator/aggregation/function/index.html#function-aggregation`), which allows you to insert custom JavaScript directly into an aggregation pipeline. This effectively negates any of the benefits of using Map-Reduce if you are using MongoDB 4.4 or above.

It's important to note that any query that breaks down revenue by differing criteria forms an important part of risk analysis. In this example, revenue totals by country can identify weak financial returns. Any investment in infrastructure in a geographic area of weak return would be considered a risk.

Let's now have a look at handling **geospatial** queries.

Handling geospatial data

In order to conduct geospatial queries, you must specify one or more pairs of coordinates. For real-world data, this often translates into latitude and longitude. Before you go too much further, however, you need to stop and carefully consider what sorts of geospatial queries you plan to perform. Although all current MongoDB geospatial queries involve one or more sets of *two-dimensional X,Y* coordinates, there are two radically different models available: **flat** and **spherical**.

We begin our discussion by covering the flat 2D model, which performs queries on legacy coordinate pairs (for example, latitude and longitude).

Working with the flat 2D model

In order to perform geospatial searches, we first need to prepare the database by creating the necessary geospatial fields from our existing latitude and longitude information. We then need to define an index on the newly added field.

Creating a new field from existing ones is another opportunity to use the aggregation framework. In earlier versions of MongoDB, it was necessary to create a JavaScript function that looped through the database collection and updated each document individually. As you can imagine, this sort of operation was enormously expensive in terms of time and resources needed. Essentially, two commands had to be performed for every document: a query followed by an update.

As the name implies, the *flat* 2D MongoDB geospatial model is used for situations where a street map is appropriate. This model is perfect for giving directions to properties, or for locating stores within a certain geographic area, and so forth.

Using aggregation to create the geospatial field

Using aggregation we can consolidate the geospatial field creation into a single command. It is important to note that the current values for latitude and longitude are stored as a data type string, and hence are unusable for the purposes of geospatial queries. Accordingly, we employ the $toDouble pipeline expression operator.

Here is the command that adds a new field, `geo_spatial_flat`, consisting of legacy coordinate pairs, from existing latitude and longitude coordinates in the `world_cities` collection:

```
db.world_cities.aggregate(
    { "$match": {} },
    { "$addFields": {
        "geo_spatial_flat" : [
            { "$toDouble" : "$longitude" },
            { "$toDouble" : "$latitude" } ]
        }
    },
    { "$out" : "world_cities" }
);
```

This command uses three stages: `$match`, `$addFields`, and `$out`:

- The `$match` stage has an empty filter, and thus matches all documents in the collection.
- The `$addFields` stage creates a new field, `geo_spatial_flat`, using the values of the existing field's longitude and latitude.
- Finally, the `$out` stage instructs MongoDB to pipe the resulting collection into the one named.

As the collection identified by the `$out` stage is the same as the source of the data in our example, the effective result is the same collection with a new field added to each document.

Here is how the new field appears:

```
File  Edit  View  Search  Terminal  Help
MongoDB jed@booksomeplace[up:74110 secs]
37>db.world_cities.findOne({},{"geo_spatial_flat":1});
{
        "_id" : ObjectId("5ebf753a0db42badd870ebfe"),
        "geo_spatial_flat" : [
                1.53414,
                42.50729
        ]
}

MongoDB jed@booksomeplace[up:74215 secs]
38>
```

Before performing any global database conversions, it is strongly advised to first run some tests on a copy of the database. Alternatively, you could simply add a new field containing the same price data, but using the `NumberDecimal` data type. You should also back up the database (see the next section in this chapter).

Please note that you need to import the same world city data in order to follow this example. Sample data is available by restoring the `/path/to/repo/sample_data/booksomeplace_world_cities_insert.js` file.

We will next have a look at indexing.

Defining an index

Before conducting a geospatial query, you need to create a 2D index on this field. Here is the command we use to create a flat 2D index on the newly created geospatial field:

```
db.world_cities.createIndex( { "geo_spatial_flat" : "2d" } );
```

Here is a screenshot of this operation:

```
File  Edit  View  Search  Terminal  Help
MongoDB jed@booksomeplace[up:74082 secs]
35>db.world_cities.createIndex( { "geo_spatial_flat" : "2d" } );
{
        "createdCollectionAutomatically" : false,
        "numIndexesBefore" : 1,
        "numIndexesAfter" : 2,
        "ok" : 1
}

MongoDB jed@booksomeplace[up:74097 secs]
36>db.world_cities.getIndexes();
[
        {
                "v" : 2,
                "key" : {
                        "_id" : 1
                },
                "name" : "_id_"
        },
        {
                "v" : 2,
                "key" : {
                        "geo_spatial_flat" : "2d"
                },
                "name" : "geo_spatial_flat_2d"
        }
]
```

It is not recommended to create `geoHayStackindex` (`https://docs.mongodb.com/manual/core/geohaystack/#geohaystack-indexes`) as this is deprecated in MongoDB 4.4. It's also worth noting that the `geoSearch` database command (`https://docs.mongodb.com/master/reference/command/geoSearch/#geosearch`) has also been deprecated in MongoDB 4.4.

Before we examine the **spherical** model, it's important to discuss GeoJSON objects.

Understanding GeoJSON objects

GeoJSON objects are needed in order to conduct geospatial queries involving a *spherical* model. These object types are modeled after RFC 7946, *The GeoJSON Format*, published in August 2016. Here is the generic syntax for creating a MongoDB GeoJSON object:

```
<field>: { type: <GeoJSON type> , coordinates: <coordinates> }
```

Please note that, depending on the GeoJSON object type, there may be more than one pair of coordinates. MongoDB supports a number of these objects, summarized here:

- `Point`:
  - **Definition**: A distinct location represented by two coordinates on an imaginary *X-Y* grid.
  - **Syntax**:

    ```
    <field>: { type: "Point", coordinates: [ x, y ] }
    ```

- `LineString`:
  - **Definition**: A set of points in a straight line starting at *x1,y1* and ending at *x2,y2*.
  - **Syntax**:

    ```
    <field>: { type: "LineString", coordinates: [[x1,y1], [x2,y2]] }
    ```

- `Polygon`:
  - **Definition**: An array of GeoJSON `LinearRing` coordinate arrays. `LinearRing` arrays are `LineString` objects that have the same beginning and ending coordinates, thus closing a loop.

- **Syntax (single ring)**:

```
<field>: { type: "Polygon", coordinates: [[x1,y1],[x2,y2],[x3, y3],
etc., [x1,y1]]}
```

- **Syntax (multiple rings)**:

```
<field>: { type: "Polygon", coordinates: [
    [[x1,y1],[x2,y2],[x3, y3], etc. ],
    [[xx1,yy1],[xx2,yy2],[xx3, yy3], etc. ] ] }
```

When using *latitude* and *longitude* for coordinates, MongoDB requires that the first coordinate must always be *longitude*, regardless of the field name.

Please refer to the following links for more information:

- **GeoJSON objects:** `https://docs.mongodb.com/manual/reference/geojson/#geojson-objects.`
- **RFC 7946,** *The GeoJSON Format,* **published August 2016:** `https://tools.ietf.org/html/rfc7946.`

Working with the spherical 2D geospatial model

The *spherical* 2D MongoDB geospatial model is used for situations where coordinates need to be located on a sphere. An obvious example would be creating queries that provide locations on planet Earth, the moon, or other planetary bodies. Here is the command that creates a GeoJSON object from existing coordinates in the `properties` collection. Again, please note the use of `$toDouble` to convert latitude and longitude values from `string` to `double`:

```
db.properties.aggregate(
    { "$match": {} },
    {
        "$addFields": {
            "address.geo_spatial":{
                "type":"Point",
                "coordinates":
                [
                    { "$toDouble" : "$address.longitude" },
                    { "$toDouble" : "$address.latitude" }
```

```
                    ]
                }
            }
        },
      { "$out" : "properties" }
    );
```

This query is identical to the one described a few sections earlier, except for the addition of geospatial fields. We are then in a position to create an index to support geospatial queries based on the 2D spherical model:

```
db.properties.createIndex( { "address.geo_spatial" : "2dsphere" } );
```

After having created a GeoJSON field from the latitude and longitude fields for the `world_cities` collection, we also need to do the same for the properties collection.

Using geospatial query operators

MongoDB geospatial query operators are used to perform queries based on geographical or spatial sets of coordinates. As noted previously, only two operators support the flat model; however, all geospatial operators support the spherical model. Here is a brief summary of the MongoDB geospatial query operators:

Operator	Spherical	Flat	Notes
$near	Y	Y	Returns documents in order of nearest to furthest from the point specified using the $geometry parameter. You can also supply two additional parameters, $minDistance and $maxDistance, in order to limit the number of documents returned. You can use either a 2D or a 2Dsphere index.
$nearSphere	Y	N	Operates exactly like $near, but uses a 2Dsphere index and only follows the spherical model.
$geoWithin	Y	Y	Accepts a $geometry argument with type and coordinates keys. The type can be either Polygon or MultiPolygon.
$geoIntersects	Y	N	This operator accepts the same arguments and keys as $geoWithin but operates on the spherical model.

As an example, let's say that we produced a list of Book Someplace properties in India using this query:

```
db.properties.aggregate(
    { "$match"   : { "address.country" : "IN" } },
    { "$project" : { "address.city" : 1, "address.geo_spatial" : 1, \
                     "_id" : 0 } },
    { "$sort"    : { "address.city" : -1 } }
);
```

We now wish to obtain a list of world cities within 100 kilometers of this property. There is no need to add the country code as `find` criteria, as we are already dealing with geospatial coordinates.

Here is the query (note that `$maxDistance` is in meters):

```
db.world_cities.find(
    { "geo_spatial_sphere" :
        { "$near" : {
            "$geometry" : {
                "type" : "Point",
                "coordinates" : [77.8956, 18.0806]
            },
            "$maxDistance" : 100000
        }
    }
},{ "name" : 1, "geo_spatial_sphere" : 1, "_id" : 0 });
```

The preceding example assumes that a `geo_spatial_sphere` field, consisting of the longitude and latitude, has been added and that a 2dsphere index has been created on this field. For a full list of operations preceding this query, have a look at `/path/to/repo/chapters/09/geo_spatial_query_examples.js`.

Here is the resulting output:

```
File Edit View Search Terminal Help
> db.world_cities.find(
...     { "geo_spatial_sphere" :
...         { "$near" : {
...             "$geometry" : {
...                 "type" : "Point",
...                 "coordinates" : [77.8956, 18.0806]
...             },
...             "$maxDistance" : 100000
...         }
...     }
... },{ "name" : 1, "geo_spatial_sphere" : 1, "_id" : 0 });
{ "name" : "Bānswāda", "geo_spatial_sphere" : [ 77.88007, 18.37725 ] }
{ "name" : "Andol", "geo_spatial_sphere" : [ 78.07713, 17.81458 ] }
{ "name" : "Medak", "geo_spatial_sphere" : [ 78.26078, 18.04531 ] }
{ "name" : "Bīdar", "geo_spatial_sphere" : [ 77.53011, 17.91331 ] }
{ "name" : "Sadāseopet", "geo_spatial_sphere" : [ 77.95263, 17.61925 ] }
{ "name" : "Zahirābād", "geo_spatial_sphere" : [ 77.60743, 17.68138 ] }
{ "name" : "Aurād", "geo_spatial_sphere" : [ 77.41761, 18.25397 ] }
{ "name" : "Kāmāreddi", "geo_spatial_sphere" : [ 78.34177, 18.32001 ] }
{ "name" : "Sangāreddi", "geo_spatial_sphere" : [ 78.08669, 17.62477 ] }
{ "name" : "Diglūr", "geo_spatial_sphere" : [ 77.57695, 18.54829 ] }
{ "name" : "Bodhan", "geo_spatial_sphere" : [ 77.88581, 18.66208 ] }
{ "name" : "Nizāmābād", "geo_spatial_sphere" : [ 78.0988, 18.67154 ] }
{ "name" : "Singāpur", "geo_spatial_sphere" : [ 78.12574, 17.46982 ] }
{ "name" : "Patancheru", "geo_spatial_sphere" : [ 78.2645, 17.53334 ] }
{ "name" : "Bhālki", "geo_spatial_sphere" : [ 77.206, 18.04348 ] }
{ "name" : "Serilingampalle", "geo_spatial_sphere" : [ 78.30196, 17.49313 ] }
{ "name" : "Vikārābād", "geo_spatial_sphere" : [ 77.90441, 17.3381 ] }
{ "name" : "Chincholi", "geo_spatial_sphere" : [ 77.41874, 17.46508 ] }
{ "name" : "Kūkatpalli", "geo_spatial_sphere" : [ 78.41376, 17.48486 ] }
{ "name" : "Quthbullapur", "geo_spatial_sphere" : [ 78.45818, 17.50107 ] }
Type "it" for more
>
```

 Please refer to the following link for more information on geospatial query operators: https://docs.mongodb.com/manual/geospatial-queries/#geospatial-query-operators.

Having seen the use of geospatial operators in a query, let's examine their use in an aggregation pipeline stage.

Examining the geospatial aggregation pipeline stage operator

The $geoNear aggregation operator supports geospatial operations within the context of the aggregation framework. The $geoNear operator can be used to form a new pipeline stage, upon which geospatial data comes into play. In this case, documents are placed in the pipeline by the indicated GeoJSON field in order of nearest to furthest from a given point.

Here is a brief summary of the more important parameters available for this operator:

Option	Notes
near	If the field used has a `2dsphere` index, you can use either a `GeoJSON` object of the `point` type or a list of legacy coordinate pairs (for example, `[ <longitude>,<latitude>])` as an argument. On the other hand, if the field only has a flat 2D index, you can only supply a legacy coordinate pair list as an argument.
spherical	When set to `True`, this parameter forces MongoDB to use a spherical model of calculation. Otherwise, if set to the default of `False`, the calculation model used depends on what type of index is available for the geospatial field.
minDistance maxDistance	Lets you set a minimum and maximum distance, which is extremely useful in limiting the number of documents in the resulting output. If you are using a geospatial field with a `GeoJSON` object, the distance is set in meters. Otherwise, when using legacy coordinate pairs, the distance can only be set in radians.
key	This parameter represents the name of the geospatial field you plan to use to calculate the distance. Although technically optional, you run into problems if you do not set a value for this parameter and the operation occurs on a collection with multiple geospatial indexes.
distanceField	The value you supply for this parameter is the name of a field that is added to the pipeline output and contains the calculated distance.

For more information on the `$geoNear` aggregation operator, refer to `https://docs.mongodb.com/master/reference/operator/aggregation/geoNear/#geonear-aggregation`.

Now that you have an understanding of working with geospatial data in MongoDB, it's time to have a look at generating financial reports.

Using aggregation to produce financial reports

Most financial reports are fairly simple in terms of the database queries and code needed to present the report. Management reports requested tend to fall into one of the following categories:

- Revenue reports
- Aging reports
- Revenue trends

All three reports form tools that help you perform **risk analysis**. Let's start with revenue reports.

NumberDecimal is a **Binary JSON (BSON)** class that allows the precise modeling of financial data. Using NumberDecimal is most useful in situations where you need extreme precision. An example would be when you are performing currency conversions when transferring money between countries.

For more information on this usage, please refer to Chapter 11, *Administering MongoDB Security*. For further information, refer to the discussion of modeling monetary data at https://docs.mongodb.com/manual/tutorial/model-monetary-data/#model-monetary-data.

Using aggregation to generate revenue reports

Revenue reports, detailing the money received, are most often requested based on certain periods of time. For example, you might be asked to produce a report that provides a summary of the revenue generated over the past year, with subtotals for each quarter.

Revenue reports are an important part of **risk analysis**. In the financial community, a *risk* can be defined as an investment that causes the company to lose money. Revenue reports, at their core, are just a total of money coming into the company broken down by time, location, or customer **demographic** information.

The criteria used in generating the report can highlight crucial areas of weakness, and thus *risk*. For example, if you generate a report by location, your company may discover that revenue is lower in certain regions than in others. If your company is investing heavily in a certain region but is generating little revenue in return, your report has identified a risk. Management can then use this information to reallocate advertising in an effort to improve revenue from the weak region. If a later revenue report reveals that the revenue has not increased given an increased advertising budget, management may decide to either seek a different sales and marketing strategy or perhaps withdraw from the region entirely.

In this example, we will generate a revenue report of customer reservations by room type. We will start by using the mongo shell to build a query.

Using the mongo shell to build a query

We will start by modeling the query, either using MongoDB Compass or directly from the `mongo` shell. In order to run this query, we need information from the `bookings` collection's `bookingInfo` field (described in Chapter 7, *Advanced MongoDB Database Design*). You can see that there is often a twist. In this case, the twist is that we must be sure to only include bookings where the payment status is confirmed.

Query criteria are grouped together using the `$and` query operator. We use a regular expression to match the target year of 2018. In addition, we add a condition that specifies that only documents where the payment status is *confirmed* are matched. It is also worth mentioning at this point that in order to reference a property of an embedded object, you should use a dot (.) as a separator. Thus, in order to match `arrivalDate`, which is a property of the object represented by the `bookingInfo` field, we use this syntax:

```
bookingInfo.arrivalDate.
```

Here is a query that produces the data needed to generate the report for 2019:

```
db.bookings.find(
    { "$and":[
        { "bookingInfo.arrivalDate":{"$regex":"^2019"},
          "bookingInfo.paymentStatus":"confirmed"
        }
    ] },
    {"totalPrice":1,"bookingInfo.arrivalDate":1}
);
```

This query, however, simply produces a block of data and places the burden of breaking down the information and producing totals on the Python script. With some thought, using the query shown previously as a base, it's easy enough to modify the query using aggregation.

The `$match` pipeline stage restricts the number of documents in the pipeline to only those where the date begins with the target year – in this case, 2019. Here is the revised query:

```
db.bookings.aggregate( [
    { $match :
        { "$and":[
            { "bookingInfo.arrivalDate":{"$regex":"^2019"},
              "bookingInfo.paymentStatus":"confirmed"
            }
        ] }
    }
}]);
```

If you take a quick look at a couple of bookings, however, you'll find that bookings are by room type, and that room types are organized into an array of `RoomType` objects. Here is an example from the `mongo` shell:

```
root@server1: /repo/chapters/09
{
        "bookingInfo" : {
                "arrivalDate" : "2018-03-21 14:07:00",
                "departureDate" : "2018-03-28 14:00:00",
                "checkoutTime" : "14:00:00",
                "refundableUntil" : "2018-03-20 14:07:00",
                "reservationStatus" : "confirmed",
                "paymentStatus" : "confirmed"
        },
        "rooms" : [
                {
                        "roomType" : "VIP",
                        "roomTypeKey" : "dd49",
                        "price" : 100,
                        "qty" : 1
                },
                {
                        "roomType" : "family",
                        "roomTypeKey" : "bb51",
                        "price" : 258,
                        "qty" : 2
                }
        ]
}
Type "it" for more
> db.bookings.find({},{"bookingInfo":1,"rooms":1,"_id":0}).pretty();
```

In order to access information by room type, we'll first need to *unwind* the array. What this pipeline operator does is create duplicate documents, one for each array element, effectively flattening the array. As an example, consider this query:

```
db.bookings.aggregate( [
    {$unwind : "$rooms"},
    {$project: {"bookingKey":1,"rooms":1,"_id":0}}
]);
```

Now have a look at the output:

```
                          root@server1: /repo/chapters/09
> db.bookings.aggregate([{$unwind : "$rooms"},{$project: {"bookingKey":1,"rooms":1,"_id":0}}]);
{ "bookingKey" : "20190710SCOT", "rooms" : { "roomType" : "premium", "roomTypeKey" : "aa44", "price" : 435, "qty" : 3 } }
{ "bookingKey" : "20180416MALO", "rooms" : { "roomType" : "VIP", "roomTypeKey" : "aa75", "price" : 156, "qty" : 1 } }
{ "bookingKey" : "20190801PRUI", "rooms" : { "roomType" : "premium", "roomTypeKey" : "bb72", "price" : 117, "qty" : 1 } }
{ "bookingKey" : "20190801PRUI", "rooms" : { "roomType" : "suite", "roomTypeKey" : "cc45", "price" : 232, "qty" : 1 } }
{ "bookingKey" : "20190801PRUI", "rooms" : { "roomType" : "standard", "roomTypeKey" : "aa66", "price" : 104, "qty" : 2 } }
{ "bookingKey" : "20181210FUEN", "rooms" : { "roomType" : "VIP", "roomTypeKey" : "bb14", "price" : 115, "qty" : 1 } }
{ "bookingKey" : "20181210FUEN", "rooms" : { "roomType" : "standard", "roomTypeKey" : "aa99", "price" : 372, "qty" : 2 } }
{ "bookingKey" : "20201106WHIT", "rooms" : { "roomType" : "family", "roomTypeKey" : "aa24", "price" : 298, "qty" : 2 } }
{ "bookingKey" : "20180907KNOX", "rooms" : { "roomType" : "standard", "roomTypeKey" : "aa53", "price" : 296, "qty" : 2 } }
{ "bookingKey" : "20180907KNOX", "rooms" : { "roomType" : "standard", "roomTypeKey" : "aa53", "price" : 296, "qty" : 2 } }
{ "bookingKey" : "20210220ESTR", "rooms" : { "roomType" : "standard", "roomTypeKey" : "aa58", "price" : 400, "qty" : 4 } }
{ "bookingKey" : "20181105HOWE", "rooms" : { "roomType" : "suite", "roomTypeKey" : "bb85", "price" : 45, "qty" : 1 } }
{ "bookingKey" : "20191005PATT", "rooms" : { "roomType" : "double", "roomTypeKey" : "dd88", "price" : 489, "qty" : 3 } }
{ "bookingKey" : "20171214SWEE", "rooms" : { "roomType" : "double", "roomTypeKey" : "aa66", "price" : 51, "qty" : 1 } }
{ "bookingKey" : "20171214SWEE", "rooms" : { "roomType" : "double", "roomTypeKey" : "aa66", "price" : 51, "qty" : 1 } }
{ "bookingKey" : "20200513NORM", "rooms" : { "roomType" : "premium", "roomTypeKey" : "aa60", "price" : 159, "qty" : 1 } }
{ "bookingKey" : "20200513NORM", "rooms" : { "roomType" : "premium", "roomTypeKey" : "aa60", "price" : 159, "qty" : 1 } }
{ "bookingKey" : "20200513NORM", "rooms" : { "roomType" : "premium", "roomTypeKey" : "cc83", "price" : 504, "qty" : 3 } }
{ "bookingKey" : "20171003NAVA", "rooms" : { "roomType" : "family", "roomTypeKey" : "cc23", "price" : 184, "qty" : 1 } }
{ "bookingKey" : "20171003NAVA", "rooms" : { "roomType" : "premium", "roomTypeKey" : "bb97", "price" : 241, "qty" : 1 } }
Type "it" for more
>
```

As you can see, the original booking documents are duplicated; however, only one room
type is represented. We can then modify the model query by adding three new stages:
$unwind, $project, and $sort, as shown here:

```
{ $unwind      : "$rooms" },
{ $project     : { "rooms" : 1 }},
{ $sort        : { "rooms.roomType" : 1 }}
```

You can see that following the $unwind pipeline stage, we then limit the fields to rooms
only using $project. After that, we sort by room type using $sort. The only thing left to
do is to group by room type and add totals. Here are the additional stages:

```
{ $addFields :
    { "total_by_type" : { $multiply : [ \
        "$rooms.price", "$rooms.qty" ] }}
},
{ $group : {
    _id : { "type" : "$rooms.roomType" },
    "count": { $sum: "$rooms.qty"},
    "total_sales": { $sum: "$total_by_type"}}
}
```

Before implementing the Python script that generates the revenue report, we must make a simple addition to the BookingService class, which is in /path/to/repo/chapters/09/src/booksomeplace/domain/booking.py module. The new method, which we will call fetchAggregate(), accepts an aggregation pipeline document as an argument and leverages the pymongo collection-level aggregate() method. Here is the new method:

```
def fetchAggregate(self, pipeline) :
    return self.collection.aggregate(pipeline)
```

We can now use the model query (just discussed) in a Python script, financial_query_sales_by_room_type.py.py, to produce the desired results. You can see the full script in the /path/to/repo/chapters/09 directory. As usual, we start with imports and create a BookingService instance:

```
import os,sys
sys.path.append(os.path.realpath("src"))
import pymongo
from config.config import Config
from booksomeplace.domain.booking import BookingService
service = BookingService(Config())
```

Next, pass the pipeline document as an argument to the newly defined domain service method, aggregate(). Here, in this code block, you can see exactly the same query as shown previously, but incorporated as a Python dictionary:

```
target_year = 2019
pipeline  = [
    { '$match' :
        { '$and':[
            { 'bookingInfo.arrivalDate':{'$regex':'^' + str(target_year)},
              'bookingInfo.paymentStatus':'confirmed'
            }
        ] }
    },
    { '$unwind'    : '$rooms' },
    { '$project'   : { 'rooms' : 1 }},
    { '$sort'      : { 'rooms.roomType' : 1 }},
    { '$addFields' :
        { 'total_by_type':{'$multiply':['$rooms.price','$rooms.qty']}}
    },
    { '$group' : {
        '_id' : { 'type' : '$rooms.roomType' },
        'count': { '$sum' : '$rooms.qty'},
        'total_sales': { '$sum' : '$total_by_type'}}
    }
```

```
]
result = service.fetchAggregate(pipeline)
```

As you can see, the `pipeline` dictionary is identical to the JSON structure used when modeling the query using the `mongo` shell.

We then loop through the query results and display the totals by room type, with a grand total at the end:

```
total = 0
count = 0
pattern_line  = "{:10}\t{:6}\t{:10.2f}"
pattern_text  = "{:10}\t{:6}\t{:10}"
print('Sales by Room Type For ' + str(target_year))
print(pattern_text.format('Type','Qty','Amount'))
print(pattern_text.format('----------','------','----------'))
for doc in result :
    label = doc['_id']['type']
    qty   = doc['count']
    amt   = doc['total_sales']
    total += amt
    count += qty
    print(pattern_line.format(label,qty,amt))

# print total
print(pattern_text.format('==========','======','=========='))
print(pattern_line.format('TOTAL',count,total))
```

Here is how the report might appear:

Let's now have a look at a class financial report: the 30-60-90 day aging report.

Generating 30-60-90 day aging reports using MongoDB aggregation

Aside from **money received** financial reports, one of the next biggest concerns for management is *how much is still owed*. Furthermore, management often requests aging reports. These reports detail the money owed broken down by time increments (for example, 30 days). One such report is referred to as a **30-60-90 day aging report**.

Aging reports are an important part of risk analysis. If the percentage of the amount owed past 90 days is excessive compared with 30 or 60 days, a risk has been identified. It could be that a financial crisis is looming in one or more areas where you do business. Another reason might be that your company's customers are unable (or perhaps unwilling!) to pay promptly. If further investigation reveals that customers are unwilling to pay promptly, this could indicate a problem with quality control, or a failure to deliver products in time (depending on the nature of your business).

In the context of Book Someplace, this type of report for a hotel booking is not very important as most bookings are made within a 30-day window. However, for the purposes of illustration, we have included in the sample data booking entries where the payment is still pending, furnishing us with enough data to generate a decent 30-60-90 day aging report.

 In this section, we will work with dates as strings. A more sophisticated way of handling date information is to use a **BSON date** object. The MongoDB aggregation framework includes a pipeline operator, `$toDate`, which allows you to convert a date string into a BSON date instance. Once converted, you are able to perform a wide variety of date-related operations, including comparisons and date arithmetic.

We can now apply this query to a Python script that uses the `pymongo.collection.find()` method. The full script is named `financial_30_60_90_day_aging_report_using_aggregation.py` and can be found at `/path/to/repo/chapters/09`.

As before, we first define the `import` statements and create an instance of the `BookingService` class from the `booksomeplace.domain.booking` module:

```
import os,sys
sys.path.append(os.path.realpath("src"))
import pymongo
from datetime import date,timedelta
from config.config import Config
from booksomeplace.domain.booking import BookingService
service = BookingService(Config())
```

Next, we stipulate the target date from which we work backward 30, 60, and 90 days:

```
year   = '2020'
target = year + '-12-01'
now    = date.fromisoformat(target)
aging  = {
    '30' : (now - timedelta(days=30)).strftime('%Y-%m-%d'),
    '60' : (now - timedelta(days=60)).strftime('%Y-%m-%d'),
    '90' : (now - timedelta(days=90)).strftime('%Y-%m-%d'),
}
```

We now formulate a Python dictionary that forms the basis of the `$project` stage in the aggregation operation. In this stage, we allow three fields to be pushed into the pipeline: `bookingInfo.arrivalDate`, `bookingInfo.paymentStatus`, and `totalPrice`. Note that this formulation allows us to effectively create an alias for the first two fields, which makes the remaining stages easier to read:

```
project = {
    "$project" : {
        "arrivalDate"   : "$bookingInfo.arrivalDate",
        "paymentStatus" : "$bookingInfo.paymentStatus",
        "totalPrice"    : 1,
    }
}
```

If you need an actual BSON date or GeoJSON objects, all you need to do is specify an additional field and use the `$convert` pipeline operator as an argument in the `$project` stage.

We then create another dictionary to represent the $match stage. In this example, we wish to include booking documents where the payment status is pending, the date is in the year 2019, and the date is less than our target date of 2019-12-01:

```
match = {
    "$match" : {
        "$and":[
            { "paymentStatus"  : "pending" },
            { "arrivalDate"    : { "$regex" : "^" + year }},
            { "arrivalDate"    : { "$lt"    : target }}
        ]
    }
}
```

Although not technically needed, we include a $sort stage so that if we do need to examine the data in detail, the dates are sorted, making it easier to confirm the report totals:

```
sort = { "$sort" : { "arrivalDate" : 1 }}
```

Finally, the most complicated stage is $bucket. In this stage, we group the dates by arrivalDate (an alias for bookingInfo.arrivalDate) and set boundaries that represent 30, 60, and 90 days. The default parameter represents the *bucket* that contains data outside of the boundaries:

```
bucket = {
    "$bucket" : {
        "groupBy"     : "$arrivalDate",
        "boundaries" : [ aging['90'], aging['60'], aging['30'], target ],
        "default"    : target,
        "output"     : {
            "totals"  : { "$sum"  : "$totalPrice" },
            "amounts" : { "$push" : "$totalPrice" },
            "dates"   : { "$push" : "$arrivalDate" }
        }
    }
}
```

The actual execution of the aggregation is absurdly simple: just two lines of code!

```
pipeline = [ project, match, sort, bucket ]
result = service.fetchAggregate(pipeline)
```

Before the main loop, we initialize some variables:

```
days  = 90;
total = 0.00
dbTot = 0.00
pattern_line = "{:20}\t{:10.2f}\t{:10.2f}"
```

We then loop through the results, consisting of one document per bucket:

```
for doc in result :
    if doc['_id'] == target :
        amt_plus = doc['totals']
        dbl_plus = sum(doc['amounts'])
    else :
        label = str(days)
        amt   = doc['totals']
        dbl   = sum(doc['amounts'])
        total += amt
        dbTot += dbl
        days  -= 30
        print(pattern_line.format(label,amt, dbl))

print(pattern_line.format('TOTAL',total,dbTot))
```

This screenshot gives you an idea of the output, with labels added for convenience:

```
File  Edit  View  Search  Terminal  Help
root@server1:/repo/chapters/09# python financial_30_60_90_day_aging_report_using_aggregation.py
..........................................................
FROM: 2020-12-01
30: 2020-11-01 | 60: 2020-10-02 | 90: 2020-09-02
..........................................................
Aging in Days          From Aggr        From Amts
-------------------    ----------       ----------
90                       24816.00         24816.00
60                       14576.00         14576.00
30                       35384.00         35384.00
===================    ==========       ==========
TOTAL                    74776.00         74776.00
..........................................................
root@server1:/repo/chapters/09#
```

> Please refer to the following link for more information on BSON date objects: https://docs.mongodb.com/manual/reference/bson-types/#date.

Next, we will have a look at generating a trend report that can assist in risk analysis.

Performing risk analysis through revenue trends

The last category we will cover here is that of revenue trends. These reports are used by management to make decisions on where to invest money in marketing, as well as where to cut funds from non-productive initiatives. Revenue trend reports can be based on actual *historic* data, projections of *future* data, or a combination of the two. When generating revenue trends for the future, it's a good idea to first plot a graph of the actual historic data. You can then apply the graph to predict the future revenue.

 The further back you go in historic data, the more accurate your projection will be.

When analyzing historic data in order to predict future data, be careful to spot and avoid **anomalies**. These are factors that are one-time events that have a significant impact on a particular trend, but that cannot be counted on in the future. An example might be a worldwide pandemic. This would obviously cause a significant drop in bookings for properties in the area affected by the outbreak. However, even though these properties might be in a hot zone, an event such as this could (hopefully!) be considered an anomaly, and the historic data that encapsulates the timespan between the outbreak occurrence and the recovery period should be weighted when predicting potential future revenue.

Any anomalies play an important part in overall risk analysis. Downward revenue trends identify a risk. If your projection is accurate, you can help your company avoid investing in a potentially disastrous risky region or demographic.

Setting up the Django infrastructure

Since the output of this report is a line chart, we start by defining a Django app for that purpose. The app is located in the `/path/to/repo/www/chapter_09/trends` folder. We then add a reference to `/path/to/repo/www/chapter_09/booksomeplace_dj/urls.py`, such that the `/trends/future` URL points to the new Django app. In the same file, `urls.py`, we add another route for a view method that responds to an AJAX request, `/trends/json`:

```
from django.urls import path
from . import views
urlpatterns = [
    path('future/', views.future, name='trends.future'),
    path('future/json/', views.futureJson, name='trends.future_json'),
]
```

Let's now define a form to capture query parameters.

Defining a form to capture query parameters

Next, in /path/to/repo/www/chapter_09/trends/forms.py, we define a form class, TrendForm, which captures the query parameters used to formulate the revenue trends report. The fields that are defined here are ultimately used to form a query filter. The trend_region field could be used to include states, provinces, or even countries, such as England in the UK. The last field, trend_factor, is used for the purposes of future revenue projection. The initial value is .10, which represents revenue growth of 10%. The step attribute is .01, and represents the increment value for this number field.

Defining view methods

We are now in a position to define view methods, entered into /path/to/repo/www/chapter_09/trends/views.py. The first method defined is future(). It defines initial form values, creates a TrendForm instance, renders it using the trend.html template, and injects the resulting HTML into the layout. The only other method is futureJson(), which responds to an AJAX request. Its main job is to make a call to the BookingService class instance, which is in the booksomeplace.domain.booking module:

```
def futureJson(request) :
    import json
    bookSvc = BookingService(Config())
    params  = {
        'trend_city'     : None,
        'trend_region'   : 'England',
        'trend_locality' : None,
        'trend_country'  : 'GB',
        'trend_factor'   : 0.10
    }
    if request.method == 'POST':
        params = request.POST
    data = bookSvc.getTrendData(params)
    return HttpResponse(json.dumps(data), content_type='application/json')
```

The response from the getTrendData() method is a list of revenue totals, both past and estimated future, which are mapped into a line chart over a span of months.

Defining the view template

After the logic used to calculate trend data, the next most difficult task is defining JavaScript code used to make AJAX requests for trend data. This is accomplished in the `/path/to/repo/www/chapter_09/templates` directory. The template filename is `trend.html`. Inside this template file, we add a `<div>` tag, where the JavaScript-generated chart resides. For the purpose of this example, we will use `Chartist.js`, an extremely simple-to-use JavaScript responsive chart generator that has no dependencies.

 `chartist.js` is the brainchild of Gion Kunz from Zurich, Switzerland. The source code is provided under a **Massachusetts Institute of Technology (MIT)** license, which is business-friendly and has no strings attached. `Chartist.js` should not be confused with `Chart.js`, which is another open source JavaScript library for generating charts. For more information on `chartist.js`, refer to `https://gionkunz.github.io/chartist-js/`.

First, we need to define an HTML element into which the chart can be written:

```
<div class="col-md-6">
    <div class="ct-chart ct-perfect-fourth"></div>
</div>
```

Next, we link in the `stylesheet`, and an override to make the *X-Y* axis labels easier to read:

```
{% block extra_head %}
<link rel="stylesheet"
href="//cdn.jsdelivr.net/chartist.js/latest/chartist.min.css">
<style>
.ct-label {
    color: black;
    font-weight: bold;
    font-size: 9pt;
}
</style>
{% endblock %}
```

We now import the JavaScript libraries for both jQuery and Chartist:

```
<script
src="//cdn.jsdelivr.net/chartist.js/latest/chartist.min.js"></script>
<script src="https://code.jquery.com/jquery-3.4.1.min.js"></script>
```

Next, we define the JavaScript `doChart()` function, which actually draws the chart:

```
<script type="text/javascript" charset="utf8">
function doChart(data)
{
    var plot = { labels:data.labels, series:data.series };
    new Chartist.Line('.ct-chart', plot);
}
```

Another JavaScript function we define is `ajaxCall()`, which makes the jQuery AJAX request. In this method, we define the request as an HTTP POST request, identify the target URL, and specify the data type as JSON. We then use jQuery to retrieve the current value of the form fields discussed earlier. Upon success, we call the `doChart()` function, which draws the chart. The `ajaxCall()` function is shown here:

```
function ajaxCall() {
    $.ajax({
        type: "POST",
        url: "/trends/future/json/",
        dataType: 'json',
        data: {
            'trend_city'    : $('#id_trend_city').val() ,
            'trend_region'  : $('#id_trend_region').val() ,
            'trend_locality': $('#id_trend_locality').val() ,
            'trend_country' : $('#id_trend_country').val() ,
            'trend_factor'  : $('#id_trend_factor').val() ,
        },
        success: function (data) { doChart(data); }
    });
}
```

Finally, we tie the `ajaxCall()` function into a document `ready()` function call, which runs `ajaxCall()` when the document loads:

```
$(document).ready(function() {
    ajaxCall();
    $("#id_trend_city").change(function() { ajaxCall(); });
    $("#id_trend_region").change(function() { ajaxCall(); });
    $("#id_trend_locality").change(function() { ajaxCall(); });
    $("#id_trend_country").change(function() { ajaxCall(); });
    $("#id_trend_factor",).change(function() { ajaxCall(); });
});
</script>
```

Subsequently, if any of the parameter field values change, `ajaxCall()` is called again to redraw the chart.

Defining a method to generate trend data

In the `BookingService` class, located in
`/path/to/repo/chapters/09/src/booksomeplace/domain/booking.py`, we define a
method, `getTrendData()`, which returns historic and future data based on query
parameters. Next, we formulate the base query filter, as well as the desired projection and
sort criteria:

```
query       = {'bookingInfo.paymentStatus':'confirmed'}
projection  = {'bookingInfo.arrivalDate':1,'totalPrice':1}
sort        = [('bookingInfo.arrivalDate',1)]
```

We now build onto the query dictionary by checking to see whether parameters are set:

```
if 'trend_city' in params and params['trend_city'] :
    query.update({'property.propertyAddr.city':
        {'$regex' :  '*' + params['trend_city'] + '*' }})
if 'trend_region' in params and params['trend_region'] :
    query.update({'property.propertyAddr.stateProvince': \
                    params['trend_region']})
if 'trend_locality' in params and params['trend_locality'] :
    query.update({'property.propertyAddr.locality': \
                    params['trend_locality']})
if 'trend_country' in params and params['trend_country'] :
    query.update({'property.propertyAddr.country': \
                    params['trend_country']})
```

We can then glue the query fragments together using the MongoDB `$and` query operator,
and then run the query:

```
final_query = {'$and':[query]}
hist_data = self.fetch(query, projection, sort)
```

Next, we initialize the variables used in the trend data-generation process. Note that the
report marks a total span of 5 years – 2 years before and 2 years after the current year:

```
today       = datetime.now()
date_format = '%Y-%m-%d %H:%M:%S'
start_year  = today.year - 2
end_year    = today.year + 2
year_span   = end_year - start_year + 1
years       = range(start_year, end_year+1)
months      = range(1,12+1)
year1_mark  = (today.year - start_year - 1) * 12
year2_mark  = (today.year - start_year - 2) * 12
```

We then prepopulate a flat list to represent a span of 60 months:

```
total = [0.00 for key in range(1,year_span*12)]
```

We now iterate through the result of the MongoDB query and build a Python `datetime` instance from the booking arrival date. We then calculate the index within the flat total list by multiplying `year_key` by 12 and adding the key for the month:

```
for doc in hist_data :
    arrDate   = datetime.strptime(\
        doc['bookingInfo']['arrivalDate'], date_format)
    price     = doc['totalPrice']
    year_key  = arrDate.year - start_year
    month_key = arrDate.month - 1
    total[(year_key*12) + month_key] += price
```

Next, we grab the trend factor from the query or assign a default value of `0.10`:

```
if 'trend_factor' in params and params['trend_factor'] :
    factor = float(params['trend_factor'])
else :
    factor = 0.10
```

Following this, we initialize lists used to represent last year's data, the difference between the last 2 years, and next year:

```
avg_diff  = []
last_year = []
next_year = []
```

We then calculate the monthly difference between the past 2 years by simply subtracting the current year from the year before, month by month. We then append this figure, multiplied by `factor`, to produce the average difference. We also append last year's data to the `last_year` list:

```
for month in range(1,13) :
    diff = total[year1_mark + month - 1] - \
            total[year2_mark + month - 1]
    avg_diff.append(diff * factor)
    last_year.append(total[year1_mark + month - 1])
```

The next bit of logic is the key to generating future trend data. We take the data from last year and apply the average difference, which has been weighted by `factor`:

```
for key in range(0,12) :
    next_year.append(last_year[key] + avg_diff[key])
```

Finally, we build lists that represent labels, and the series, which consists of 2 years of data: last year's totals and the estimated future revenue. These are returned as a single dictionary, `data`:

```
months = ['J','F','M', 'A', 'M', 'J', 'J', 'A', 'S', 'O', 'N', 'D']
data = {'labels' : months + months, 'series':[last_year + next_year]}
return data
```

The full code file can be viewed at `/path/to/repo/chapters/09/src/booksomeplace/domain/booking.py`.

Viewing the results

Turning to our website, if we access `http://chap09.booksomeplace.local/trends/future/`, we can see the initial data that represents property bookings in England with a factor of 10%. Here is a screenshot of our work:

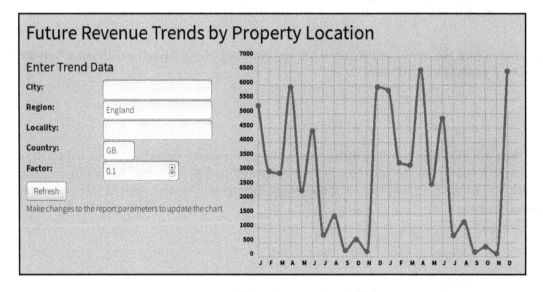

If we make a change to any of the fields, the chart changes. Other things that could be done at this point are the following:

- Adding additional trend data parameters
- Adding multiple lines representing different countries for other criteria for comparison
- Dressing up the chart using colors and redefining the labels

As you have now learned, the three types of financial reports – revenue reports, aging reports, and revenue trend reports – are all important tools in the overall process of risk analysis. Trend reports especially benefit from a graphical representation.

Summary

In this chapter, you first learned how to use **MongoDB Compass** to connect to the database and to model complex queries. You then learned how to copy the resulting JSON into a Python script to run the query. Next, you learned about the **MongoDB aggregation framework**, consisting of a series of **pipeline stage operators**, which allow you to perform various aggregation operations such as grouping, filtering, and so forth. You also learned about **pipeline operators**, which can perform operations on the data in the pipe, including conversions, summation, and even adding new fields. You were then shown a brief section on Map-Reduce functions, which allow you to perform operations very similar to that of the aggregation framework but using JavaScript functions instead.

The next section showed you how to work with MongoDB geospatial data. You learned about GeoJSON objects, and the two types of indexes and models supported: **flat** or **spherical**. You then stepped through a sample query that listed the major cities that are within a specified kilometer radius to one of the Book Someplace properties.

Finally, in the last section, you learned about key financial reports, including revenue reports and 30-60-90 day aging reports, and generating future revenue trend charts. In order to create an interactive chart, you were shown how to integrate jQuery by making AJAX requests, which used a domain service class to return JSON data used to build the chart. It's extremely important to note that all of the report types that we have covered form an important part of the overall process of risk analysis.

In the next chapter, you will learn how to work with complex documents across multiple collections.

4

Section 4: Replication, Sharding, and Security in a Financial Environment

Section 4 introduces *BigLittle Micro Finance Ltd.*, a financial start-up. The core business involves matching lenders with borrowers. This leads to complex relationships across documents that need to be addressed. Being a financial organization, they need to implement security. Another area of concern is to implement replication to ensure their data is continuously available. Finally, as they grow in size, the company expands to Asia, setting up an office in Mumbai, leading them to implement a sharded cluster.

This section contains the following chapters:

10
Working with Complex Documents Across Collections

This chapter starts with a brief overview of the scenario used for the remaining chapters of the book. After a brief discussion on how to handle monetary data, the focus of this chapter switches to working with multiple collections. Developers who are used to formulating SQL JOIN statements that connect multiple tables are shown how to do the equivalent in MongoDB. You'll also learn how to process programming operations that require multiple updates across collections. The technique can be tricky, and this chapter outlines some of the potential pitfalls novices to MongoDB might experience. Lastly, *GridFS* technology is introduced as a way of handling large files and storing documents directly in the database.

The topics covered in this chapter include the following:

- Introducing BigLittle Micro Finance Ltd.
- Handling monetary data
- Referencing documents across collections
- Performing secondary updates
- Avoiding potential pitfalls
- Storing large files using GridFS

Technical requirements

Minimum recommended hardware:

- Desktop PC or laptop
- 2 GB free disk space
- 4 GB of RAM
- 500 Kbps or faster internet connection

Software requirements:

- OS (Linux or Mac): Docker, Docker Compose, Git (optional)
- OS (Windows): Docker for Windows and Git for Windows
- Python 3.x, PyMongo driver, and Apache (already installed in the Docker container used for the book)
- MongoDB Compass (needs to be installed on your host computer; instructions are in the chapter)

Installation of the required software and how to restore the code repository for the book is explained in Chapter 2, *Setting Up MongoDB 4.x*. To run the code examples in this chapter, open a Terminal window (Command Prompt) and enter these commands:

```
cd /path/to/repo
docker-compose build
docker-compose up -d
docker exec -it learn-mongo-server-1 /bin/bash
```

To run the demo website described in this chapter, once the Docker container from Chapter 3, *Essential MongoDB Administration Techniques* to Chapter 12, *Developing in a Secured Environment* has been configured and has been started, there are two more things you need to do:

1. Add the following entry to the local hosts file on your computer:

   ```
   172.16.0.11    chap10.booksomeplace.local
   ```

 The local hosts file is located at C:\windows\system32\drivers\etc\hosts on Windows, and for Linux and Mac, /etc/hosts.

2. Open a browser on your local computer (outside of the Docker container) to this URL:

   ```
   http://chap09.booksomeplace.local/
   ```

When you are finished working with the examples covered in this chapter, return to the Terminal window (Command Prompt) and stop Docker as follows:

```
cd /path/to/repo
docker-compose down
```

The code used in this chapter can be found in the book's GitHub repository at `https://github.com/PacktPublishing/Learn-MongoDB-4.x/tree/master/chapters/10` as well as `https://github.com/PacktPublishing/Learn-MongoDB-4.x/tree/master/www/chapter_10`.

Introducing BigLittle Micro Finance Ltd.

For our last use case scenario we introduce *BigLittle Micro Finance Ltd.* This is a fictitious online company that links borrowers with lenders. Although technically a bank, BigLittle Micro Finance Ltd. specializes in *micro loans*: loan amounts that do not exceed 20,000 USD. Accordingly, in order to make money, the company needs to have a much larger customer base than an ordinary commercial bank would. Borrowers could be either individuals or small businesses. Lenders, likewise, can be individuals or small businesses (or even small banks).

For the purposes of this scenario, assume the website is based on Django for reasons stated in `Chapter 7`, *Advanced MongoDB Database Design*, in the section entitled *Using MongoDB with Django*. Unlike the previous scenario (for example, *Book Someplace Inc.*), however, in the remaining chapters of this book, we do not show full Django code. Rather we provide the pertinent code examples and leave it to you to fit the code into a full Django implementation. Also, please bear in mind that the full code can be seen in the repository associated with this book.

First, let's have a look at how monetary data should be handled.

Handling monetary data in MongoDB

Normal floating-point data representations work well for most situations. In the case of financial operations, however, extreme precision is required. As an example, take the subject of *currency conversion*. If a customer is transferring money between **United States Dollars(USD)** and **Indian Rupees (INR)**, for example, adding precision produces radically different results as the sums of money exchanged increase.

Here is an example. For the sake of argument, let's say that the actual exchange rate between USD and INR is the following:

```
ACTUAL: 1 USD = 70.64954999 INR
```

However, if the program application only extends precision to 4 decimal places, the rounded rate used might be this:

```
ROUNDED: 1 USD = 70.6495
```

The difference when transferring 1,000,000 USD starts to become significant:

```
Actual:  70649549.99
Rounded: 70649500.00
```

Even in the case of small amounts being exchanged, over time the differences could start to add up to significant amounts. Also, imagine the effect precision might have when calculating loan payments where the annual and effective percentage rates are stated without high precision! For the purposes of precision, MongoDB includes the `NumberDecimal` data type, which is based upon the `BSON Decimal128` data type. This is preferred when dealing with financial data as extreme precision is required.

The `BSON Decimal128` type (`http://bsonspec.org/spec.html`) is based on the IEEE 754-2008 standard (`https://standards.ieee.org/standard/754-2008.html`). It is a floating-point numeric representation that extends to 128 bits. Note that this standard has now been superseded by IEEE 754-2019 (`https://standards.ieee.org/standard/754-2019.html`) but has not yet been incorporated into BSON. Also see `https://docs.mongodb.com/manual/tutorial/model-monetary-data/#using-the-decimal-bson-type`.

Although you might consider using `NumberDecimal (Decimal128)` format (`https://docs.mongodb.com/manual/core/shell-types/#numberdecimal`) to store financial data, Python and Django are not able to handle it. Accordingly, you must perform a conversion in your code from `Decimal128` to the Python class `decimal.Decimal` before using it in arithmetic operations. The conversion process is covered later in this chapter.

Let's now have a brief look at the data structures needed for BigLittle Micro Finance Ltd.

User collection data structures

As seen with *Book Someplace, Inc.*, we break down each collection document structure into a series of smaller classes. Accordingly, the following data structures are exactly as shown in Chapter 7, *Advanced MongoDB Data Design*:

- Name
- Location
- Contact
- OtherContact
- OtherInfo
- LoginInfo

If you would like to review the complete data structures for these classes, please refer to
/path/to/repo/chapters/07/document_structure_definitions.
js.

The BigLittle user collection document structure includes a userType field that identifies whether the user is a borrower or a lender. Further, these two identifiers are present in the common collection under the userType key to maintain uniformity. Here is the formal JSON data structure for the user collection:

```
User = {
    "userKey"       : <string>,
    "userType"      : Common::userType,
    "amountDue"     : <NumberDecimal>,
    "amountPaid"    : <NumberDecimal>,
    "name"          : <Name>,
    "address"       : <Location>,
    "contact"       : <Contact>,
    "otherContact"  : <OtherContact>,
    "otherInfo"     : <OtherInfo>,
    "login"         : <LoginInfo>
}
```

In our scenario, in order to provide convenience to loan administrators, lenders, and borrowers, we have added two additional fields to the user collection: amountPaid and amountDue. In the case of a borrower, amountPaid represents how much money has been paid to the lender, including overpayments. The amountDue field represents a sum of the amountDue fields in the loan payments list (shown next).

In the case of a lender, `amountPaid` represents a sum of the amounts received from users who borrowed from this lender. The `amountDue` field represents the sum of all amounts due from borrowers for the lender.

Loan collection data structure

The heart of this business is the `loan` collection. Obviously, each loan document records not only the currency, amount, and interest, but the payment schedule as well. Accordingly, we need to first define a `Payment` class, as shown here. Each payment records the payment amount due, the amount paid, the due date, the payment received date, and the status. The status consists of a set of key words recorded in the `common` collection `payStatus` key:

```
Payment = {
    "amountDue"   : <NumberDecimal>,
    "amountPaid"  : <NumberDecimal>,
    "dueDate"     : <Date>,
    "recvDate"    : <Date>,
    "status"      : Common::payStatus
}
```

The `LoanInfo` class describes the amount borrowed (the principal), annual interest rate, effective interest rate, currency, and monthly payment amount:

```
LoanInfo = {
    "principal"      : <NumberDecimal>,
    "numPayments"    : <int>,
    "annualRate"     : <NumberDecimal>,
    "effectiveRate"  : <NumberDecimal>,
    "currency"       : Common::currency,
    "monthlyPymt"    : <NumberDecimal>
}
```

For the purposes of this book, we make the following assumptions:

- The annual interest rate is expressed as a percentage.
- The effective interest is calculated as *Annual Interest / 100 / 12* (that is, the number of months in a year).
- Payments are always monthly: there is no provision for quarterly nor annual payments.
- There is no support for amortized loans, negative interest loans, nor balloon payments.

Please note that the bullet points include many financial terms. If you are not familiar with these, the quickest way to get up to speed is to either do a Google search or consult Wikipedia. A study of finances is beyond the scope of this book.

The formula for calculating monthly payments is as follows:

```
monthlyPymt = principal *
    ( effectiveRate / (1 - (1 + effectiveRate) ** -numPayments ))
```

Finally, the `loan` collection document structure simply links the borrower, lender, and payment schedule, as well as loan information. In addition, the loan document contains a payment schedule in the form of a list of payment documents, as well as a field representing any overpayments:

```
Loan = {
    "loanKey"      : <string>,
    "borrowerKey"  : <string>
    "lenderKey"    : <string>,
    "loanInfo"     : <LoanInfo>,
    "overpayment"  : <NumberDecimal>
    "payments"     : [<Payment>,<Payment>,etc.]
}
```

Next, we'll tackle common values used by BigLittle Micro Finance Ltd.

Common collections

Finally, as with our other scenarios, we introduce a `common` collection that serves to house choices used in drop-down menus. We also use this to enforce uniformity when collecting data. Here are the currently suggested definitions for *BigLittle*:

```
Common = [
    { "title"        : ['Mr','Ms','Dr',etc.] },
    { "userType"     : ['borrower','lender'] },
    { "phoneType"    : ['home', 'work', 'mobile', 'fax'] },
    { "socMediaType" : ['google','twitter','facebook','skype',etc.] },
    { "genderType"   : [{'M' : 'Male'},{'F' : 'Female'},{'X' : 'Other'}] },
    { "currency"     : ['AUD','CAD','EUR','GBP','INR','NZD','SGD','USD'] },
    { "payStatus"    : ['scheduled','received','overdue'] },
]
```

Now that you have an idea of the BigLittle data structures, as well as a general understanding of the `NumberDecimal` class, let's look at how financial data might be stored in the case of BigLittle Micro Finance Ltd.

Storing financial data

As with *Book Someplace, Inc.*, our previous scenario, we use Django to build a web infrastructure. The source code can be seen at `/path/to/repo/chapters/10`. The web infrastructure is at `/path/to/repo/www/chapter_10`. In this chapter, we do not dive into the details of Django, but rather draw out the pertinent bits of code needed to illustrate the point.

In order to generate a loan proposal, we need to develop a domain service class that gives us access to the database `loans` collection. For this purpose, we created the `biglittle.domain.loan.LoanService` class. In this class, the key method generates a loan proposal, as shown here:

```python
def generateProposal(self, principal, numPayments, annualRate, \
        currency, borrowerKey, lenderKey, \
        lenderName, lenderBusiness) :
    effective_rate  = annualRate / 100 / 12;
    monthly_payment = principal * \
        ( effective_rate / (1 - (1 + effective_rate) ** -numPayments ))
    loanInfo = {
        'principal'     : principal,
        'numPayments'   : numPayments,
        'annualRate'    : annualRate,
        'effectiveRate' : effective_rate * 1000,
        'currency'      : currency,
        'monthlyPymt'   : monthly_payment
    }
    loan = {
        'borrowerKey'    : borrowerKey,
        'lenderKey'      : lenderKey,
        'lenderName'     : lenderName,
        'lenderBusiness' : lenderBusiness,
        'overpayment'    : 0.00,
        'loanInfo'       : loanInfo,
        'payments'       : []
    }
    return loan
```

The `save()` method does the actual database insertion, shown here. Note that we first convert the data format of the financial data fields to `BSON Decimal128` before storage:

```
def save(self, loan) :
    loan = self.generateLoanKey(loan)
    loan.convertDecimalToBson()
    return self.collection.insert(loan)
```

As mentioned earlier, we need to use the Python class `decimal.Decimal` for processing, but wish to store our information in MongoDB using the `BSON Decimal128` (MongoDB `NumberDecimal`) data format. The conversion process could be used any time data needs to be stored and retrieved, however, this could lead to redundant code. In this case, the best solution is to have the entity class perform the data conversion. Accordingly, from the top of the loan entity class, found in the module `/path/to/repo/chapters/10/src/biglittle/entity/loan.py`, we import the two data format classes:

```
from decimal import Decimal
from bson.decimal128 import Decimal128
```

We then add two methods to the `biglittle.entity.loan.Loan` class: `convertDecimalToBson()` and `convertBsonToDecimal()`. The `convertDecimalToBson()` method is shown first. As you can see, the conversion process simply takes the current value of the field and supplies a string representation to the constructor of `bson.decimal.Decimal128`:

```
def convertDecimalToBson(self) :
    self['overpayment'] = Decimal128(str(self['overpayment']))
    if 'payments' in self :
        for doc in self['payments'] :
            doc['amountPaid'] = Decimal128(str(doc['amountPaid']))
            doc['amountDue']  = Decimal128(str(doc['amountDue']))
    loanInfo = self['loanInfo']
    loanInfo['principal'] = Decimal128(str(loanInfo['principal']))
    loanInfo['annualRate'] = Decimal128(str(loanInfo['annualRate']))
    loanInfo['effectiveRate'] = \
        Decimal128(str(loanInfo['effectiveRate']))
    loanInfo['monthlyPymt'] = Decimal128(str(loanInfo['monthlyPymt']))
    return True
```

The reverse process, converting from `Decimal128` to `Decimal`, is in one sense easier, and in another sense more difficult. The process is easier in that there is a method, `Decimal128.to_decimal()`, that produces a decimal instance. It is more difficult in that we need to first make sure that all financial data fields are of type `Decimal128`. This is accomplished by first calling `convertDecimalToBson()`, ensuring all financial data is in BSON `Decimal128` format. The code for `convertBsonToDecimal()` is shown here:

```
def convertBsonToDecimal(self) :
    self.convertDecimalToBson()
    self['overpayment'] = self['overpayment'].to_decimal()
    if 'payments' in self :
        for doc in self['payments'] :
            doc['amountPaid'] = doc['amountPaid'].to_decimal()
            doc['amountDue'] = doc['amountDue'].to_decimal()
    loanInfo = self['loanInfo']
    loanInfo['principal']     = loanInfo['principal'].to_decimal()
    loanInfo['annualRate']    = loanInfo['annualRate'].to_decimal()
    loanInfo['effectiveRate'] = loanInfo['effectiveRate'].to_decimal()
    loanInfo['monthlyPymt']   = loanInfo['monthlyPymt'].to_decimal()
    return True
```

Now that you have an idea how to handle monetary data, it's time to examine how to connect documents across multiple collections.

Referencing documents across collections

As we have discussed throughout this book, one major objective in your initial database design should be to design a database document structure such that there is no need to cross-reference data between collections. The temptation, especially for those of us steeped in traditional RDBMS database techniques, is to *normalize* the database structure and then define a series of *relationships* that are contingent upon a series of interlocking *foreign keys*. This is the beginning of many problems ... for legacy RDBMS developers! For those of us fortunate enough to be operating in the MongoDB NoSQL world, such nightmares are a thing of the past, given a proper design. Even with a proper database design, however, you might still find occasions requiring a reference to documents across collections in MongoDB.

Let's first have a look at a number of techniques commonly used to reference documents across collections.

Techniques for referencing documents across collections

There are several techniques available to MongoDB developers. The more popular ones include the following:

- The building block approach
- Embedded documents
- Document references

A less common cross-reference technique would be to use *database references (DBRefs)*. For on-the-fly cross-references, another extremely useful tool is to use the $lookup pipeline aggregation operator. Although the building block approach has actually already been used in the scenarios describing *Book Someplace*, we present a short discussion of this approach next.

The building block approach

In this approach, the developer defines a set of small classes (that is, *building blocks*) that represent objects needed in your application that must ultimately be stored. You then define a collection of document structures as a super-set of these small classes. We have already seen an example of this approach in the design for *Book Someplace*. For example, we designed an object location, which includes fields such as streetAddress, city, postalCode, and more. This class was then used as part of the document structure for both the customers and the properties collections.

The advantage of this approach is that the size of each document is smaller and more manageable. The disadvantage is that you might need to perform a secondary lookup if additional information from the other collection is needed.

As an example, recalling the *Book Someplace* scenario, when we create an entry in the bookings collection, we include the following small objects containing customer information: Name, Location, and Contact. But what if we need to produce a booking demographic report that also includes the customer's gender? Using the building block approach, we would have two choices: expand what is included in each booking document to also include the OtherInfo class (includes gender), or perform a secondary lookup and get the full document from the customers collection.

Embedded documents

This approach is very similar to the one we've just described, except, in this case, the *entire document* from one collection is embedded directly inside the document from another collection. When using the *building block* approach, only a sub-class from one collection is inserted into the other collection.

The advantage of the *embedded document* approach is that there is absolutely no need for a further lookup should additional information from the other collection be needed. The disadvantage of this approach is that we now have duplicate data, which in turn also means potentially wasted space as we probably do not need all of this information all of the time.

As an example, in the case of *Book Someplace*, for each booking, instead of using the *building block* approach and storing just fragments of the customer, we could instead embed the *entire document* for each customer making a booking.

For more information, consult the MongoDB documentation on embedded documents: `https://docs.mongodb.com/manual/tutorial/` `model-embedded-one-to-many-relationships-between-documents/` `#model-one-to-many-relationships-with-embedded-documents`.

Document references

The third approach, most commonly used by MongoDB developers, is to use *document references*. In this approach, instead of storing a partial document (that is, a *building block*), or instead of storing an entire document (*embedded document*), you only store the equivalent of a *foreign key*, which then allows you to perform a secondary lookup to retrieve the needed information.

Although you could conceivably store the `_id` (*ObjectID* instance) field as the reference, it is considered a best practice in the MongoDB world to have a developer-generated unique key of some sort, which can be used to perform the lookup.

As an example, when storing loan information, instead of storing partial or complete borrower and lender information, just store the `userKey` fields. This can then be used in conjunction with `find()` to retrieve lender or borrower information. You will note that this is the approach taken with BigLittle Micro Finance Ltd. When creating a loan document, the `userKey` field for the lender is stored in `lenderKey`, and the `userKey` value for the borrower is stored in `borrowerKey`.

In order to extract information from related documents on the fly, the most effective approach is to use the MongoDB aggregation framework and create a new pipeline stage using the `$lookup` stage operator (discussed later in this chapter).

 For more information, consult the MongoDB documentation on *document references*: https://docs.mongodb.com/manual/tutorial/model-referenced-one-to-many-relationships-between-documents/#model-one-to-many-relationships-with-document-references.

Database references (DBRefs)

Yet another approach, used less frequently, is to use a *DBRef*. Technically speaking, there is actually no difference between creating a DBRef as compared to simply adding a *document reference* (see the previous sub-topic). In both cases, an additional lookup is required. The main difference between using a DBRef compared to using a *document reference* is that the DBRef is a uniform, formalized way of stating the reference.

DBRef is actually not a data type in and of itself. Rather, it's a formal JSON structure that completely identifies the document in the target collection. Here are the fields required to insert a properly recognized DBRef:

- `$ref`: The name of the target collection
- `$id`: The value of the `_id` field in the document to be referenced in the target collection

 In addition, a third field, `$db`, containing the name of the target collection's database, can also be added when defining a DBRef, however, this is not supported by all programming language drivers.

The generic syntax for creating a DBRef would appear as follows:

```
var = { "$ref" : <collection name>, "$id" : ObjectId(<value of _id field>) }
```

Alternatively, you can directly reference the DBref and create the instance by supplying these two arguments to the constructor:

```
var = DBRef(<collection name>, ObjectId(<value of _id field>))
```

Using DBRefs does not automatically create document associations. Retrieving full documents across collections involves multiple queries. For this reason, the MongoDB documentation actually recommends that you create your own unique document references because using DBRefs involves extra unnecessary steps. If you wish to retrieve full document information across collections in a single operation, use the $lookup aggregation pipeline operator (discussed next).

Now let's have a quick look at how you can use the *aggregation framework* to perform operations across collections.

$lookup aggregation pipeline stage

One last document cross-reference approach we present is to use the aggregation pipeline $lookup stage operator (https://docs.mongodb.com/manual/reference/operator/aggregation/lookup/#lookup-aggregation). This operator performs the equivalent of an SQL LEFT OUTER JOIN. The prerequisite for this operation is to have a field that is common between the two collections referenced. The generic syntax is as follows:

```
db.<this_collection>.aggregate([
    { $lookup: {
        from:         <target_collection_name>,
        localField:   <common_field_in_this_collection>,
        foreignField: <common_field_in_target_collection>,
        as:           <output_array_field> } }
]);
```

Two other options are available for more complex processing. let allows you to assign a value to a variable for temporary use in the pipeline. pipeline lets you run pipeline operations on the incoming data stream from the target collection. A practical example follows in the next several sections.

Practical example using document references

As we have already covered usage examples for *building-block* and *embedded* documents in earlier chapters, let's now focus on how to use and create a document that relates to other documents using a common reference. As an example, let's look at the process of creating a Loan document. As you recall from the *Loan collection data structure* section in this chapter, the Loan document includes two references to the users collection: one for the borrower and another for the lender.

The `users` collection includes a unique field named `userKey`. When we create a loan document, all we need to do is to add the value of the `userKey` property for the lender, and the `userKey` value for the borrower.

The final loan document might appear as follows:

```
{
    "_id" : ObjectId("5da950636f165e472c37a98d"),
    "loanKey"      : "LAMOHOLM1595_CHRIMARQ4459_20200303",
    "borrowerKey" : "LAMOHOLM1595",
    "lenderKey"    : "CHRIMARQ4459",
    "overpayment" : NumberDecimal("0.00"),
    "payments"     : [ <not shown> ],
    "loanInfo" : {
        "principal" : NumberDecimal("19700.0000000000"),
        "numPayments" : 24,
        "annualRate" : NumberDecimal("1.26000000000000"),
        "effectiveRate" : NumberDecimal("0.00105000000000000"),
        "monthlyPymt" : NumberDecimal("831.650110710560")
}}
```

Let's first look at the process of collecting loan information from a potential borrower and generating loan proposals.

Gathering information to generate a loan proposal

In our scenario, we build the loan document after collecting information from the potential borrower using a Django form. Here is the code, located in `/path/to/repo/www/chapter_10/loan/views.py`. At the beginning of the code file are the `import` statements:

```
from django.template.loader import render_to_string
from django.http import HttpResponse
from config.config import Config
from db.mongodb.connection import Connection
from biglittle.domain.common import CommonService
from biglittle.domain.loan import LoanService
from biglittle.domain.user import UserService
from biglittle.entity.loan import Loan
from biglittle.entity.user import User
from utils.utils import Utils
```

The method that presents the form to the fictitious borrower is index(). We first initialize the important variables:

```
def index(request) :
    defCurr     = 'USD'
    proposals   = {}
    have_any    = False
    principal   = 0.00
    numPayments = 0
    currency    = defCurr
    config      = Config()
    comSvc      = CommonService(config)
    userSvc     = UserService(config, 'User')
    maxProps    = 3
    utils       = Utils()
    borrowerKey = 'default'
```

We then use the biglittle.domain.common.CommonService class to retrieve a list of payment lengths and currencies that are offered by our fictitious lenders:

```
payments   = comSvc.fetchByKey('paymentLen')
currencies = comSvc.fetchByKey('currency')
```

Next, we check to see if the request is an HTTP POST. If so, we gather the parameters needed to calculate loan payments:

```
if request.method == 'POST':
    if 'borrower' in request.POST :
        borrowerKey = request.POST['borrower']
    if 'principal' in request.POST :
        principal = float(request.POST['principal'])
    if 'num_payments' in request.POST :
        numPayments = int(request.POST['num_payments'])
    if 'currency' in request.POST :
        currency = request.POST['currency']
```

If the amount entered for *principal* is greater than zero, we proceed with the loan calculation. Please note that the pickRandomLenders() method shown here is a simulation. In actual practice, the web app would send a notification to potential lenders. The web app would then present proposals to the borrower for final selection. Although we use a custom *cache* class to store proposals for the next cycle, we could just as easily have used the Django *session* mechanism:

```
if principal > 0 :
    utils    = Utils()
    have_any = True
    loanSvc  = LoanService(config, Loan())
```

```
lenders    = userSvc.pickRandomLenders(maxProps)
have_any   = True
proposals = loanSvc.generateMany(float(principal), \
        int(numPayments), currency, borrowerKey, lenders)
utils.write_cache(borrowerKey, proposals)
```

Finally, as with the web apps shown for our previous scenario, *Book Someplace, Inc.*, we render nested templates – first `loan.html`, and then `layout.html`, and return an `HttpResponse`:

```
params = {
    'payments'    : payments,
    'currencies'  : currencies,
    'proposals'   : proposals,
    'have_any'    : have_any,
    'borrowers'   : borrowers,
    'principal'   : principal,
    'numPayments' : numPayments,
    'currency'    : currency,
    'borrower'    : str(borrowerKey)
}
main    = render_to_string('loan.html', params)
home    = render_to_string('layout.html', {'contents' : main})
return HttpResponse(home)
```

The loan generation method is located in the `/path/to/repo/chapters/10/src/biglittle/domain/loan.py` module. Again, as mentioned above, this is a simulation where we pick annual rates at random. In actual practice, the web app would solicit loan proposals from lenders who would offer an appropriate annual rate:

```
def generateMany(self, principal, numPayments, currency, \
                borrowerKey, lenders) :
    proposals = {}
    for item in lenders :
        import random
        annualRate = random.randint(1000,20000) / 1000
        doc = self.generateProposal(principal, numPayments, \
            annualRate, currency, borrowerKey, item['key'], \
            item['name'], item['business'])
        proposals.update({ item['key'] : doc })
    return proposals
```

Please note that the actual loan generation process, generateProposal(), was discussed in detail in the *Handling monetary data in MongoDB* section at the beginning of this chapter, and is not duplicated here.

Now, it's time to have a look at the code in which a borrower accepts a loan proposal.

Creating a loan document using document references

Once a loan proposal has been accepted by the borrower, we are able to put together a *Loan* document. This is where we put document references to use. The code that captures the borrower's selection is also located in views.py, in the accept() method. We start by checking to see if the request is an HTTP POST. We also check to make sure the accept button was pressed, and that the key for lender was included in the post, shown here:

```
def accept(request) :
    message = 'Sorry! Unable to process your loan request'
    utils = Utils()
    if request.method == 'POST':
        if 'accept' in request.POST :
            if 'lender' in request.POST :
                lenderKey = request.POST['lender']
                borrowerKey = request.POST['borrower']
```

We retrieve the proposals from our simple cache class and loop through them until we match the lender key, after which we use the loan domain service to save the document:

```
                proposals = utils.read_cache(borrowerKey)
                if lenderKey in proposals :
                    loan = proposals[lenderKey]
                    config  = Config()
                    loanSvc = LoanService(config, Loan())
                    loanSvc.save(loan)
```

All that is left is to return to the previous web page – simply a matter of calling the index() method:

```
        return index(request)
```

Let's now turn our attention to a practical example using the $lookup aggregation pipeline operator.

Practical example using the $lookup aggregation pipeline operator

For the sake of illustration, let's assume that management wishes to be able to generate a report of lender names and addresses sorted by the lender's country, and then by the total amounts loaned, from largest to smallest. Lender information is found in the `users` collection, however, loan amount information is in `loans`; accordingly, the aggregation framework is needed.

Connecting two collections using $lookup

The first aggregation pipeline stage uses the `$lookup` pipeline operator to join the `users` and `loans` collections. As with other complex examples discussed in this book, we first model the query using the mongo shell. Here is how the query might appear:

```
db.users.aggregate([
    { $match      : { "userType" : "lender" }},
    { $project    : { "userKey" : 1, "businessName" : 1, "address" : 1 }},
    { $lookup     : { "from" : "loans", "localField" : "userKey",
                      "foreignField" : "lenderKey", "as" : "loans" }},
    { $addFields : { "total" : { "$sum" : "$loans.loanInfo.principal" }}},
    { $sort       : { "address.country" : 1, "total" : -1 }}
]);
```

Now let's break this down into stages:

Stage	Notes
$match	Adds documents to the pipeline from the `users` collection whose `userType` is `lender`.
$project	Strips out all fields in the pipeline other than `userKey`, `businessName`, and `address`.
$lookup	Adds documents from the `loans` collection into a new array field called `loans`.
$addFields	Adds a new field, `total`, which represents a sum of the *principal* amounts in the `loans` array.
$sort	Sorts all documents in the pipeline first by *country* and then by `total` in descending order.

We do not have space to show the entire result set, however, here is how a single document might appear after launching the query just shown:

```
{
    "_id" : ObjectId("5d9d3d008b1c565ae0d2813e"),
    "userKey" : "CASSWHEE6724",
    "businessName" : "Illuminati Trust",
    "address" : {
        "streetAddress" : "5787 Red Mountain Boulevard",
        "geoSpatial" : [ "103.7", "17.2167"]
        # not all fields shown
    },
    "loans" : [ {
            "_id" : ObjectId("5da5981b72d8437f7436c419"),
            "borrowerKey" : "SUNNCHAM4815",
            "lenderKey" : "CASSWHEE6724",
            "loanInfo" : {
                "principal" : NumberDecimal("12800.0000000000"),
                # not all fields shown
    } } ],
    "total" : NumberDecimal("12800.0000000000")
}
```

Having successfully modeled the query using the mongo shell, it's time to formulate it into Python code.

Producing Python code from the JSON query

For the purposes of this illustration, we create a new Django app called `maintenance` located in `/path/to/repo/www/chapter_10/maintenance`. The logic in `views.py`, shown here, is extremely simple. The main work is done by the user domain service. The resulting iteration is then rendered in the `totals.html` template and injected into the main `layout.html` template.

As usual, the module starts with a set of `import` statements:

```
from django.template.loader import render_to_string
from django.http import HttpResponse
from config.config import Config
from db.mongodb.connection import Connection
from biglittle.domain.user import UserService
```

We then define a method, `totals()`, which runs the report and sends the results to the view template:

```
def totals(request) :
    config  = Config()
    userSvc = UserService(config, 'User')
    totals  = userSvc.fetchTotalsByLender()
    params  = { 'totals'    : totals }
    main    = render_to_string('totals.html', params)
    home    = render_to_string('layout.html', {'contents' : main})
    return HttpResponse(home)
```

Likewise, because most of the work is passed on to the MongoDB aggregation framework, the new method, `fetchTotalsByLender()`, in the `biglittle.domain.user.UserService` class is also quite simple. Here is the code:

```
def fetchTotalsByLender(self) :
    pipe = [
        { "$match"    : { "userType" : "lender" }},
        { "$project" : { "userKey":1, "businessName":1, "address":1 }},
        { "$lookup"  : { "from" : "loans", "localField" : "userKey",
                         "foreignField" : "lenderKey", "as" : "loans" }},
        { "$addFields" : {"total":{"$sum":"$loans.loanInfo.principal"}}},
        { "$sort" : { "address.country" : 1, "total" : -1 }}
    ]
    return self.collection.aggregate(pipe)
```

Finally, the view template logic is shown next. Notice that we use the Django template language to loop through the list of `totals`. We use `loop.counter0` to create a double-width table to conserve screen space:

```
<h2>Total Amount Lent By Lender</h2>
<hr>
<table border=1>
    <tr><th>Business</th><th>Country</th><th>Total Lent</th>
        <th>Business</th><th>Country</th><th>Total Lent</th></tr>
    {% for doc in totals %}
        {% if doc.total %}
            {% if forloop.counter0|divisibleby:2 %}<tr>{% endif %}
```

```
            <td class="td-left">{{ doc.businessName }}</td>
            <td class="td-left">{{ doc.address.country }}</td>
            <td class="td-right">{{ doc.total }}</td>
        {% if not forloop.counter0|divisibleby:2 %}</tr>{% endif %}
    {% endif %}
  {% endfor %}
</table>
```

The output appears similar to the following screenshot:

BigLittle Micro Finance Ltd					View release notes for Django 2.2
		Total Amount Lent By Lender			
Business	**Country**	**Total Lent**	**Business**	**Country**	**Total Lent**
Nonstop Trust Inc	AU	18700.00000000000	Illuminati Partners	AU	8100.00000000000
Comfort Partners Company	AU	4500.00000000000	Industrious Holdings Company	BY	7400.00000000000
Leisure Associates LLC	CA	22500.00000000000	Comfort Industries Ltd	CA	19000.0000000000
Bizarro Trust Ltd	CA	18000.00000000000	Powerhouse Associates Ltd	CA	16600.0000000000
Lazy Bird Holdings Inc	CA	14200.0000000000	Lazy Bird Partners Inc	CA	10700.0000000000
Serious Associates	CZ	6600.00000000000			
Industrious Holdings	GB	81200.0000000000	Lazy Bird Partners LLC	GB	10600.0000000000
Lazy Bird Trust Ltd	GB	7700.00000000000	Specialty Associates LLC	GT	7200.00000000000
Comfort Associates Company	IE	19600.0000000000	Lazy Bird Partners Company	IE	13000.0000000000
Accomodation Associates Inc	IN	16000.0000000000	Ninety Nine Percenter Associates	IN	10400.00000000000
Leisure Associates Inc	IN	6400.00000000000	Bizarro Partners Ltd	IN	3900.00000000000
Leisure Associates	IN	1800.00000000000	Friendly Associates LLC	LI	5200.00000000000
Industrious Industries LLC	MK	35800.00000000000			
Powerhouse Business Company	PL	33600.0000000000			
Powerhouse Partners Inc	PR	11100.0000000000	Accomodation Trust Ltd	RE	3600.00000000000
Friendly Trust LLC	SE	13900.0000000000			
Round the Clock Business Ltd	TH	13100.0000000000	Illuminati Trust	TH	12800.0000000000
Nonstop Holdings	UA	22200.00000000000	Serious Industries Inc	US	18200.00000000000
Friendly Associates	US	14700.0000000000	Powerhouse Associates LLC	US	10800.0000000000
Industrious Partners LLC	YT	18800.0000000000	Specialty Holdings Inc	ZA	21000.00000000000

To test the working code using the Docker-based test environment, enter this URL in the browser of your host computer:

```
http://chap10.biglittle.local/maintenance/totals/
```

In the next section, we take a look at situations where you need to update documents across collections.

Performing secondary updates

In the case of *BigLittle Micro Finance,* when a borrower makes a loan payment, the `status` field for that payment needs to be updated to reflect payments being received. However, given our example document structure, we also need to update the `amountDue` and `amountPaid` fields in the user document. This is referred to as a *secondary update.*

This is a step easily overlooked when developing financial applications. What can often happen is that the total amount paid as recorded in the `users` collection could fall out of sync with the sum of the `amount` field values in the `loans` collection. Accordingly, it's not a bad idea when performing secondary updates to either provide an immediate double-check and note any discrepancies in a log, or alternatively, provide management with a discrepancies report allowing administrators to perform the double-check themselves. Let's first look at the process of accepting a loan payment in our sample scenario.

Accepting a loan payment

For the purposes of this illustration, we make the following assumptions:

- Loan payments are entered by Big Little Micro Finance Ltd. loan administrators.
- If the loan amount is greater then the amount due, the overpayment is added to an `overpayment` field in the loan document and dealt with later.
- If the loan amount is less than the amount due, the amount paid is added to an `overpayment` field in the loan document, and the payment is considered overdue.
- Each borrower can have only one outstanding loan at a time. When the loan has been paid off, we assume, for the purposes of this illustration, that the loan information is moved into a yet to be defined `history` collection.

In order to accept a loan payment, logic needs to be devised in the Django `views.py` module of the newly created maintenance app. In addition, we need to add the appropriate methods to the loan domain service for *BigLittle.* Finally, a form template needs to be created that lets a loan administrator choose the borrower and process a loan payment. Also, in the `urls.py` file found in `/path/to/repo/www/chapter_10/maintenance`, we define two routes: one that lets a loan admin choose the borrower, and another to accept the payment. The contents of this file are shown here:

```
from django.urls import path
from . import views
urlpatterns = [
    path('', views.index, name='maintenance.index'),
```

```
        path('totals/', views.totals, name='maintenance.totals'),
        path('payments/', views.payments, name='maintenance.payments'),
    ]
```

Now we can examine the view logic.

Django view logic for choosing the borrower

The view logic for this illustration is found in
`/path/to/repo/www/chapter_10/maintenance/views.py`. As with other classes, we
first import the necessary Python and Django classes. In addition, we import the
necessary *BigLittle* entity and domain service classes:

```
from django.template.loader import render_to_string
from django.http import HttpResponse
from config.config import Config
from db.mongodb.connection import Connection
from biglittle.domain.user import UserService
from biglittle.domain.loan import LoanService
from biglittle.entity.loan import Loan, LoanInfo, Payment
from biglittle.entity.user import User
```

In the `index()` method, the loan admin chooses the borrower whose payment is processed.
We first initialize key variables:

```
def index(request) :
    loan          = Loan()
    amtPaid       = 0.00
    amtDue        = 0.00
    message       = ''
    borrower      = None
    borrowerKey   = None
    borrowerName  = ''
    payments      = []
    overpayment   = 0.00
```

After that, we initialize domain services and fetch a list of borrower keys and names:

```
    config    = Config()
    userSvc   = UserService(config, 'User')
    borrowers = userSvc.fetchBorrowerKeysAndNames()
```

Finally, we render the results using the inner template, `maintenance.html`, subsequently injected into the layout:

```
params = {
        'loan'          : loan,
        'payments'      : payments,
        'borrowers'     : borrowers,
        'borrowerName'  : borrowerName,
        'borrowerKey'   : borrowerKey,
        'amountDue'     : amtDue,
        'overpayment'   : overpayment,
        'message'       : message
}
main    = render_to_string('maintenance.html', params)
home    = render_to_string('layout.html', {'contents' : main})
return HttpResponse(home)
```

Next, we look at the view logic for processing a payment.

Django view logic for payment processing

As with the `index()` method, we first initialize the same variables as, ultimately, we'll be sending results to the same template for rendering:

```
def payments(request) :
    loan            = Loan()
    amtPaid         = 0.00
    amtDue          = 0.00
    message         = ''
    borrower        = None
    borrowerKey     = None
    borrowerName    = ''
    overpayment     = 0.00
```

We then initialize `payments` as a list with a single blank `Payment` instance:

```
payment  = Payment()
payments = [payment.populate()]
```

Next, in a similar manner to the index() method, we get instances of domain service classes representing users and loans. With the user domain service, we retrieve a list of borrower keys and names:

```
config   = Config()
userSvc  = UserService(config, 'User')
loanSvc  = LoanService(config, 'Loan')
borrowers = userSvc.fetchBorrowerKeysAndNames()
```

We then check to see if the HTTP request method was POST. If so, we check for the existence of borrowerKey and amount_paid values:

```
if request.method == 'POST':
    if 'borrowerKey' in request.POST :
        borrowerKey = request.POST['borrowerKey']
    if 'amount_paid' in request.POST :
        amtPaid = float(request.POST['amount_paid'])
```

If we have a borrowerKey, it is used to retrieve borrower information. If valid, we proceed to retrieve loan information using the validated borrowerKey:

```
if borrowerKey :
    borrower = userSvc.fetchUserByBorrowerKey(borrowerKey)
if borrower :
    borrowerName = borrower.getFullName()
    loan = loanSvc.fetchLoanByBorrowerKey(borrowerKey)
```

A valid loan instance is returned. If the amount paid by the borrower in this processing pass is greater than zero, we process the loan and retrieve the modified loan document:

```
if loan :
    if amtPaid > 0 :
        if loanSvc.processPayment(borrowerKey, amtPaid, loan) :
            message = 'Payment processed'
            loan = loanSvc.fetchLoanByBorrowerKey(borrowerKey)
        else :
            message = 'Problem processing payment'
```

Inside the block created by the `if loan` statement, we add logic to check to see if we have a valid loan document. We also extract a list of payments and the overpayment amount. Because Python and Django do not directly support the `BSON Decimal128` format, we first convert all financial information to `Decimal` before rendering:

```
loan.convertBsonToDecimal()
payments = loan.getPayments()
loanInfo = loan.getLoanInfo()
amtDue = loanInfo.getMonthlyPayment()
overpayment = loan.get('overpayment')
```

We then present all this information to the view rendering process. The parameters are initialized, and a the sub document `maintenance.html` is rendered first. It is then injected into `layout.html` to produce the final view:

```
params = {
     'loan'          : loan,
     'payments'      : payments,
     'borrowers'     : borrowers,
     'borrowerName'  : borrowerName,
     'borrowerKey'   : borrowerKey,
     'amountDue'     : amtDue,
     'overpayment'   : overpayment,
     'message'       : message
}
main    = render_to_string('maintenance.html', params)
home    = render_to_string('layout.html', {'contents' : main})
return HttpResponse(home)
```

Let's now have a look at the domain service methods referenced in the view logic.

Payment processing domain service methods

`fetchBorrowerKeysAndNames()` is the first method referenced in the view logic. This method is found in the `biglittle.domain.user.UserService` class. Its purpose is to return a list of documents containing the values of the `_id`, `userKey`, and `name` fields. This list is subsequently used in the view template to provide a drop-down list from which the loan administrator can choose. Here is that method:

```
def fetchBorrowerKeysAndNames(self) :
    keysAndNames = []
    lookup = self.collection.find({'userType':'borrower'})
    for borrower in lookup :
        doc = {
            'id'    : borrower.getId(),
```

```
                'key'   : borrower.getKey(),
                'name'  : borrower.getFullName()
        }
        keysAndNames.append(doc)
    return keysAndNames
```

Another method used during payment acceptance is `fetchUserByBorrowerKey()`, found in the same class. Its purpose is to return a `biglittle.entity.user.User` instance from which the user's name is extracted. It is a simple lookup method, shown here:

```
def fetchUserByBorrowerKey(self, borrowerKey) :
    return self.collection.find_one({ \
        "userKey" : borrowerKey, "userType" : "borrower" })
```

In a similar manner, once the borrower has been identified, and the loan payment amount has been recorded, the `fetchLoanByBorrowerKey()` method from the `LoanService` class, found in the `biglittle.domain.loan.py` module is called. Again, it's a simple lookup, this time by the borrower key. The outstanding feature of this method is that we first convert all financial information from MongoDB `NumberDecimal` (BSON `Decimal128`) to Python `decimal.Decimal`, for ease of processing. Here is that method:

```
def fetchLoanByBorrowerKey(self,  borrowerKey) :
    loan = self.collection.find_one({"borrowerKey":borrowerKey})
    loan.convertBsonToDecimal()
    return loan
```

As you may recall from our list of assumptions discussed at the start of this subsection, we assume any given borrower can have at most one outstanding loan. Thus, if we do a lookup by borrower key, we are assured of retrieving a unique loan document.

Finally, we have the `processPayment()` method, located in the `LoanService` class, which applies the payment, updates the status, and records any overpayment information. As with most complex methods, we start by initializing key variables. We also split out `loanInfo` from the loan document in order to retrieve the expected payment amount due:

```
def processPayment(self, borrowerKey, amtPaid, loan) :
    from decimal import Decimal
    from bson.decimal128 import Decimal128
    config = Config()
    result = False
    overpayment = 0.00
    loanInfo = loan.getLoanInfo()
    amtDue = loanInfo.getMonthlyPayment()
```

Next, we check the data type of `amtPaid` and convert it to `decimal.Decimal` if needed:

```
if not isinstance(amtPaid, Decimal) :
    amtPaid = Decimal(amtPaid)
```

The heart of this method loops through the payments list, looking for the first document in the list for which the `amountPaid` field is zero. If found, we check to see if the amount paid is less than the amount due. If so, we apply that amount to `overpayment`, but do not otherwise touch the loan payment list:

```
for doc in loan['payments'] :
    if doc['amountPaid'] == 0 :
        if amtPaid < amtDue :
            overpayment = amtPaid
```

Still in the loop, if the amount paid is equal to or greater than the amount due, we set the `amountPaid` field in the payments list equal to the amount due. Any excess amount is added to the `overpayment` field. `status` is updated to `received`, and the date the payment was accepted is stored in `recvDate`. We also break out of the loop as there is no need for further processing:

```
        else :
            overpayment = amtPaid - amtDue
            doc['amountPaid'] = doc['amountDue']
            doc['status'] = 'received'
            from time import gmtime, strftime
            now = strftime('%Y-%m-%d', gmtime())
            doc['recvdate'] = now
        break
```

Now out of the payments loop, we update the `overpayment` field, adding any excess payments:

```
currentOver = loan.get('overpayment')
loan.set('overpayment', currentOver + overpayment)
```

Lastly, we convert all decimal values to `NumberDecimal`, set the filter to `borrowerKey`, and perform an update:

```
loan.convertDecimalToBson()
filt = { "borrowerKey" : borrowerKey }
result = self.collection.replace_one(filt,loan)
return result
```

In this case, we perform the update using the `replace_one()` method. The reason this method is much faster in this case is because we are working at the *entity* level: we have a complete document ready to hand. If we were only updating one or a few fields, the `update_one()` method would be preferred.

Now, let's have a quick look at the loan maintenance template.

Loan payment view template

The loan maintenance view template is located at `/path/to/repo/www/chapter_10/templates/maintenance.html`. The first item of interest is how to present the drop-down list of borrowers. The view calls a method, `fetchBorrowerKeysAndNames()`, from the `UserService` class, subsequently sent to the template. We are then able to build the HTML `SELECT` element using the Django template language, as shown here:

```
<select name="borrowerKey">
{% if borrowerKey and borrowerName %}
    <option value="{{ borrowerKey }}">{{ borrowerName }}</option>
{% endif %}
{% for user in borrowers %}
    <option value="{{ user.key }}">{{ user.name }}</option>
{% endfor %}
</select>
```

We also provide the complete payment schedule for the borrower in the same template. This is so that the loan administrator can see which payments have been received. The view extracts the payments list from the loan document and sends it to the template, where we again use the Django template language to provide the list:

```
<h3>Payment Schedule</h3>
{% if payments %}
<table>
    <tr><th>Amount Paid</th><th>Due Date</th><th>Status</th></tr>
    {% for doc in payments %}
    <tr><td>{{ doc.amountPaid|floatformat:2 }}</td>
        <td>{{ doc.dueDate }}</td><td>{{ doc.status }}</td></tr>
    {% endfor %}
</table>
{% endif %}
```

Now that you have an idea of how payment processing works, it's time to look at the secondary updates that need to occur.

Defining code to process secondary updates

In order to avoid the overhead involved in creating *normalized* data structures with lots of foreign keys and relationships, MongoDB DevOps may choose to store redundant information across collections. The main danger here is that if a change occurs in one collection, you, the developer, may need to ensure that a secondary update occurs in the collection with related information.

Accordingly, when a loan payment is accepted, the following secondary updates need to occur in the `users` collection:

- Add the amount paid to the `amountPaid` field for that borrower.
- Subtract the amount paid from the `amountDue` field.

Before we get started on the implementation, we need to first digress and discuss using *listeners* and *events*.

The publish/subscribe design pattern

What is now known as the *Publish-Subscribe* (or affectionately PubSub) software design pattern was originally proposed by Birman and Joseph in a paper entitled *Exploiting Virtual Synchrony in Distributed Systems*, presented to the 11th Association for Computing Machinery (ACM) symposium on operating system principles. To boil down a rather complex algorithm into simple terms, a *publisher* is a software class that sends a *message* before or after an event of significance takes place. A common example of such an event would be a database modification. A *subscriber* is a software class that *listens* to messages produced by its associated publisher. The subscriber examines the message, and may or may not take action, at its own determination. In our example scenario, we create a publisher class that sends a message to subscribers when a loan payment is received. The subscriber performs the secondary update on the relevant borrower document.

 Beginning with MongoDB 4.0, the concept of multi-document ACID transactions were introduced. The PubSub concept can potentially introduce a point of failure for the overall transaction. If the two documents must be updated, they should be wrapped in a transaction.

At this point, you might ask yourself: why not just update the borrower document right away? The answer is that such an operation would violate an important software principal known as *separation of concerns*. Each software class and each method within that class should have a single logical focus. In our example, it's the job of the `LoanService` class to service the `loans` collection, and no other.

If we start to have one domain service interfere with the responsibilities of another domain service, soon we end up with what eventually came to be known as spaghetti code, made famous by a *Dilbert* cartoon published on June 10, 1994.

Important terms for you to be aware of include the following:

- **Topic**: A label used to represent an event of significance.
- **Listener**: Callable logic that performs an action when sent a message. A function, class method, or class defining __call__() is considered callable.
- **Subscribe**: The process of associating a listener with a topic.
- **Message**: A call from the publisher that includes the topic and any associated parameters. The parameters are sent to all listeners subscribing to this topic.

Rather than having to invent a custom Publish-Subscribe class, in this illustration, we take advantage of a very simple PubSub implementation released in 2015, called *Python Simple PubSub*. It is a simple matter to define a *Publisher* class that provides a wrapper for *PubSub*. The code can be found at /path/to/repo/chapters/10/src/biglittle/events/publisher.py. The main thing to note here is that the pubsub.pub.sendMessage() method takes its second argument in **kwargs format. Accordingly, you need to be careful to always use whatever key you define. In this example, the name of the key is arg:

```python
import pubsub
class Publisher :
    def attach(self, topic, listener) :
        pubsub.subscribe(listener, topic)
    def trigger(self, topic, obj) :
        pubsub.sendMessage(topic, arg=obj)
```

Please refer to the following links for more information:

- *Exploiting Virtual Synchrony in Distributed Systems:* https://dl. acm.org/citation.cfm?id=41457.37515coll=portaldl=ACM.
- Dilbert spaghetti logic cartoon: https://dilbert.com/strip/ 1994-06-10.
- Python Simple PubSub: https://pypi.org/project/pubsub/.
- **kwargs* format: https://docs.python.org/3/glossary. html#term-argument.

We'll look at creating listeners for loan events next.

Defining listeners for loan events

Since the secondary update needs to occur on the `users` collection, it's appropriate to define the listener in the `UserService` class already defined. We add a new method, `updateBorrowerListener()`, which is later subscribed as a listener. Note that the argument to the listener is `arg`, in order to match that defined in our `Publisher` wrapper class described in the section just before.

We first import the Python classes to convert between `Decimal` and `Decimal128`:

```
def updateBorrowerListener(self, arg) :
    from decimal import Decimal
    from bson.decimal128 import Decimal128
```

We then initialize two sets of two values representing amounts due, and amounts paid, as defined by information from the `users` collection and also the `loans` collection:

```
loan = arg['loan']
amtPaid = arg['amtPaid']
amtDueFromLoan = Decimal(0.00)
amtPaidFromLoan = Decimal(0.00)
amtDueFromUser = Decimal(0.00)
amtPaidFromUser = Decimal(0.00)
```

Next, we retrieve a `User` entity class instance based on the borrower key stored in the loan entity. If not found, an entry is made in the discrepancy log:

```
config = Config()
log_fh = open(config.getConfig('discrepancy_log'), 'a')
borrowerKey = loan.get('borrowerKey')
borrower = self.fetchUserByBorrowerKey(borrowerKey)
if not borrower :
    message = now + ' : User ' + borrowerKey + ' not found'
    log_fh.write(message + "\n")
```

We can now retrieve and update the information on the amount due and paid as recorded in the `User` document:

```
else :
    amtDueFromUser = borrower.get('amountDue').to_decimal()
    amtPaidFromUser = borrower.get('amountPaid').to_decimal()
    amtDueFromUser -= amtPaid
    amtPaidFromUser += amtPaid
```

The secondary update is then performed using the pymongo `update_one()` method:

```
filt = {'userKey' : borrower.getKey()}
updateDoc = { '$set' : {
    'amountDue' : Decimal128(str(amtDueFromUser)),
    'amountPaid' : Decimal128(str(amtPaidFromUser))
}}
self.collection.update_one(filt, updateDoc)
```

Let's now look at how the publisher notifies the subscriber.

Event notification

Setting up the publisher to notify the subscriber involves two steps:

1. Associating the listener with the topic
2. Sending a message to the subscriber

The first step is achieved in the `views.py` module of the Django maintenance app. In the `accept()` method, the following is added immediately after variable initialization:

```
from biglittle.events.publisher import Publisher
publisher = Publisher()
config    = Config()
userSvc   = UserService(config, 'User', publisher)
loanSvc   = LoanService(config, 'Loan', publisher)
publisher.attach(publisher.EVENT_LOAN_UPDATE_BORROWER, \
                 userSvc.updateBorrowerListener)
```

In the `biglittle.domain.base.Base` class, we modify the `__init__()` method to accept the publisher as an argument:

```
class Base :
    def __init__(self, config, result_class = None, publisher = None) :
        self.publisher = publisher
        # other logic in this method now shown
```

In the `processPayment()` method of the `biglittle.domain.loan.LoanService` class (described earlier), we add a call to the publisher to notify (that is, send a message to) the subscriber. Note that we convert the loan entity back to `Decimal` for processing:

```
filt = { 'borrowerKey' : borrowerKey }
result = self.collection.replace_one(filt, loan)
if result :
    loan.convertBsonToDecimal()
    arg = { 'loan' : loan, 'amtPaid' : amtPaid }
    self.publisher.trigger( \
        self.publisher.EVENT_LOAN_UPDATE_BORROWER, arg)
return result
```

Next, we look at common problems encountered when working with documents across collections.

Avoiding cross-collection problems

Many of the *show stoppers* that are prevalent in the legacy **relational database management system (RDBMS)** arena are simply not an issue when using MongoDB due to differences in architecture. There are a few major issues in the area of cross-collection processing of which you need to be aware:

- Ensuring uniqueness
- Orphaned documents
- Value synchronization

Let's start this discussion by examining ways to ensure *uniqueness* when using MongoDB.

Ensuring uniqueness

The first question that naturally comes to mind is: *what is uniqueness?* In a traditional RDBMS database design, you must provide a mechanism whereby each database table row is different from every other row. If you had two RDMBS table rows, X and Y, which were identical, it would be impossible to guarantee access to X or to Y. Further, it would be a pointless waste of space to maintain duplicate data in such a manner. Accordingly, in the RDBMS world, when designing your database, you need to identify one or more database columns as part of the *primary key*.

Using ObjectId as a unique key

In the case of MongoDB, the equivalent of an RDBMS primary key would be the autogenerated `ObjectId`. You can reference this value via the alias `_id`. Unlike RDBMS *sequences* or *auto-increment* fields, however, the MongoDB `_id` field includes both sequential as well as random elements. It consists of 12 bytes broken down as follows:

- A 4-byte value representing the number of seconds since midnight on January 1, 1970 (Unix epoch)
- A 5-byte random value
- A 3-byte counter (seeded by a random value)

The `_id` field could potentially be used when conducting cross-collection operations (see the discussion on *DBRefs* earlier in this chapter). There is a method, `ObjectId.valueOf()`, that returns a hexadecimal string of 24 characters. However, within your application code, or when performing an operation using the aggregation framework, as this field is actually an object (for example, `ObjectId`), you would need to perform further conversions before being able to use it directly. Accordingly, it is highly recommended that within your application you create a custom unique and meaningful key that can be used to cross-reference documents between collections.

Defining a custom unique key

There are no specific rules, regulations, nor protocols that apply to define a unique custom key, however, there are two criteria that stand out. The key you choose should be the following:

- Unique
- Meaningful

Usually, in order to achieve uniqueness, some sort of random strategy needs to be employed. Using a value derived from the year, month, day, hours, minutes, and seconds might be of use. You can also use bits from other fields within the document to build a unique key that is meaningful.

As an example, in the BigLittle Micro Finance Ltd. scenario, each borrower or lender has information stored in the `users` collection. Each document, in addition to the automatically generated `ObjectId`, has a unique `userKey` field. As you can see from the following screenshot, the key consists of the first 4 letters of first and last names with a random 4-digit number:

```
File Edit View Search Terminal Help
> db.users.find({},{"userKey":1,"name.first":1,"name.last":1});
{ "_id" : ObjectId("5ec4d4d0ec2b5adab4754de0"), "userKey" : "CHARROSS3456", "name" : { "first" : "Charisse", "last" : "Ross" } }
{ "_id" : ObjectId("5ec4d4d0ec2b5adab4754de1"), "userKey" : "CLYDROJA7788", "name" : { "first" : "Clyde", "last" : "Rojas" } }
{ "_id" : ObjectId("5ec4d4d0ec2b5adab4754de2"), "userKey" : "ERICPOLL6194", "name" : { "first" : "Ericka", "last" : "Pollard" } }
{ "_id" : ObjectId("5ec4d4d0ec2b5adab4754de3"), "userKey" : "MYRTMCCA8673", "name" : { "first" : "Myrtie", "last" : "Mccall" } }
{ "_id" : ObjectId("5ec4d4d0ec2b5adab4754de4"), "userKey" : "CELIBLAC8153", "name" : { "first" : "Celina", "last" : "Black" } }
{ "_id" : ObjectId("5ec4d4d0ec2b5adab4754de5"), "userKey" : "KEIKSIER2074", "name" : { "first" : "Keiko", "last" : "Sierra" } }
{ "_id" : ObjectId("5ec4d4d0ec2b5adab4754de6"), "userKey" : "KATHHOWE6236", "name" : { "first" : "Kathe", "last" : "Howell" } }
{ "_id" : ObjectId("5ec4d4d0ec2b5adab4754de7"), "userKey" : "CLAUMUNO0734", "name" : { "first" : "Claude", "last" : "Munoz" } }
{ "_id" : ObjectId("5ec4d4d0ec2b5adab4754de8"), "userKey" : "MILOWHIT5333", "name" : { "first" : "Milo", "last" : "Whitney" } }
{ "_id" : ObjectId("5ec4d4d0ec2b5adab4754de9"), "userKey" : "JARRKIM6948", "name" : { "first" : "Jarred", "last" : "Kim" } }
{ "_id" : ObjectId("5ec4d4d0ec2b5adab4754dea"), "userKey" : "HERMGILM2578", "name" : { "first" : "Herman", "last" : "Gilmore" } }
{ "_id" : ObjectId("5ec4d4d0ec2b5adab4754deb"), "userKey" : "ANASCHUN7156", "name" : { "first" : "Anastacia", "last" : "Chung" } }
{ "_id" : ObjectId("5ec4d4d0ec2b5adab4754dec"), "userKey" : "ODESLEAC7490", "name" : { "first" : "Odessa", "last" : "Leach" } }
{ "_id" : ObjectId("5ec4d4d0ec2b5adab4754ded"), "userKey" : "ADELBLAK2695", "name" : { "first" : "Adelia", "last" : "Blake" } }
{ "_id" : ObjectId("5ec4d4d0ec2b5adab4754dee"), "userKey" : "GERTVELA6336", "name" : { "first" : "Gertrude", "last" : "Velasquez" } }
{ "_id" : ObjectId("5ec4d4d0ec2b5adab4754def"), "userKey" : "FLETBARR1534", "name" : { "first" : "Fletcher", "last" : "Barron" } }
{ "_id" : ObjectId("5ec4d4d0ec2b5adab4754df0"), "userKey" : "YONBARR0402", "name" : { "first" : "Yon", "last" : "Barrett" } }
{ "_id" : ObjectId("5ec4d4d0ec2b5adab4754df1"), "userKey" : "SHAWTRAN4223", "name" : { "first" : "Shawna", "last" : "Tran" } }
{ "_id" : ObjectId("5ec4d4d0ec2b5adab4754df2"), "userKey" : "LUCISAND0674", "name" : { "first" : "Luciana", "last" : "Sandoval" } }
{ "_id" : ObjectId("5ec4d4d0ec2b5adab4754df3"), "userKey" : "RANDSTEV0697", "name" : { "first" : "Randell", "last" : "Stevenson" } }
Type "it" for more
>
```

 Note that the `ObjectId` field does have a certain sequential aspect as these values were imported in bulk, thus the initial Unix epoch is almost the same for each one.

Now that you have an idea of how uniqueness might be defined, let's have a look at how it can be enforced.

Enforcing uniqueness

Once you have settled on an algorithm for choosing a unique key, is there a way to *guarantee uniqueness*? One very useful and effective solution is to create a unique index on the key field. This has two benefits:

- MongoDB prevents your application from adding a document with a duplicate key.
- Searches involving this key becomes faster and more efficient.

 The general rule for creating indexes in MongoDB is to create indexes on fields frequently used in queries. Do not go overboard on creating indexes, however, as they introduce extra overhead to maintain them.

Using the *mongo* shell, you can create an index on a field as follows:

```
db.<name_of_collection>.createIndex(keys, options);
```

Let's jump right into an example involving the `users` collection and the `userKey` field mentioned earlier in this subsection. Here is the query that creates a unique index on the `userKey` field in ascending order:

```
db.users.createIndex( { "userKey" : 1 }, { "unique" : true } );
```

If we then attempt to insert a document with a duplicate `userKey`, MongoDB refuses to perform the operation, thus enforcing uniqueness. Here is a screenshot of this attempt:

```
File Edit View Search Terminal Help
> db.users.createIndex({"userKey":1},{"unique":true});
{
        "createdCollectionAutomatically" : false,
        "numIndexesBefore" : 1,
        "numIndexesAfter" : 2,
        "ok" : 1
}
> db.users.insertOne({"userKey":"CHARROSS3456","name":{"first":"Fred","last":"Flintstone"}});
2020-05-21T05:11:26.543+0000 E  QUERY    [js] WriteError({
        "index" : 0,
        "code" : 11000,
        "errmsg" : "E11000 duplicate key error collection: biglittle.users index: userKey_1 dup key: { userKey: \"CHARROSS3456\" }",
        "op" : {
                "_id" : ObjectId("5ec60d7e0776678639c02dce"),
                "userKey" : "CHARROSS3456",
                "name" : {
                        "first" : "Fred",
                        "last" : "Flintstone"
                }
        }
}) :
WriteError({
        "index" : 0,
        "code" : 11000,
        "errmsg" : "E11000 duplicate key error collection: biglittle.users index: userKey_1 dup key: { userKey: \"CHARROSS3456\" }",
        "op" : {
                "_id" : ObjectId("5ec60d7e0776678639c02dce"),
                "userKey" : "CHARROSS3456",
                "name" : {
                        "first" : "Fred",
                        "last" : "Flintstone"
                }
        }
})
WriteError@src/mongo/shell/bulk_api.js:458:48
mergeBatchResults@src/mongo/shell/bulk_api.js:855:49
executeBatch@src/mongo/shell/bulk_api.js:919:13
Bulk/this.execute@src/mongo/shell/bulk_api.js:1163:21
DBCollection.prototype.insertOne@src/mongo/shell/crud_api.js:264:9
@(shell):1:1
>
```

 Please refer to the following link for more information on MongoDB indexes: `https://docs.mongodb.com/manual/indexes/#indexes`.

Next, we'll look at the subject of *orphans*.

Avoiding creating orphans across collections

The first question to be addressed is: *what is an orphan*? In the context of RDBMS, a database table row becomes an *orphan* when its parent relation is deleted. As an example, suppose you have one table for customers and another table for phone numbers. Each customer can have multiple phone numbers. In the phone number table, you would have a *foreign key* that refers back to the customer table. If you then delete a given customer, all the related rows in the phone numbers table become orphans.

As we mentioned before, MongoDB is not *relational*, so such a situation would not arise. A lot depends on good MongoDB database design. Let's look at two approaches that avoid creating orphans: embedded arrays and documents.

Using embedded arrays to avoid orphans

Using the example in the previous subsection, instead of creating a separate table for phone numbers, in MongoDB, simply include the phone numbers as an embedded array. Here is an example from the `users` collection. Note how we store multiple email addresses and phone numbers as embedded arrays within the `otherContact` field:

```
db.users.findOne({},{"userKey":1,"otherContact":1,"_id":0});
{
    "userKey" : "CHARROSS3456",
    "otherContact" : {
        "emails" : [
            "cross100@swisscom.com",
            "cross100@megafon.com"
        ],
        "phoneNumbers" : [
            "100-688-6884",
            "100-431-1074"
        ]
    }
}
```

Thus, when a user document is deleted, all of the related information is deleted at the same time, completely avoiding the issue of orphaned information.

Using embedded documents to avoid orphans

In a similar manner, as discussed earlier in this chapter, instead of using DBRefs, or unique keys to reference documents across collections, you can simply embed the entire document, or a subset of that document, where the information is needed. Going back to the *Book Someplace, Inc.* scenario as an example, a `Booking` document includes fragments of both `Customer` and `Property`. To refresh your memory, have a look at the `Booking`, `CustBooking`, and `PropBooking` document structures:

```
Booking = {
    "bookingKey"    : <string>,
    "customer"      : <CustBooking>,
    "property"      : <PropBooking>,
    "bookingInfo"   : <BookingInfo>,
    "rooms"         : [<RoomBooking>,<RoomBooking>,etc.],
    "totalPrice"    : <float>
}
CustBooking = {
    "customerKey"     : <string>,
    "customerName"    : <Name>,
    "customerAddr"    : <Location>,
    "customerContact" : <Contact>,
    "custParty"       : [<CustParty>,<CustParty>,etc.]
}
PropBooking = {
    "propertyKey"     : <string>,
    "propertyName"    : <string>,
    "propertyAddr"    : <Location>
    "propertyContact" : <Contact>
}
```

 Full definitions of these document structures can be found at `/path/to/repo/chapters/07/document_structure_definitions.js`.

This approach has its own disadvantages. If a customer document representing a customer who has made a booking is deleted, no orphaned documents are created. However, we now have a strange situation where the booking document contains a value for `customerKey`, but the corresponding customer document no longer exists! So, in a certain sense, although there are no orphaned documents, we now have an orphaned *field*. One way this situation can be avoided is to move deleted documents into an `archive` collection. You could then perform a secondary update on all booking documents associated with the removed customer, adding a new field: `archived`.

We'll now look at considerations for keeping values across collections properly synchronized.

Keeping values properly synchronized

Continuing the discussion from the previous subsection, we once again look at the database structure for *Book Someplace, Inc*. Another issue with the embedded document approach is what happens when the original customer document is updated? Now you have two places where customer information is stored: once in the `customers` collection, another in the `bookings` collection. The solution to this problem depends on the company's philosophy with regard to bookings. On the one hand, you could argue that at the time of the booking the customer information was correct, and must be maintained as is for historic reasons. On the other hand, you could argue that the latest information should always be present in `bookings` regardless of the situation that existed at the time of the booking. In the latter case, following a customer update, you would need to perform a series of secondary updates on all relevant bookings.

Returning to the BigLittle Micro Finance Ltd. scenario, the situation gets slightly more complicated because financial amounts are involved. Accordingly, it's highly important that your programming code includes a double-check for accuracy. In this case, we need to add programming logic that produces a sum of the `amountDue` and `amountPaid` fields in the `loan.payments` list and compare it with the same fields in the associated user document. If the information does not match, a discrepancies log file entry must be made. What would normally follow would be a tedious manual process of going through all loan payments, beyond the scope of this book.

In order to implement this secondary accuracy check, we add additional logic to the `updateBorrowerListener()` method in the `UserService` class already defined. We draw upon our `config.Config()` class to provide the name and location of the discrepancies log file. In addition, we pull today's date for logging purposes. This logic would then appear immediately after variables are initialized:

```
def updateBorrowerListener(self, arg) :
    # imports (not shown)
    # init vars (not shown)
    config = Config()
    log_fh = open(config.getConfig('discrepancy_log'), 'a')
    from time import gmtime, strftime
    now = strftime('%Y-%m-%d', gmtime())
    # other code already shown earlier in this chapter
```

Subsequently, just after we perform the update, we add logic to double-check the accuracy of the updated information. First, we calculate the amount due and paid from the list of payments:

```
# earlier code not shown
self.collection.update_one(filt, updateDoc)
for doc in loan.getPayments() :
    amtDueFromLoan += doc['amountDue']
    amtPaidFromLoan += doc['amountPaid']
```

If the amount due or paid does not match, discrepancy log entries are made. Also, of course, it's good practice to close the log file, which ends this method:

```
if amtDueFromUser != amtDueFromLoan :
    log_fh.write(now + ':Amount due discrepancy ' + \
        borrowerKey+"\n")
    log_fh.write('--"users" collection: ' + \
        str(amtDueFromUser)+"\n")
    log_fh.write('--"loans" collection: ' + \
        str(amtDueFromLoan)+"\n")
if amtPaidFromUser != amtPaidFromLoan :
    log_fh.write(now + ' :Amount paid discrepancy ' + \
        borroweKey+"\n")
    log_fh.write('--"users" collection: ' + \
        str(amtPaidFromUser)+"\n")
    log_fh.write('--"loans" collection: ' + \
        str(amtPaidFromLoan)+"\n")
log_fh.close()
```

In many cases, you can have your app perform updates to make sure the information is in sync in the case of a discrepancy. In a financial application, however, this is not recommended as information that is out of sync is a warning sign that either there is a logic error somewhere in the application (that is, secondary updates are not being performed properly), or an indicator that amounts are not being reported correctly, in which case manual investigation and intervention are needed.

MongoDB provides a feature known as *document validation* (https://docs.mongodb.com/v3.2/core/document-validation/#document-validation), which greatly minimizes the risk of creating orphaned documents as well as facilitating keeping information synchronized.

In the next section, we introduce a feature related to storing images in the database called *GridFS*.

Uploading files into GridFS

GridFS (`https://docs.mongodb.com/manual/core/gridfs/#gridfs`) is a technology included with MongoDB that allows you to store files directly in the database. GridFS is a virtual filesystem mainly designed to handle files larger than 16 MB in size. Each document is broken down into *chunks* of 255 KB except for the last chunk, which could be less than 255 KB. By default, two collections assigned to an arbitrary *bucket* are used. The `chunks` collection stores the actual contents of the file. The `files` collection stores file metadata. Within your database, you can designate as many *buckets* as needed.

Let's start by discussing why you might want to incorporate GridFS into your applications.

Why use GridFS?

One big advantage is that once stored, MongoDB offers the same advantages to large files as it does to any other type of data: replication to ensure high availability, sharding to facilitate scalability, and the ability to access a file regardless of the host operating system. Accordingly, if the server's operating system places a limit on the number of files in a single directory, you can use GridFS to store the files and get around OS limitations.

Another big advantage is that due to the way GridFS breaks files into 255 KB chunks, reading extremely large files is not as memory intensive as when you try to directly access the same file stored on the server's filesystem. Finally, using GridFS is a great way to keep files synchronized across servers.

It is recommended that you do not use GridFS if all the files you need to store are less than 16 MB in size. If this is the case, and you still want to take advantage of using MongoDB, simply store the file directly in a normal MongoDB document. The maximum size of a BSON document is 16 MB, so if all your files are smaller, using GridFS is a waste of resources.

Also, it is not recommended to use GridFS if the entire document needs to be updated atomically. As an example, if you are maintaining legal documents, each revision needs to be stored as a separate document with changes clearly marked. In this case, either store the document using the server's filesystem or clone the document and store clearly marked revisions.

How to use GridFS from the command line

GridFS is available from the command line via the `mongofiles` utility. The generic syntax is as follows:

```
mongofiles <options> <commands> <filename>
```

Most of the *options* are identical to those of the *mongo* shell and include command-line flags to identify a URI style connection string, host, port, and authentication information. You can also specify the IP version (for example, `--ipv6`) and SSL/TLS secure connectivity options.

> The complete list of options is found here: `https://docs.mongodb.com/manual/reference/program/mongofiles/#options`. For more information on options, please review the subsection entitled *Mongo Shell Command Line Switches* in `Chapter 3`, *Essential MongoDB Administration Techniques*.

GridFS command-line options

Some options are not relevant to the database connection. These additional options, peculiar to *GridFS*, are summarized here:

Option	Notes
`--local=<filename>`	Use this option if you want the filename recorded in GridFS to be different from the local filename on the server's OS. In place of `<filename>`, enter the actual name of the file on the local server filesystem. At the end of the entire command string, enter the filename as you wish it to appear in GridFS.
`--replace`	When you use the `put` command (described in the next subsection) to store a file, existing files of the same name are not overwritten. Instead, a new entry is created. If you want the currently stored file to be completely replaced, use this option.
`--prefix=<string>`	Use this option to specify a different *bucket*. The default bucket is `fs`. Think of a bucket as being like a subdirectory.
`--db=<database>`	This option specifies which database to use for GridFS storage.
`--uri="<string>"`	Just as with the *mongo* shell, you can consolidate command-line options into a single URI-style string.

In addition to *options*, the `mongofiles` utility has commands, discussed next.

GridFS commands

In the following table is a summary of the more important commands used with the `mongofiles` utility:

Command	Notes
list \<filter\>	This command operates much like the Linux `ls` or the Windows `dir` commands. If you specify one or more characters in the `filter`, you get a `list` of filenames starting with those characters. Using Linux as an example, this command is similar to `ls xyz*` (returns a list of files starting with `xyz`).
search \<filter\>	This command is the same as `list` except that a list of filenames matching any part of `filter` are displayed. Using Linux as an example, this command is similar to `ls *xyz*` (returns a `list` of filenames containing `xyz`).
put \<filename\>	Copies a file from the local filesystem to GridFS. By default, GridFS uses the local filename. If you need to have the GridFS filename be different, use the `--local` option. Use the `--replace` option to overwrite an existing GridFS of the same name.
get \<filename\>	The opposite of `put`. Copies a file from GridFS into the local filesystem.
get_id "\<_id\>"	The same as `get` except that the file to be copied is identified by its `_id` field.
delete \<filename\>	Erases a file from GridFS storage.
delete_id "\<_id\>"	The same as `delete` except that the file to be deleted from GridFS is identified by its `_id` field.

Now let's look at a usage example.

Command-line usage examples

In this example, we copy a number of large files drawn from the open source GeoNames project representing world city data and postal codes. The data provided by the project, authored by Marc Wick, is licensed under a Creative Commons Attributions v4.0 license. In the book repository, there is a script, `/path/to/repo/sample_data/geonames_downloads.sh`, that imports and unzips two large GeoNames project sample data files. Here is that download script:

```
#!/bin/bash
export
STATS_URL="https://download.geonames.org/export/dump/allCountries.zip"
export ZIP_URL="https://download.geonames.org/export/zip/allCountries.zip"
wget -O /tmp/stats.zip $STATS_URL
wget -O /tmp/postcodes.zip $ZIP_URL
```

```
unzip /tmp/stats.zip -d /tmp
mv /tmp/allCountries.txt /tmp/allCountriesStats.txt
unzip /tmp/postcodes.zip -d /tmp
mv /tmp/allCountries.txt /tmp/allCountriesPostcodes.txt
rm /tmp/*.zip
ls -l /tmp/all*
```

Here is a list of the files to be copied into GridFS:

```
File  Edit  View  Search  Terminal  Help
root@server1:/repo/sample_data# ls -l /tmp/all*
-rw-r--r-- 1 root root  119200479 May 21 03:01 /tmp/allCountriesPostcodes.txt
-rw-r--r-- 1 root root 1551551316 May 21 02:48 /tmp/allCountriesStats.txt
root@server1:/repo/sample_data#
```

As you can see, most files are well over 16 MB in size. We use the `put` command to store the files in `GridFS` in the `biglittle` database on localhost. Here is an example using the `--local` option. In this example, we want the filename in GridFS to appear as `/postalCodes/world_postal_codes.txt`. We can then use `list` to view results (`search` could also be used):

```
File  Edit  View  Search  Terminal  Help
root@server1:/repo/sample_data# mongofiles --db=biglittle --local=/tmp/allCountriesPostcodes.txt
 put /postalCodes/world_cities_codes.txt
2020-05-21T05:52:41.238+0000    connected to: mongodb://localhost/
2020-05-21T05:52:42.212+0000    added gridFile: /postalCodes/world_cities_codes.txt

root@server1:/repo/sample_data# mongofiles --db=biglittle list
2020-05-21T05:53:31.038+0000    connected to: mongodb://localhost/
/postalCodes/world_cities_codes.txt     119200479
root@server1:/repo/sample_data#
```

From the *mongo* shell, you can see the new **fs** collection. `fs.chunks` contains the actual contents of the files. `fs.files` contains file metadata, as shown here:

```
File  Edit  View  Search  Terminal  Help
> show collections;
common
fs.chunks
fs.files
loans
users
world_cities
> db.fs.files.find().pretty();
{
        "_id" : ObjectId("5ec617294270dd1f9cea992c"),
        "length" : NumberLong(119200479),
        "chunkSize" : 261120,
        "uploadDate" : ISODate("2020-05-21T05:52:42.210Z"),
        "filename" : "/postalCodes/world_cities_codes.txt",
        "metadata" : {

        }
}
{
        "_id" : ObjectId("5ec6182e08597ebba8913889"),
        "length" : NumberLong(1551551316),
        "chunkSize" : 261120,
        "uploadDate" : ISODate("2020-05-21T05:57:17.948Z"),
        "filename" : "/statistics/world_cities_stats.txt",
        "metadata" : {

        }
}
>
```

Please refer to the following link for more information:
GeoNames project: http://www.geonames.org/ . Author: Marc Wick:
marc@geonames.org. Creative Commons Attributions v4.0 license:
https://creativecommons.org/licenses/by/4.0/.

Next, let's look at how *GridFS* is supported by the PyMongo driver.

PyMongo GridFS support

The PyMongo driver provides a package, `gridfs`, that provides support for GridFS from a programming context. The core class is `gridfs.GridFS`. There are a number of supporting packages, however, in order to retain our focus on the essentials, we concentrate here only on the core class. GridFS class methods mirror the commands discussed in the previous subsection. The following table provides a summary of the more important methods:

gridfs.GridFS.*method*	Notes
`find(*args, **kwargs)`	This method operates in a similar manner to `pymongo.collection.find()`. `*args` is a JSON document with key/value pairs. The key can be anything in the `fs.files` collection. The most common key is `filename`. `**kwargs` is a series of key-value keypairs including the keys `filter`, `skip`, `limit`, `no_cursor_timeout` and `sort`. `find_one()` takes the same arguments except only a single file is returned.
`list()`	Operates much like the *mongofiles* `list` command. It returns a list of of files in GridFS.
`put(data, **kwargs)`	Operates like the *mongofiles* `put` command. It writes a file to GridFS. Data can take the form of a bytes data type or an object (for example, a file handle) that provides a `read()` method. `**kwargs` can be any of the options mentioned in the documentation for `gridfs.grid_file.GridIn`.
`get(file_id)`	Much like the *mongofiles* `get` command, returns the file based on its `_id` field value. The return value is the `gridfs.grid_file.GridOut` instance (much like a file handle). The `GridOut` object provides methods that allow you to get metadata on the file. Of special interest is its `read (NNN)` method, giving access to the file contents, NNN number of bytes at a time. There is also a `readLine()` method, which reads a single line, and `readChunk()`, which reads out the file in chunks (preset to 255KB).
`exists(file_id)`	Returns a Boolean set to `True` if the given file `_id` field value exists; `False` otherwise.
`delete(file_id)`	Deletes a file with a given value of the `_id` field.

Please refer to the following links for more information:

- `GridFS` module: `https://api.mongodb.com/python/current/api/gridfs/index.html#module-gridfs`.
- `gridfs.grid_file.GridIn`: `https://api.mongodb.com/python/current/api/gridfs/grid_file.html#gridfs.grid_file.GridIn`.

Now it's time to have a look at a practical example using this class in our sample scenario.

Example using GridFS to store loan documents

We need a way to upload loan documents, so we add a file upload HTML element to the `maintenance.html` template. We also add a loop that displays the currently uploaded documents for this borrower:

```
<tr><th class="th-right">Loan Documents Stored</th>
    <td class="td-left">
        {% for name in loan_docs %} <br>{{ name }}  {% endfor %}
    </td>
</tr>
<tr>
    <th class="th-right">Upload Loan Document</th>
    <td class="td-left"><input type="file" name="loan_doc" /></td>
</tr>
```

In the view logic (`/path/to/repo/www/chapter_10/maintenance/views.py`), we add the following to the `payments()` method. If there is a `borrowerKey` value available, we import `GridFS` and `os`, and retrieve borrower information (discussed earlier):

```
if borrowerKey :
    from gridfs import GridFS
    import os
    borrower = userSvc.fetchUserByBorrowerKey(borrowerKey)
```

We then check to see whether any files have been uploaded. If so, we prepend the value of the `borrowerKey` property and store it in GridFS:

```
grid = GridFS(loanSvc.getDatabase())
if request.FILES and 'loan_doc' in request.FILES :
    fn = request.FILES['loan_doc']
    newFn = borrowerKey + '/' + os.path.basename(fn.name)
    grid.put(fn, filename=newFn)
    message = 'File uploaded'
```

We then get a list of only filenames, subsequently sent to the view template for rendering:

```
loan_docs = []
temp = grid.list()
for name in temp :
    if name.find(borrowerKey) >= 0 :
        loan_docs.append(os.path.basename(name))
```

Now that you have an idea of how to use GridFS from both the command line and from a program script, it's time to jump to the chapter summary.

Summary

In this chapter, you learned about the MongoDB `NumberDecimal` class, based upon the `BSON Decimal128` class, used to store financial numbers with high precision. You also learned that not all programming languages provide direct support for this class, which means a conversion might be necessary in order to perform processing.

Next, you learned how to handle database operations across collections. Several approaches were covered, including *building block, embedded document, database reference,* and DBRef approaches. You also learned how to produce a single iteration of documents across collections using the `$lookup` aggregation pipeline operator. After this, you learned about secondary updates and where they might be performed, as well as common pitfalls when working with documents across collections. These include problems associated with ensuring uniqueness, orphaned documents or fields, and keeping values synchronized.

Finally, at the end of this chapter, you learned about GridFS, a technology provided by MongoDB that lets you store large documents directly in the database. You learned that GridFS files can be accessed using the `mongofiles` command from the command line, or via a programming driver. You then learned how to use the `gridfs.GridFS` class provided by the pymongo driver to store and then access files.

In the next chapter, continuing with the BigLittle Micro Finance Ltd. scenario, you'll learn how to secure connections to the database at both the user level as well as at the SSL/TLS level.

11
Administering MongoDB Security

Security is an important consideration for any organization, especially companies conducting financial transactions over the internet, and also companies dealing with sensitive user data such as health service providers, counselling services, law firms, and so forth. This chapter focuses on the administration needed to secure a MongoDB database. First, you are shown how to secure the database by creating database users and how to implement role-based access control. After that, you learn how to secure the MongoDB database communications by implementing transport layer (SSL/TLS) security based upon X.509 certificates.

The following topics are covered in this chapter:

- Enabling database security
- Understanding role-based access control
- Exploring database user scenarios
- Setting up transport layer security using X.509 certificates
- Additional security considerations

Technical requirements

The minimum recommended hardware is as follows:

- Desktop PC or laptop
- 2 GB free disk space
- 4 GB of RAM
- 500 Kbps or faster internet connection.

The software requirements are as follows:

- OS (Linux or Mac): Docker, Docker Compose, Git (optional)
- OS (Windows): Docker for Windows and Git for Windows
- Python 3.x, PyMongo driver, and Apache (already installed in the Docker container used for the book)
- SSL certificate (use the script described later in this chapter to install a test certificate)

Installation of the required software and how to restore the code repository for the book is explained in `Chapter 2`, *Setting Up MongoDB 4.x*. To run the code examples in this chapter, open a terminal window (Command Prompt) and enter these commands:

```
cd /path/to/repo
docker-compose build
docker-compose up -d
docker exec -it learn-mongo-server-1 /bin/bash
```

When you are finished working with the examples covered in this chapter, return to the terminal window (Command Prompt) and stop Docker as follows:

```
cd /path/to/repo
docker-compose down
```

The code used in this chapter can be found in the book's GitHub repository at `https://github.com/PacktPublishing/Learn-MongoDB-4.x/tree/master/chapters/11`.

Enabling database security

The first step in securing a MongoDB database is to enable security. By default, a MongoDB database is without transport layer security mechanisms and without any role-based access controls. This is by design, otherwise how would you be able to access the database in order to make it secure? Also, not having to deal with authentication is convenient for initial application development and database document design. If you implement an unsecured database in a production environment, you might end up being unemployed!

Before moving into a discussion on how to enable security, it is important to gain an understanding of the MongoDB authentication process, and what forms of authentication are available.

For an excellent overview of the steps needed to properly secure a MongoDB database, in the MongoDB documentation, see the security checklist (`https://docs.mongodb.com/manual/administration/security-checklist/#security-checklist`).

Starting with MongoDB version 3.6, by default, the `mongod` or `mongos` instance binds only to *localhost* (IP address `127.0.0.1`). Although this does not represent an alternative to implementing a proper security policy, it does provide a limited degree of protection in that any attempt to connect to the database from outside the local server is rejected.

Understanding MongoDB authentication

MongoDB authentication involves a number of different factors: the authentication database, the authentication mechanism, and the database user. The *authentication database* is the place where authentication information is stored. The *authentication mechanism* concerns how the authentication is to take place. Finally, the database user brings together one or more *roles*, the authentication mechanism, and the password. Before getting into the details, however, it's important to understand the difference between *authentication* and *authorization*.

Authentication versus authorization

An important distinction we need to make right away is to describe the difference between *authentication* and *authorization*:

- **Authentication**: Authentication proves the *identity* of an entity attempting to access the database. The entity could be a DevOp connecting through the *mongo* shell, or an application performing a database read or write operation. As an analogy, when traveling from one country to another, a traveller is asked to display a passport that contains identifying information before being allowed to enter the country.
- **Authorization**: Authorization determines what actions the entity is allowed to perform once authentication is granted. Thus, authentication precedes authorization: MongoDB needs to know who you are before determining your ability to perform the requested action. As an analogy, once the traveler successfully crosses the border, that person's *role* determines what activities they can perform. Thus, for example, if the traveller is an airline pilot, and the border crossing is in an airport, the traveler is able to use the airline staff entrance, and take advantage of the airline staff lounge.

In this section, we discuss *authentication*. In the next major section in this chapter, we discuss *authorization*, which, in MongoDB, takes the form of *role-based access control*. Now let's turn our attention to the *authentication database*.

Understanding the authentication database

The *authentication database* (`https://docs.mongodb.com/manual/core/security-users/#user-authentication-database`) can be any MongoDB database in which you have defined users and roles. The authentication database chosen for a given user depends on what level of access that user is to be assigned. If the user has privileges in many different databases, it doesn't make sense to define the same user multiple times in each database. Instead, it makes more sense to define the user in a single central database, and to specify the central database as the *authentication database* when attempting access.

On the other hand, it's quite possible that you wish to confine the activities of a particular database user to one single database. In this case, you could create the user in that one database, and assign the appropriate privileges. We now move our attention to the different *authentication mechanisms* available in MongoDB.

Authentication mechanisms

There are four primary *authentication mechanisms* (https://docs.mongodb.com/manual/core/authentication-mechanisms/#authentication-mechanisms) currently supported by MongoDB. **Salted Challenge Response Authentication Mechanism** (**SCRAM**) and X.509 Certificate Authentication are supported in all versions of MongoDB. Two additional mechanisms, Keberos and **Lightweight Directory Access Protocol** (**LDAP**), are available only in the *Enterprise* edition. Let's look at them individually:

- **SCRAM**: SCRAM is the default authentication mechanism for MongoDB. It is based upon RFC 5802 and RFC 7677, providing a secure way to support usernames and passwords. To authenticate using this mechanism, simply provide a valid username and password and indicate the authentication database.

- **X.509 certificate authentication**: The X.509 certificate authentication process, defined by RFC 6818, involves generating public key infrastructure certificates used in an exchange between client and server. Although it requires more work to set up, this implementation avoids the necessity of always having to provide a username and password.

- **Kerberos authentication**: In order to use this authentication method, you must upgrade to the MongoDB *Enterprise Edition*. The authentication scheme relies upon defining Kerberos *realms* and a server that hosts the **Key Distribution Center** (**KDC**). When an authentication request is made, the KDC grants a *ticket* if the credentials are valid. It is important to note that Kerberos is at the heart of Microsoft Active Directory authentication. The KDC is implemented as a Windows domain service. For each Kerberos realm user (referred to as a *principal*) needing access to the database, you must also create a matching MongoDB user. In addition, you need to define each mongod and/or mongos instance as a Kerberos principal. There is also an open source KDC implementation included in the ApacheDS project.

- **LDAP authentication**: The MongoDB *Enterprise Edition* also offers authentication support via an LDAP server. The LDAP server maintains its own database of users; however, as with Kerberos, you need to define a matching MongoDB user. There are several open source LDAP implementations, including OpenLDAP, OpenDJ (originally developed by Sun MicroSystems), and the 389 Directory Server (RedHat). The ApacheDS project also provides an LDAP service. In addition, there are a number of commercial implementations that respond to LDAP requests, including Microsoft Active Directory and NetIQ eDirectory (evolved from Novell Directory Services).

Documentation references: SCRAM (`https://docs.mongodb.com/manual/core/security-scram/#scram`), RFC 5802 (`https://tools.ietf.org/html/rfc5802`) and RFC 7677 (`https://tools.ietf.org/html/rfc7677`), RFC 6818 (`https://tools.ietf.org/html/rfc6818`), Microsoft Active Directory authentication (`https://docs.microsoft.com/en-us/windows/win32/secauthn/key-distribution-center`), the ApacheDS project (`http://directory.apache.org/apacheds/`), OpenLDAP (`http://www.openldap.org/`), OpenDJ (`https://backstage.forgerock.com/docs/opendj/3.5`), 389 Directory Server (`https://directory.fedoraproject.org/`), the ApacheDS project (`http://directory.apache.org/apacheds/`), Microsoft Active Directory (`https://docs.microsoft.com/en-us/windows-server/identity/ad-ds/get-started/virtual-dc/active-directory-domain-services-overview`), and NetIQ eDirectory (`https://www.microfocus.com/en-us/products/netiq-edirectory/overview`).

As the focus in this book is on the MongoDB *Community Edition*, in this section we cover implementation only of the first two mechanisms. In order to enable database security, you need to know how to create a database user, discussed next.

 The database in the Docker container used for this book is not secured. You must perform the security actions described in this chapter in order to reproduce the results shown in the code examples.

Creating a database user

When you create a database user, you bring together both authentication and authorization. The key command used for this purpose is a database command, `createUser()`. Here is the generic syntax:

```
db.createUser( { user_document }, { writeConcern_document } )
```

Both the `user` and `writeConcern` documents are JSON documents with a number of parameters. The following table summarizes `user_document` parameters.

Parameter	Required	Notes
`user`	Yes	Username in the form of a text string.
`pwd`	Yes	Password in the form of a text string. When the password is actually stored in the database, it is first converted into a BCRYPT hash. MongoDB also allows you to insert a `passwordPrompt ()` function in place of a text string. In this case, you are first prompted for the password before the insertion operation proceeds.
`roles`	Yes	A list of one or more roles to be assigned to this user.
`customData`	No	A JSON document with any additional information to store for this user (such as department, title, phone number, and so on).
`authenticationRestrictions`	No	A list consisting of a list of IP addresses and/or **Classless Inter-Domain Routing (CIDR)** ranges from which the user is allowed to connect. The restriction can apply directly to the client, in which case you need the `clientSource` sub-parameter, Otherwise, by using the `serverAddress` sub-parameter, the restriction applies to the server this database user uses as a connection.
`mechanisms`	No	Use this parameter to specify which SCRAM authentication can be used. Choices can be either `SCRAM-SHA-1` or `SCRAM-SHA-256`. Note that these choices are in effect regardless of the `authenticationMechanisms` setting in the `mongod` config file.
`passwordDigestor`	No	This setting controls where the password hash is created when a plain text password is received. The default is `server`. If using *SCRAM-SHA-1* you can specify either `client` or `server`.

The second document is optional and can be used to specify the behavior of *write concerns*. Here is the prototype for the `writeConcern` document:

```
{ w: <int | string>, j: <boolean>, wtimeout: <number of milliseconds> }
```

Here is a brief summary of these three arguments visible in the `writeConcern` document:

- `w : <int | string>`: w stands for *write acknowledgement*. Specify a value of 1 if you wish to receive an acknowledgement that the write operation has been successfully passed to a `mongod` instance. A value of 0, on the other hand, tells MongoDB an acknowledgement is not required. MongoDB still returns information even if there is an error or socket exception.

  If you are passing a write request to a replica set, any value above 1 signals that many servers must acknowledge. So, for example, a value of 3 indicates that the *primary* and two *secondaries* must acknowledge the write operation.

  If the value for `w` is not a string, it must be either *majority* or a custom write concern name. In this case, the value assigned to `w` would be the number of servers in the custom write concern that is named, or the number of servers considered to be a majority within the given replica set.

- `j : <boolean>`: j stands for *journal*. If set to `true`, the number of servers indicated by the `w` option must acknowledge the write operation has been written to their on-disk journals.

- `wtimeout : <int milliseconds>`: wtimeout stands for *write timeout*. The value specified is in milliseconds. If the write operation is not successful within this time limit, an error condition is raised.

For more information on replica sets, see `Chapter 13`, *Deploying a Replica Set*. For replica set programming considerations, including *read preferences* and *write concerns*, see `Chapter 14`, *Replica Set Runtime Management and Development*.

Please refer to the following links for more information:
authenticationMechanisms (`https://docs.mongodb.com/manual/reference/parameters/#param.authenticationMechanisms`)
write concerns (`https://docs.mongodb.com/manual/reference/write-concern/#write-concern`)
writeConcern document: (`https://docs.mongodb.com/manual/reference/write-concern/#write-concern-specification`).

Let's now have a look at how to enable security in a MongoDB database.

Enabling database security

In order to enable database security, you need to start the mongod instance *without* security, create at least one user in the authentication database with the userAdminAnyDatabase role and restart mongod with security enabled. You can then authenticate as the new admin user and create additional users and make assignments as needed.

Here are the detailed steps you should follow to enable database security:

1. Stop the local mongod instance if running. Restart it *without* security. This can be done by changing the security.authorization parameter in the mongod.conf file (mongod.cfg in Windows) to disabled. Alternatively, start mongod manually from Command Prompt using the --noauth option. If using Command Prompt, you also need to supply the --dbpath argument, needed to tell MongoDB where its database files are located.

   Here is an example on an Ubuntu 18.04 server:

```
File  Edit  View  Search  Terminal  Help
ked@ked-VirtualBox:~$ sudo mongod --noauth --dbpath /var/lib/mongodb
[sudo] password for ked:
{"t":{"$date":"2020-05-24T07:03:43.044+01:00"},"s":"I",  "c":"CONTROL",  "id":23285,   "ctx":"main","msg":
"Automatically disabling TLS 1.0, to force-enable TLS 1.0 specify --sslDisabledProtocols 'none'"}
{"t":{"$date":"2020-05-24T07:03:43.053+01:00"},"s":"W",  "c":"ASIO",     "id":22601,   "ctx":"main","msg":
"No TransportLayer configured during NetworkInterface startup"}
{"t":{"$date":"2020-05-24T07:03:43.053+01:00"},"s":"I",  "c":"NETWORK",  "id":4648601, "ctx":"main","msg":
"Implicit TCP FastOpen unavailable. If TCP FastOpen is required, set tcpFastOpenServer, tcpFastOpenClient,
 and tcpFastOpenQueueSize."}
{"t":{"$date":"2020-05-24T07:03:43.054+01:00"},"s":"I",  "c":"STORAGE",  "id":4615611, "ctx":"initandliste
n","msg":"MongoDB starting","attr":{"pid":2346,"port":27017,"dbPath":"/var/lib/mongodb","architecture":"64
-bit","host":"ked-VirtualBox"}}
{"t":{"$date":"2020-05-24T07:03:43.060+01:00"},"s":"I",  "c":"CONTROL",  "id":23403,   "ctx":"initandliste
n","msg":"Build Info","attr":{"buildInfo":{"version":"4.4.0-rc5","gitVersion":"bbdf0a11d1c61be0760a829e827
99129beac7be0","openSSLVersion":"OpenSSL 1.1.1  11 Sep 2018","modules":[],"allocator":"tcmalloc","environm
```

2. You can now get into the *mongo* shell and create a new admin user. Here is an example of the command to use:

```
superMan = {
 user : "superMan",
 pwd : "password",
 roles : [ { role : "userAdminAnyDatabase", db : "admin" },
 "readWriteAnyDatabase" ]
}
db.createUser(superMan);
```

A discussion of *role-based access control* is in the next section. Please be aware that the `userAdminAnyDatabase` role, along with the `readWriteAnyDatabase` privilege, creates a *super user* who can access any database, run any command, and in particular, make rights assignments. This username and password must be carefully guarded to avoid a security breach. See `https://docs.mongodb.com/manual/reference/built-in-roles/#userAdminAnyDatabase`.

3. You can now shut down the database using the `db.adminCommand({shutdown:1});` command. In the next step, you restart it with authentication enabled.

Note that when you are in the shell and you shut down the database, the shell reports connection errors. Do not be alarmed, this is to be expected.

The following screenshot shows what you might see at this point:

```
File Edit View Search Terminal Help
> db.adminCommand({shutdown:1});
{"t":{"$date":"2020-05-24T06:38:53.194Z"},"s":"I",  "c":"NETWORK",  "id":20125,  "ctx":"js","msg":
"DBClientConnection failed to receive message from {getServerAddress} - {swm_getStatus}","attr":{"g
etServerAddress":"127.0.0.1:27017","swm_getStatus":"HostUnreachable: Connection closed by peer"}}
uncaught exception: Error: error doing query: failed: network error while attempting to run command
 'shutdown' on host '127.0.0.1:27017'  :
DB.prototype.runCommand@src/mongo/shell/db.js:169:19
DB.prototype.adminCommand@src/mongo/shell/db.js:187:12
@(shell):1:1
{"t":{"$date":"2020-05-24T06:38:53.205Z"},"s":"I",  "c":"NETWORK",  "id":20120,  "ctx":"js","msg":
"Trying to reconnnect","attr":{"connString":"127.0.0.1:27017 failed"}}
{"t":{"$date":"2020-05-24T06:38:53.206Z"},"s":"I",  "c":"NETWORK",  "id":20125,  "ctx":"js","msg":
"DBClientConnection failed to receive message from {getServerAddress} - {swm_getStatus}","attr":{"g
etServerAddress":"127.0.0.1:27017","swm_getStatus":"HostUnreachable: Connection reset by peer"}}
{"t":{"$date":"2020-05-24T06:38:53.206Z"},"s":"I",  "c":"NETWORK",  "id":20121,  "ctx":"js","msg":
"Reconnect attempt failed","attr":{"connString":"127.0.0.1:27017 failed","reason":""}}
>
```

4. To enable authentication and access control permanently, before restarting `mongod`, set the `security.authorization` config setting to `enabled`. Here is how the `mongod` config file might now appear:

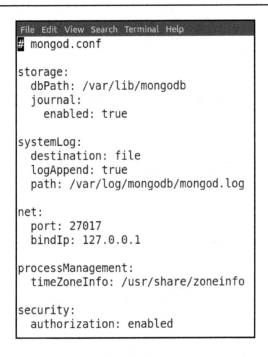

```
File Edit View Search Terminal Help
# mongod.conf

storage:
  dbPath: /var/lib/mongodb
  journal:
    enabled: true

systemLog:
  destination: file
  logAppend: true
  path: /var/log/mongodb/mongod.log

net:
  port: 27017
  bindIp: 127.0.0.1

processManagement:
  timeZoneInfo: /usr/share/zoneinfo

security:
  authorization: enabled
```

5. You can now restart `mongod`. Check the server status to confirm its operation. The screen shows a bad status:

```
File Edit View Search Terminal Help
● mongod.service - MongoDB Database Server
   Loaded: loaded (/lib/systemd/system/mongod.service; disabled; vendor preset: enabled)
   Active: failed (Result: exit-code) since Sun 2020-05-24 07:28:51 BST; 7s ago
     Docs: https://docs.mongodb.org/manual
  Process: 2963 ExecStart=/usr/bin/mongod --config /etc/mongod.conf (code=exited, status=14)
 Main PID: 2963 (code=exited, status=14)

May 24 07:28:50 ked-VirtualBox systemd[1]: mongod.service: Main process exited, code=killed, sta
May 24 07:28:50 ked-VirtualBox systemd[1]: mongod.service: Failed with result 'signal'.
May 24 07:28:50 ked-VirtualBox systemd[1]: Stopped MongoDB Database Server.
May 24 07:28:50 ked-VirtualBox systemd[1]: Started MongoDB Database Server.
May 24 07:28:51 ked-VirtualBox systemd[1]: mongod.service: Main process exited, code=exited, sta
May 24 07:28:51 ked-VirtualBox systemd[1]: mongod.service: Failed with result 'exit-code'.
lines 1-13/13 (END)
```

6. If you encounter errors, it's possible while starting `mongod` manually as the root user that the file permissions in the `dbpath` were reset. An example is shown here:

```
File  Edit  View  Search  Terminal  Help
-rw------- 1 mongodb mongodb    20480 Dec 24 09:42 index-5-7500462118803848817.wt
-rw------- 1 mongodb mongodb    12288 May 24 07:28 index-6-6735730018938818569.wt
-rw------- 1 mongodb mongodb    20480 Nov 19  2019 index-7-3426497892779560040.wt
drwx------ 2 mongodb mongodb     4096 May 24 07:17 journal
-rw------- 1 mongodb mongodb    36864 May 24 07:28 _mdb_catalog.wt
-rw------- 1 mongodb mongodb        5 May 24 07:17 mongod.lock
-rw------- 1 mongodb mongodb    36864 May 24 07:28 sizeStorer.wt
-rw------- 1 mongodb mongodb      114 Aug 26  2019 storage.bson
-rw------- 1 mongodb mongodb       44 Aug 26  2019 WiredTiger
-rw------- 1 mongodb mongodb     4096 May 24 07:28 WiredTigerHS.wt
-rw------- 1 mongodb mongodb       21 Aug 26  2019 WiredTiger.lock
-rw------- 1 root    root       1259 May 24 07:22 WiredTiger.turtle
-rw------- 1 mongodb mongodb   106496 May 24 07:28 WiredTiger.wt
ked@ked-VirtualBox:~/learn-mongodb$
```

Looking at the screenshot, note that of the last four entries, `WiredTiger.turtle` has reverted to `root`. All files in the directory where `mongod` expects to find its database files must be set to the `mongodb` user and group.

7. To find the location of this file, look in the `mongod` config file for the `storage.dbpath` setting. Use the appropriate operating system commands to reset ownership to the `mongodb` user and group.

8. Once the `mongod` instance is up and running with security, you can access the *mongo* shell by adding the `-u` or `--username` and `-p` or `--password` options. Also add the `--authenticationDatabase` directive to indicate the source database for authentication. In the example shown here, the authentication database is `admin`. Here is an example of accessing the *mongo* shell as the new super user:

```
File  Edit  View  Search  Terminal  Help
ked@ked-VirtualBox:~$ sudo service mongod restart
ked@ked-VirtualBox:~$ mongo --quiet -u superMan -p password
{"t":{"$date":"2020-05-24T06:17:35.844Z"},"s":"I",  "c":"NETWORK", "id":4648601, "ctx":
"main","msg":"Implicit TCP FastOpen unavailable. If TCP FastOpen is required, set tcpFas
tOpenServer, tcpFastOpenClient, and tcpFastOpenQueueSize."}
> show dbs;
admin            0.000GB
biglittle        0.000GB
config           0.000GB
local            0.000GB
sweetscomplete   0.002GB
test             0.000GB
>
```

An excellent tutorial on enabling database security is available on the MongoDB documentation page on enabling access control (`https://docs.mongodb.com/manual/tutorial/enable-authentication/#enable-access-control`).

You now have a good idea on what needs to be done to enable database security. Next, you will learn how to configure access control based on roles.

Understanding role-based access control

MongoDB *role-based access control* (`https://docs.mongodb.com/manual/core/authorization/#role-based-access-control`) involves three main factors: *role, resource,* and *rights*. In the documentation on security, you see *rights* referred to as *actions, privileges,* and also *privilege actions*. To form a mental picture of a *role*, picture managing a server. The person who creates users and assigns filesystem rights assumes the *administrator* role. However, in a small company, this person could also manage the accounting department, and thus also assumes the role of *accounting manager*.

A *resource*, in the context of MongoDB security, is most often a database or collection. Resources could also be a server *cluster* or a built-in resource such as `anyResource`. The most complicated to understand are rights, thus we begin our discussion with privilege actions.

Understanding privilege actions (rights)

MongoDB privilege actions fall into two main categories: *operational* and *maintenance*. Rights in the operational category are documented under *Query and Write Actions*. Maintenance rights are documented starting with the *Database Management Actions* section. See the documentation on privilege actions (`https://docs.mongodb.com/manual/reference/privilege-actions/`) for more information.

It is extremely important to understand that rights assignments affect *database commands*. Each *mongo* shell method is actually a *wrapper* for the deeper infrastructural database command set. As an example, the database command `find` serves as a basis for the `db.collection.find()` and `db.collection.findOne()` *mongo* shell methods, and so forth. For more information, see the documentation on *MongoDB Database Commands* (`https://docs.mongodb.com/manual/reference/command/#database-commands`).

We first look at operational rights.

Operational rights

Operational rights are the set of rights most often assigned by DevOps, and are the ones most often used when writing application code. The four primary rights in this category are summarized in the following table. As you scan this table, please bear in mind that if a given database command is affected, it has a ripple effect on any *mongo* shell methods or programming language driver leveraging the given database command.

Right	Database Command(s) Affected
find	Allows the user assigned this right the ability to query the database. Database commands enabled by this assignment include `find`, `geoSearch`, and `aggregate`. Other less obvious commands also affected include `count`, `distinct`, `getLastError`, `mapReduce`, `listCollections`, and `listIndexes`, to name a few.
insert	When this right is assigned, the database user is able to add documents to a collection using the `insert` or `create` database commands. It is interesting to note that the `aggregation` database command is also affected if the `$out` pipeline stage operator is used.
remove	The primary database commands affected by this right `remove` and `findAndModify`. However, *mongo* collection-level shell methods such as `mapReduce()` and `aggregate()` are also affected if directed to write results to a collection if the collection already exists and replacements are made.
update	This affects the `update` and `findAndModify` database commands. The mongo shell `db.collection.mapReduce()` method is also affected when outputting results to a collection.

We now turn our attention to maintenance rights.

Maintenance rights

The list of maintenance rights is much more extensive than the operational rights listed in the previous section. In order to conserve space, a summary of the more important rights is presented here:

Right	User With This Assignment Can ...
changePassword	Change passwords of users.
createCollection, createIndex, createRole, createUser	Create collections, indexes, roles, or users. The user with createIndex can create indexes, the user with createUser can create users, and so on.
dropCollection, dropRole, dropUser	Drop a collection, role, or user.
grantRole, revokeRole	Assign roles (grantRole) or take a role away (revokeRole). The grantRole right is a very dangerous ability to assign. Please exercise caution and do not assign this right indiscriminately!
replsetConfigure	Make changes to the configuration of a replica set.
addShard, removeShard	Add or remove shards from a sharded cluster.
dropConnection, dropDatabase, dropIndex	Drop a connection, database, or index.
shutdown	Shut down a database.
listSession	List sessions of a *cluster* resource.
setFreeMonitoring	Enable or disable free monitoring on a *cluster* resource.
collStats, indexStats, dbStats	Gather statistics on collections, indexes, or databases.
listDatabases, listCollections, listIndexes	Get a list of databases, collections, or indexes.

Please note that this list is by no means comprehensive. We have covered barely a third of all possible actions. It would be well worth your while to consult the documentation on privilege actions (`https://docs.mongodb.com/manual/reference/privilege-actions/#privilege-actions`) for more information.

 Although there is no provision for creating your own privileges, you can create custom roles using the *mongo* shell `db.createRole()` method. You can assign any set of privileges to the custom role as needed. Any user assigned this role automatically inherits all privilege rights assigned.

The next subsection describes a series of built-in roles consisting of pre-defined sets of the rights described in the previous tables.

Understanding built-in roles

MongoDB provides *built-in roles* (`https://docs.mongodb.com/manual/reference/built-in-roles/#built-in-roles`) that represent preset combinations of rights. Some of the built-in roles are available on a specific database, others apply to any database. Management roles are available to enable individual users to perform backup and restore as well as replica sets and/or sharded clusters. Let's now examine built-in roles that apply to a specific database.

Assigning built-in roles applying to a single database

The `read` and `readWrite` built-in roles can be applied to the database in use or to the database designated specifically when creating a user. Here is a summary of these two roles:

- `read`: The `read` built-in role grants the following privilege actions (rights) to the database: `changeStream`, `collStats`, `dbHash`, `dbStats`, `find`, `killCursors`, `listIndexes`, and `listCollections`. It's important to note that there are minor differences between implementations in MongoDB versions 4.0 to 4.04 in that granting the `read` built-in role grants the `find` right. This right, in turn, enables the user to run `listDatabases`. As this might not be desirable, it might be necessary to deny the `listDatabases` right manually. For MongoDB versions 4.05 and above, an additional permission, `authorizedDatabases`, can be added.

- `readWrite`: The `readWrite` built-in role grants all the rights of the `read` built-in role, with the additional ability to write to the assigned database as well.

Effectively, these rights are granted: `collStats`, `convertToCapped`, `createCollection`, `dbHash`, `dbStats`, `dropCollection`, `createIndex`, `dropIndex`, `find`, `insert`, `killCursors`, `listIndexes`, `listCollections`, `remove`, `renameCollectionSameDB`, and `update`.

Let's look at an example using database-specific rights.

Creating a user with rights on a single database

For our first scenario, let's say that the management appoints an admin to run financial reports on the *biglittle* database. However, the database server also hosts another database to which this admin should *not* have rights. Here is an example of how the rights might be assigned:

```
bgReader = {
  user   : "bgReader",
  pwd    : "password",
  roles : [ { role:"read", db:"biglittle" } ],
  mechanisms: [ "SCRAM-SHA-256" ]
}
db.createUser(bgReader);
```

This user can now authenticate to the server using the *mongo* shell, specifying the –u and –p parameters. In addition the user needs to identify the authentication source using the ––`authenticationDatabase` parameter. As you can see from the screenshot shown here, the new user, bgReader, can access the *biglittle* database and issue the findOne() command:

```
File Edit View Search Terminal Help
ked@ked-VirtualBox:~$ mongo --quiet -u bgReader -p password --authenticationDatabase=biglittle
{"t":{"$date":"2020-05-24T06:18:31.717Z"},"s":"I",  "c":"NETWORK",  "id":4648601, "ctx":"main","
msg":"Implicit TCP FastOpen unavailable. If TCP FastOpen is required, set tcpFastOpenServer, tcp
FastOpenClient, and tcpFastOpenQueueSize."}
> use biglittle;
switched to db biglittle
> db.loans.findOne({},{"payments":0});
{
        "_id" : ObjectId("5dc4fd4722a7c0328fb3dcf1"),
        "loanKey" : "CLYDROJA7788_RUFULOZA8328_20190801",
        "borrowerKey" : "CLYDROJA7788",
        "lenderKey" : "RUFULOZA8328",
        "loanInfo" : {
                "principal" : NumberDecimal("18400.0000000000"),
                "numPayments" : 60,
                "annualRate" : NumberDecimal("6.77000000000000"),
                "effectiveRate" : NumberDecimal("0.00564166666666670"),
                "monthlyPymt" : NumberDecimal("362.348698431970")
        },
        "overpayment" : NumberDecimal("0")
}
>
```

However, if the same user attempts to insert something into the database, the operation fails and an authentication error appears, similar to that shown in the next screenshot:

```
File Edit View Search Terminal Help
> db.test.insertOne({"entry_date":"2020-06-01"});
uncaught exception: WriteCommandError({
        "ok" : 0,
        "errmsg" : "not authorized on biglittle to execute command { insert: \"test\", ordered:
true, lsid: { id: UUID(\"0a077bb7-99b9-45bf-8abc-4cda5034511d\") }, $db: \"biglittle\" }",
        "code" : 13,
        "codeName" : "Unauthorized"
}) :
WriteCommandError({
        "ok" : 0,
        "errmsg" : "not authorized on biglittle to execute command { insert: \"test\", ordered:
true, lsid: { id: UUID(\"0a077bb7-99b9-45bf-8abc-4cda5034511d\") }, $db: \"biglittle\" }",
        "code" : 13,
        "codeName" : "Unauthorized"
})
WriteCommandError@src/mongo/shell/bulk_api.js:417:48
executeBatch@src/mongo/shell/bulk_api.js:915:23
Bulk/this.execute@src/mongo/shell/bulk_api.js:1163:21
DBCollection.prototype.insertOne@src/mongo/shell/crud_api.js:264:9
@(shell):1:1
>
```

Likewise, if the `bgReader` user attempts any operation on another database, unless assigned rights to that database specifically, the operation fails. It might become too cumbersome to have to visit every single database, and create the same user over and over again. Accordingly, MongoDB also provides roles that apply to *all* databases, which we examine next.

Assigning built-in roles applying to all databases

The `read` and `readWrite` built-in roles can be applied to the database in use or to the database designated specifically when creating a user. Here is a summary of these two roles:

- `readAnyDatabase`: The `readAnyDatabase` built-in role grants the same rights as the `read` database right described earlier to *all* databases except for `local` and `config`, which are reserved internally used databases. If the user also needs rights to either of these, use the *admin* database, and make a specific `read` assignment to either one as appropriate.
- `readWriteAnyDatabase`: The `readWriteAnyDatabase` built-in role grants the same rights as the `readWrite` database right to *all* databases except for `local` and `config`, which are reserved internally used databases. If the user also needs rights to either of these, use the *admin* database, and make a specific `readWrite` assignment to either one as appropriate.

The next subsection looks at how to create a user with all database rights.

Creating a user with rights on all databases

For our next scenario, let's say that management wants a senior admin to have the ability to not only run financial reports on any hosted databases, but have the ability to perform create, update, and delete operations as well. Here is an example of how the rights might be assigned:

```
use admin;
rwAll = {
  user   : "rwAll",
  pwd    : "password",
  roles  : [ "readWriteAnyDatabase" ],
  mechanisms: [ "SCRAM-SHA-256" ]
}
db.createUser(rwAll);
```

This user can now authenticate to the server using the *mongo* shell, specifying the −u and −p parameters. As you can see from the following screenshot, the `rwAll` user can perform read and write operations on both the *sweetscomplete* and *biglittle* databases:

```
File  Edit  View  Search  Terminal  Help
ked@ked-VirtualBox:~$ mongo --quiet -u rwAll -p password
{"t":{"$date":"2020-05-24T05:33:20.762Z"},"s":"I",  "c":"NETWORK",  "id":4648601, "ctx":"main"
,"msg":"Implicit TCP FastOpen unavailable. If TCP FastOpen is required, set tcpFastOpenServer,
 tcpFastOpenClient, and tcpFastOpenQueueSize."}
> use sweetscomplete;
switched to db sweetscomplete
> db.test.insertOne({"entry_date" : "2020-06-01"});
{
        "acknowledged" : true,
        "insertedId" : ObjectId("5eca072564c988ab5175ec28")
}
>
> use biglittle;
switched to db biglittle
> db.users.findOne({},{"name":1});
{
        "_id" : ObjectId("5dc4fd41541e1aa5a62f7bbb"),
        "name" : {
                "title" : "Ms",
                "first" : "Charisse",
                "middle" : "X",
                "last" : "Ross",
                "suffix" : null
        }
}
>
```

Administrative built-in roles are examined in the next subsection.

Administrative built-in roles

The remaining set of built-in roles falls into the general category of administration. Here is a summary of the built-in most commonly assigned administrative roles:

- `dbAdmin`: This role gives a database user the ability to administer the database by gathering statistics and listing, creating, and dropping the assigned database, and all of its collections and indexes.
- `dbAdminAnyDatabase`: This is the same as `dbAdmin` except that it applies to all databases.
- `userAdmin`: The `userAdmin` role gives a database user the ability to manage users for the current database, including the ability to reset passwords, as well as creating, dropping, or modifying roles and database users.
- `userAdminAnyDatabase`: This is the same as `userAdmin` except that it applies to all databases.
- `dbOwner`: This is a composite role that combines `readWrite`, `dbAdmin`, and `userAdmin` for a single database.
- `clusterManager`: Gives a database user to list, add, or remove members of a sharded cluster or replica set.
- `clusterMonitor`: This role allows the assigned user to get statistics and performance metrics, on a read-only basis, from databases residing on sharded clusters or replica sets.
- `hostManager`: Grants a database user the rights associated with individual database server management, including database shutdown, re-synchronization, and rogue cursor management (such as complex queries gone astray).
- `backup`: As the name implies, this allows this database user to back up the database.
- `restore`: This gives the ability to restore the database from a backup.
- `root`: This role is a combination of the `readWriteAnyDatabase`, `dbAdminAnyDatabase`, `userAdminAnyDatabase`, `clusterAdmin`, `restore`, and `backup` roles. Exercise extreme caution when assigning this role!

Please remember that if the server is part of a replica set, the entire notion of backup and restore needs to be handled with extreme caution. In fact, if replication is properly managed, backup and restore is, in a sense, being performed continuously in real time, alleviating the need for a formal backup and restore operation.

Next, we look at what's involved in enabling SCRAM authentication.

Creating a user with SCRAM-based authentication

As mentioned in the previous subsection, SCRAM is the default authentication mechanism. The only thing you need to do is to identify the type of hash when creating a database user. When defining a new database user, add a `mechanisms` key, and set the value to either `SCRAM-SHA-1` or `SCRAM-SHA-256`. The SHA-1 hashing algorithm is no longer considered secure; however, you may need this option for backwards compatibility with older hashed passwords. The SHA-256 algorithm is recommended.

If using `SCRAM-SHA-1`, you can also set a value of `scramIterationCount`, which has a minimum value of `5000` and a default value of `10000`. The higher the value, the more secure is the hash. The higher the value, the longer it takes to calculate, so you'll need to arrive at a good balance between security and performance. The default is recommended.

Likewise, if using `SCRAM-SHA-256`, you can set a value of `scramSHA256IterationCount`, which has a minimum value of `5000` and a default value of `20000`. Here is an example creating a user using `SCRAM-SHA-256` with an iteration count of `10000`:

```
db.adminCommand( { setParameter: 1, scramSHA256IterationCount: 10000 } );
scramUser = {
  user  : "scramUser",
  pwd   : "password",
  roles : [ "readAnyDatabase" ],
  mechanisms: [ "SCRAM-SHA-256" ]
}
db.createUser(scramUser);
// set iteration back to default
db.adminCommand( { setParameter: 1, scramSHA256IterationCount: 15000 } );
```

Important: `db.adminCommand()` affects the entire database. In order to set the iteration count back to the default, you need to issue another `db.adminCommand()` as shown in the code sample in this subsection on creating a SCRAM user.

SHA documentation references: SHA-1 (`https://tools.ietf.org/html/rfc3174`), SHA-256 (`https://tools.ietf.org/html/rfc4634`)
MongoDB SCRAM references:
scramIterationCount (`https://docs.mongodb.com/manual/reference/parameters/#param.scramIterationCount`),
scramSHA256IterationCount (`https://docs.mongodb.com/manual/reference/parameters/#param.scramSHA256IterationCount`).

At this point, you have an idea of how MongoDB role-based access control works. The next topic is setting up secure communications using SSL/TLS and X.509 certificates.

Setting up transport layer security with X.509 certificates

In this section, you'll learn how to configure a `mongod` (MongoDB server daemon) instance to communicate with clients and peers in a secure manner. In this section, we discuss how to secure communications to and from a `mongod` instance. We also discuss securing communications between the *mongo* shell and a `mongod` instance. The sections that follow cover *who* can connect to the MongoDB database, and what that person (or *role*) is allowed to do.

As your database grows in size and you implement a *sharded cluster*, you can secure the `mongos` used to route requests to shards in exactly the same manner described in this section.

Before we get going on the details, however, we need to first examine what is meant by *transport layer*.

Understanding TLS, SSL and the transport layer

The *transport layer* in TCP/IP networks sits below the *application layer* (that is, TCP sits just below HTTP) and above the *internet* layer (that is, TCP sits above **Internet Protocol (IP)**). The two possibilities are either **Transmission Control Protocol (TCP)**, defined by RFC 793, or **User Datagram Protocol (UDP)**, defined by RFC 768. The main difference between these two is that TCP is slower, but provides *reliable transport*, in that there is a packet sequencing control aspect not present in UDP. UDP, on the other hand, although not considered *reliable*, is faster.

Secure Sockets Layer (SSL) was first introduced in the early 1990s by Netscape as a way to secure communications between browser and server (see RFC 8446 for more details). Its goal was to provide *privacy* and *data integrity* to secure transport (such as TCP). Version 1.0 was never published, but version 2.0 was first made available in 1995, followed a year later by version 3.0, defined by RFC 6101.

TCP/IP references: TCP (`https://tools.ietf.org/html/rfc793`) and UDP (`https://tools.ietf.org/html/rfc768`). RFC 6101 defines SSL version 3 (`https://tools.ietf.org/html/rfc6101`). RFC 8446 describes TLS 1.3 (`https://tools.ietf.org/html/rfc8446`).

Security vulnerabilities soon cropped up, however, and work began on a replacement for SSL, which became known simply as **Transport Layer Security** (**TLS**), starting with version 1.0 in 1999. TLS versions 1.0 and 1.1 are now considered compromised, however, and are expected to be deprecated in 2020. It is recommended that you use at a minimum version 1.2, and later versions if supported.

Starting with MongoDB version 4.0, support for TLS 1.0 is disabled. Also, as of MongoDB 4.0, native operating system libraries are used for TLS support. Thus, if your OS supports TLS 1.3, MongoDB automatically supports it as well. The libraries used are as follows:

- **Linux/BSD** uses *OpenSSL*.
- **Windows** uses *Schannel* (Secure Channel).
- **Mac** uses *Secure Transport*.

MongoDB provides automatic support for a number of extremely secure TLS algorithms, including **Ephemeral Elliptic Curve Diffie-Hellman** (**ECDHE**) and **Ephemeral Diffie-Hellman** (**DHE**). If you need to provide specific settings in order to influence these algorithms, a command-line parameter, `opensslCipherConfig`, is available. A complete discussion of all possible settings is well beyond the scope of this book.

For an excellent review of the foundation of end-to-end secure communications, read *End-to-End Arguments in System Design* by *Salzar, Reed,* and *Clark* (`http://web.mit.edu/Saltzer/www/publications/endtoend/endtoend.pdf`), made available by the MIT Computer Science Laboratory in 1980.

For more information on the latest version of TLS (version 1.3), see RFC 8446 (`https://tools.ietf.org/html/rfc8446`). RFC 8422 describes ECDHE: `https://tools.ietf.org/html/rfc8422`. RFC 8446 describes DHE: `https://tools.ietf.org/html/rfc8446`.

Documentation on *opensslCipherConfig* can be found here: `https://docs.mongodb.com/manual/reference/parameters/index.html#param.opensslCipherConfig`. Finally, for information on *TLS/SSL Ciphers* please see `https://docs.mongodb.com/manual/core/security-transport-encryption/#tls-ssl-ciphers`.

Now let's take a quick look at what is needed in terms of certificates.

Certificates needed for secure MongoDB communications

In order to secure communications between `mongod` or `mongos` instances, you need a *public key certificate* (also called an *X.509 certificate*). This is also often misleadingly referred to as an *SSL certificate*. It's best to use a certificate that is *signed* by a trusted **certificate authority (CA)**. However, if you are setting up a MongoDB database just for testing purposes, or if you are only developing inside a company network, a *self-signed* certificate might be sufficient.

 Self-signed certificates do not validate the server's identity, which opens the communication to man-in-the-middle attacks.

To generate a self-signed certificate, use the server's operating system SSL library. Here is an example script, found at `/path/to/repo/chapters/11/install_ssl_cert.sh`, that installs test certificates for both client and server:

```bash
#!/bin/bash
echo "Generating SSL certificates ..."
export RAND_DIGITS=`date |cut -c 18-20`
export EMAIL_ADDR="doug@unlikelysource.com"
mkdir /etc/.certs
cd /etc/.certs
echo $RAND_DIGITS >file.srl
touch /root/.rnd
openssl req -out ca.pem -new -x509 -days 3650 -subj \
    "/C=TH/ST=Surin/O=BigLittle/CN=root/emailAddress=$EMAIL_ADDR" \
    -passout pass:password
openssl genrsa -out server.key 2048
openssl req -key server.key -new -out server.req -subj  \
    "/C=TH/ST=Surin/O=BigLittle/CN=$HOSTNAME/emailAddress=$EMAIL_ADDR"
openssl x509 -req -in server.req -CA ca.pem -CAkey privkey.pem \
    -CAserial file.srl -out server.crt -days 3650 -passin pass:password
cat server.key server.crt > server.pem
openssl verify -CAfile ca.pem server.pem
openssl genrsa -out client.key 2048
openssl req -key client.key -new -out client.req -subj \
    "/C=TH/ST=Surin/O=BigLittle/CN=client1/emailAddress=$EMAIL_ADDR"
openssl x509 -req -in client.req -CA ca.pem -CAkey privkey.pem \
    -CAserial file.srl -out client.crt -days 3650 -passin pass:password
```

```
cat client.key client.crt > client.pem
openssl verify -CAfile ca.pem client.pem
chmod -R -v 444 /etc/.certs
```

For development purposes, an excellent article on how to create self-signed certificates can be found in *Appendix A* of the MongoDB documentation entitled *OpenSSL CA Certificate for Testing* (https://docs.mongodb.com/manual/appendix/security/appendixA-openssl-ca/).

Although self-signed certificates are fine for the development environment, it's vital that production servers have a certificate signed by a recognized and trusted *certificate authority*. There is an excellent, free, open source certificate authority known as *Let's Encrypt* (https://letsencrypt.org/) that is extremely widely used. The main downside to Let's Encrypt is that the certificates they generate are generally only good for 90 days, meaning you must put into place some mechanism to renew the certificate. The Let's Encrypt website offers more information on how this can be done.

Other alternatives would be to conduct a search for companies that offer SSL certificates for sale. Preferably your own web hosting provider might offer commercial SSL certificates that can be installed on your website. Let's now have a look at how to configure a mongod server instance to communicate using TLS.

Configuring mongod for TLS/SSL

Once you have a valid SSL certificate installed on your server, whether self-signed or generated by a recognized signing authority, you are in a position to secure the mongod server instance. If you are using self-signed certificates, please be aware that although communications to and from the server are encrypted, no identity validation takes place. This means the communications chain can potentially be subject to man-in-the-middle attacks.

Before you begin, please consider that if your database operates behind a firewall, and all of your mongod server instances communicate solely inside the company **Local Area Network (LAN)**, implementing TLS/SSL secure communications might introduce unnecessary performance degradation. By their very nature, TLS/SSL communications are slower as an extra handshake is needed to initiate the connection, and a cumbersome encryption/decryption process must take place for each communication.

No matter what operating system is used, you can use **Privacy-enhanced Electronic Mail (PEM)** *key files* to establish secure communications. For Windows or Mac servers, you have the option to configure secured communications using the operating system *certificate store*. In addition, by adding an additional configuration parameter, the server can force the client to present a certificate for identity verification.

In order to configure a `mongod` or `mongos` instance to use TLS, add the appropriate `net.tls` settings to the MongoDB `config` file (such as `/etc/mongod.conf`). For an overview, see the documentation page on *Procedures Using* `net.tls` *Settings*.

In MongoDB 4.0 and earlier, `net.tls` settings are not available. Instead, you need to use `net.ssl`. For more information on this set of configuration settings, providing identical functionality, please see the documentation on *Procedures Using* `net.tls` *Settings* (`https://docs.mongodb.com/manual/tutorial/configure-ssl/#procedures-using-net-tls-settings`).
RFC 7468 describes **Privacy-enhanced Electronic Mail (PEM)** key files: `https://tools.ietf.org/html/rfc7468`.

Please note that as an alternative to setting parameters in the MongoDB `config` file, you can also use the equivalent command-line switches. The following table summarizes useful `net.tls` settings and their equivalent command-line switches:

net.tls.<setting>	Command Line	Notes
mode	`--tlsMode`	Possible values include `disabled`, `requireTLS`, `allowTLS`, and `preferTLS`. `requireTLS` refuses any connection that is not using TLS. Use `requireTLS` if connections default to insecure, but where there might be some TLS connections involved. Use `preferTLS` if connections are secure by default, but where there might be some non-TLS connections involved. If your setup does not use TLS at all, use `disabled` (default).
certificateKeyFile	`--tlsCertificateKeyFile`	The value is a string representing the full path to the PEM file containing both the certificate and key. Cannot be used with the *certificateSelector* option.
certificateSelector	`--tlsCertificateSelector`	This option is only available on a Windows or Mac server and allows MongoDB to select the certificate from the local certificate store. Values under this option include `subject`, a string representing the name associated with the certificate in the store, and `thumbprint`, a hex string representing SHA-1 hash of the public key. See the next subsection for more details on this option. You cannot use this with the `certificateKeyFile` option.

certificateKeyFilePassword	`--tlsCertificateKeyFilePassword`	The value associated with this option is a string representing the password if the certificate is encrypted. If the certificate is encrypted and this option is not present, you are prompted for the password when the mongod (or mongos) instance first starts on Linux installations. If the certificate is encrypted and MongoDB is running on a Windows or Mac server, this option *must* be set. It's important to note that this password is removed from any MongoDB log files.
CAFile	`--tlsCAFile`	Use this option if you need to specify the root certificate chain from the Certificate Authority being used.
allowInvalidCertificates	`--tlsAllowInvalidCertificates`	If an invalid certificate is presented by the client, and this setting is t rue, a TLS connection is established, but client X.509-based authentication fails.
allowInvalidHostnames	`--tlsAllowInvalidHostnames`	If this setting is t rue, MongoDB disables verifying that the client hostname matches the one specified in the certificate.
disabledProtocols	`--tlsDisabledProtocols`	Options accepted by this parameter include none, TLS1_0, TLS1_1, TLS1_2, and TLS1_3. You can specify multiple protocols to disable by separating them with commas. MongoDB 4 automatically disables TLS 1.0 if TLS 1.1 or greater is available. You can only specify protocols that your OS and your version of MongoDB support. If you specify none, all protocols are allowed, including TLS 1.0 (now considered insecure).

Examples of these settings follow in the next subsections. Now let's examine how to use PEM key files.

For a complete list of net.tls settings, see the documentation on net.tls options (https://docs.mongodb.com/manual/reference/configuration-options/#net-tls-options). For a list of TLS command-line switches, see the documentation on TLS options (https://docs.mongodb.com/manual/reference/program/mongod/#tls-options).

Establishing secure communications using PEM key files

Once you have access to valid server certificate and key files, they are generally combined into a single file referred to as a PEM file. Unfortunately, there are a number of variants of the PEM format, so you'll have to consult the documentation for the server's operating system. As an example, assuming you ran the install_ssl_cert.sh script described earlier, let's add the following to the existing MongoDB config file:

```
net:
    port: 27017
```

```
bindIp: 0.0.0.0
tls:
    mode: requireTLS
    certificateKeyFile: /etc/.certs/server.pem
    CAFile: /etc/.certs/ca.pem
```

Before you restart the `mongod` instance, be sure to perform a proper shutdown as follows:

```
mongo admin
> db.shutdownServer();
> exit
```

Notice that after we restart the `mongod` instance, when we try to connect using a client that does not connect using TLS, the connection is rejected, as seen in the screenshot:

```
                        root@server1: /repo/chapters/11
root@server1:/repo/chapters/11# mongod -f /etc/mongod.conf &
[1] 462
root@server1:/repo/chapters/11# mongo
MongoDB shell version v4.4.0-rc13
connecting to: mongodb://127.0.0.1:27017/?compressors=disabled&gssapiServiceName=mongodb
Error: network error while attempting to run command 'isMaster' on host '127.0.0.1:27017'  :
connect@src/mongo/shell/mongo.js:362:17
@(connect):2:6
exception: connect failed
exiting with code 1
root@server1:/repo/chapters/11# ▮
```

If we subsequently modify the MongoDB `config` file, and change the setting for `mode` to `allowTLS`, a connection is allowed.

> For an excellent discussion on how to obtain a PEM file, please see the *StackOverflow* article entitled *How to get .pem file from .key and .crt files?* (https://stackoverflow.com/questions/991758/how-to-get-pem-file-from-key-and-crt-files).

Let's now look at configuring the client to enforce an end-to-end TLS connection.

Creating a client connection using TLS

Now that you know how to configure the database server to communicate using a secure TLS connection, it's time to have a look at how the client can connect over a TLS connection. For the purposes of this discussion, we focus on connecting using the *mongo* shell. The next chapter, `Chapter 12`, *Developing in a Secured Environment*, covers how to connect over a TLS connection using the PyMongo client and an X.509 certificate.

For command line *mongo* shell connections over a secure TLS connection, command-line switches are summarized in the following table:

Switch	Arguments	Notes
--tls	--	Causes the *mongo* shell to request a TLS handshake from the target `mongod` or `mongos` instance.
--tlsCertificateKeyFile	string	The value represents the full path to the PEM file containing the *client* certificate and key.
--tlsCertificateKeyFilePassword	string	The value associated with this option is a string representing the password if the client certificate is encrypted. If the certificate is encrypted and this option is not present, you are prompted for the password from the command line.
--tlsCAFile	string	Use this option if you need to specify the full path and filename for the Certificate Authority file on the server used to verify the client certificate.
--tlsCertificateSelector	string	This option is only available on a Windows or Mac server and allows the *mongo* shell to select the certificate from the local certificate store.

Here is an example connection using the test certificates described in *Appendix C* of the MongoDB security documentation in the article entitled *OpenSSL Client Certificates for Testing*. This example assumes the `hostname` is `server.biglittle.local`, the CA file is in the `/etc/.certs` directory, and the client certificate is in the home directory:

```
mongo --tls --tlsCertificateKeyFile /etc/.certs/client.pem  \
      --tlsCAFile /etc/.certs/ca.pem --host server1 --quiet
```

Here is a screenshot of the result:

```
root@server1: /repo/chapters/11
root@server1:/repo/chapters/11# mongo --tls --tlsCertificateKeyFile /etc/.certs/client.pem
  --tlsCAFile /etc/.certs/ca.pem --host server1 --quiet
> show dbs;
admin            0.000GB
biglittle        0.003GB
booksomeplace    0.005GB
config           0.000GB
local            0.000GB
sweetscomplete   0.002GB
>
```

If you still have role-based access control enabled, for the preceding examples to work you also need to add parameters to represent the username and password. To take the last example shown, here is how the client connection might be formulated:

```
mongo --tls --tlsCertificateKeyFile
/etc/.certs/client.pem  --tlsCAFile /etc/.certs/ca.pem
--host server1 -u superMan -p password
```

Summary

In this chapter, you learned how to secure a MongoDB database. This involves enabling security and providing a mechanism for both authentication and authorization. You learned how to create a database user and what database rights (or action privileges) are available, as well as built-in roles. You learned about built-in roles that affect a single database, or all databases, and also roles involved in maintenance. In the last section, you learned how to enable secure communications to and from a MongoDB database using TLS and X.509 certificates.

In the next chapter, you'll put this knowledge to use by writing program applications that require security.

12

Developing in a Secured Environment

In this chapter, you will learn how to write applications that access a database in which role-based access control has been implemented. In addition, this chapter shows you how to develop an application where the client-to-database communications are secured at the transport layer using x.509 certificates.

Building on the work started in the previous chapter, Chapter 11, *Administering MongoDB Security*, in this chapter, you will learn how to configure a Python application to use different database users with different sets of permissions. In addition, you will learn how to configure a Python application to communicate using an encrypted TLS connection using x.509 certificates.

The topics covered in this chapter include the following:

- Developing applications with minimal rights
- Creating applications with expanded rights
- Configuring applications to communicate using TLS

Technical requirements

The minimum recommended hardware is as follows:

- A desktop PC or laptop
- 2 GB free disk space
- 4 GB of RAM
- 500 KBPS or faster internet connection.

The software requirements are as follows:

- **OS (Linux or macOS)**: Docker, Docker Compose, and Git (optional)
- **OS (Windows)**: Docker for Windows and Git for Windows
- Python 3.x, a PyMongo driver, and Apache (already installed in the Docker container used for this book)
- An SSL certificate (already installed in the Docker container used for this book)

The installation of the required software and how to restore the code repository for the book is explained in Chapter 2, *Setting Up MongoDB 4.x*. To run the code examples in this chapter, open a Terminal window (Command Prompt) and enter these commands:

```
cd /path/to/repo
docker-compose build
docker-compose up -d
docker exec -it learn-mongo-server-1 /bin/bash
```

When you are finished working with the examples covered in this chapter, return to the Terminal window (Command Prompt) and stop Docker, as follows:

```
cd /path/to/repo
docker-compose down
```

The code used in this chapter can be found in the book's GitHub repository at https://github.com/PacktPublishing/Learn-MongoDB-4.x/tree/master/chapters/12.

Developing applications with minimal rights

One of the golden security rules of web application development is the principle of *least privilege*. According to the United States Department of Homeland Security, Cyber Infrastructure Division, this principle is defined as follows:

> *"Only the minimum necessary rights should be assigned to a subject that requests access to a resource and should be in effect for the shortest duration necessary (remember to relinquish privileges). Granting permissions to a user beyond the scope of the necessary rights of an action can allow that user to obtain or change information in unwanted ways. Therefore, careful delegation of access rights can limit attackers from damaging a system."*

> *– Taken from an article entitled Least Privilege as seen on the US-CERT website* (`https://www.us-cert.gov/bsi/articles/knowledge/principles/least-privilege`).

In the context of a MongoDB database application, this means assigning a role to a database user with the least amount of rights needed to perform the task. In this section, we will discuss how to set up the MongoDB database and configure an application to perform queries with the minimum amount of rights.

Creating a database user with read rights

When an application conducts queries only, there is no need to write anything to the database. Accordingly, we need to find a privilege action (that is, *right*) allowing a user to read from the database. The most likely right would be **find**. This right allows us to run the `pymongo.collection.find()` and `pymongo.collection.aggregate()` methods. This is perfect for an application that only needs to return the results of a query.

We do not assign this right directly to a user, however; instead, we select a **role** that includes this right, and assign the role to a database user. With this in mind, we will now consider the **read** built-in role. This role includes the **find** action privilege (or *right*), and also allows related operations, such as the ability to list collections and manage cursors (for example, the results of a query).

The database found in the Docker container used for this book is not secured in order to facilitate exercises in earlier chapters. In order for the examples in this chapter to work, you need to consult the previous chapter, `Chapter 11`, *Administering MongoDB Security*, and create the database the users referenced here and assign rights.

In order to create this user using the `mongo` shell, proceed as follows:

1. Log in as a database user who has the ability to create users and assign rights:

```
File  Edit  View  Search  Terminal  Help
ked@ked-VirtualBox:~$ mongo --quiet -u superMan -p
{"t":{"$date":"2020-05-29T00:56:43.606Z"},"s":"I",  "c":"NETWORK",  "id":4648601, "ctx":"main","msg
":"Implicit TCP FastOpen unavailable. If TCP FastOpen is required, set tcpFastOpenServer, tcpFastOp
enClient, and tcpFastOpenQueueSize."}
Enter password:
> show dbs;
admin            0.000GB
biglittle        0.000GB
config           0.000GB
local            0.000GB
sweetscomplete   0.002GB
test             0.000GB
>
```

2. Create a user document with the appropriate security settings.
3. In this example, the user document specifies a username of `biglittle_reader`. If you do not wish to directly enter the password, you can insert the JavaScript `passwordPrompt()` helper method, causing the `mongo` shell to prompt for a password when the document is accepted. The role assigned is `read`, only for the `biglittle` database. The password challenge uses an SHA-256 hash. You will also note that the user is restricted to `localhost` access only. This presumes that the application using this user is also running on the same server:

```
doc = {
    user: "biglittle_reader",
    pwd: passwordPrompt(),
    roles: [ { role: "read", db: "biglittle" } ],
    authenticationRestrictions: [ {
        clientSource: ["127.0.0.1"],
        serverAddress: ["127.0.0.1"]
    }],
    mechanisms: [ "SCRAM-SHA-256" ],
    passwordDigestor: "server"
}
```

4. When the `mongo` shell creates the JSON document described in the preceding code block, a password prompt appears when the *Enter* key is pressed:

```
File  Edit  View  Search  Terminal  Help
> doc = {
...    user: "biglittle_reader",
...    pwd: passwordPrompt(),
...    roles: [ { role: "read", db: "biglittle" } ],
...    authenticationRestrictions: [
...       {
...          clientSource: ["127.0.0.1"],
...          serverAddress: ["127.0.0.1"]
...       }
...    ],
...    mechanisms: [ "SCRAM-SHA-256" ],
...    passwordDigestor: "server"
... }
Enter password: ▮
```

5. You can then issue the db.createUser() command using the document
   described previously:

```
File  Edit  View  Search  Terminal  Help
> db.createUser(doc);
Successfully added user: {
        "user" : "biglittle_reader",
        "roles" : [
                {
                        "role" : "read",
                        "db" : "biglittle"
                }
        ],
        "authenticationRestrictions" : [
                {
                        "clientSource" : [
                                "127.0.0.1"
                        ],
                        "serverAddress" : [
                                "127.0.0.1"
                        ]
                }
        ],
        "mechanisms" : [
                "SCRAM-SHA-256"
        ],
        "passwordDigestor" : "server"
}
> ▮
```

The *find* database privilege: `https://docs.mongodb.com/manual/reference/privilege-actions/#find`.
The *read* built-in role: `https://docs.mongodb.com/manual/reference/built-in-roles/#read`.

Next, we will look at configuring the application to use the newly created database user.

Configuring the application for a database user

To configure an application to use the newly created database user, proceed as follows:

1. We start by creating a simple application that reads the name and address for the first user in the `biglittle` database:

```
import pprint
from pymongo import MongoClient
client = MongoClient('localhost');
find_result =
client.biglittle.users.find_one({},{"name":1,"address":1})
pprint.pprint(find_result)
```

2. As the database is now configured to require authentication, we expect this application to fail, as seen in the following screenshot:

```
File Edit View Search Terminal Help
ked@ked-VirtualBox:~/learn-mongodb/chapters/12$ python3 test_read_no_auth.py
Traceback (most recent call last):
  File "test_read_no_auth.py", line 11, in <module>
    find_result = client.biglittle.users.find_one({},{"name":1,"address":1})
  File "/home/ked/.local/lib/python3.6/site-packages/pymongo/collection.py", line 1273, in find_one
    for result in cursor.limit(-1):
  File "/home/ked/.local/lib/python3.6/site-packages/pymongo/cursor.py", line 1156, in next
    if len(self.__data) or self._refresh():
  File "/home/ked/.local/lib/python3.6/site-packages/pymongo/cursor.py", line 1073, in _refresh
    self.__send_message(q)
  File "/home/ked/.local/lib/python3.6/site-packages/pymongo/cursor.py", line 955, in __send_message
    address=self.__address)
  File "/home/ked/.local/lib/python3.6/site-packages/pymongo/mongo_client.py", line 1346, in _run_operation_with_response
    exhaust=exhaust)
  File "/home/ked/.local/lib/python3.6/site-packages/pymongo/mongo_client.py", line 1464, in _retryable_read
    return func(session, server, sock_info, slave_ok)
  File "/home/ked/.local/lib/python3.6/site-packages/pymongo/mongo_client.py", line 1340, in _cmd
    unpack_res)
  File "/home/ked/.local/lib/python3.6/site-packages/pymongo/server.py", line 136, in run_operation_with_response
    _check_command_response(first)
  File "/home/ked/.local/lib/python3.6/site-packages/pymongo/helpers.py", line 159, in _check_command_response
    raise OperationFailure(msg % errmsg, code, response)
pymongo.errors.OperationFailure: command find requires authentication
ked@ked-VirtualBox:~/learn-mongodb/chapters/12$
```

3. We then modify the initial client connection to include a reference to the database user with least privileges:

```
import pprint
from pymongo import MongoClient
client = MongoClient(
    'localhost',
    username='biglittle_reader',
    password='password',
    authSource='biglittle',
    authMechanism='SCRAM-SHA-256');
find_result =
client.biglittle.users.find_one({},{"name":1,"address":1})
pprint.pprint(find_result)
```

4. When this program is executed, the desired query is successfully executed, as shown in the following screenshot:

```
File  Edit  View  Search  Terminal  Help
ked@ked-VirtualBox:~/learn-mongodb/chapters/12$ python3 test_read_with_auth.py

Find Result:
{'_id': ObjectId('5dc4fd41541e1aa5a62f7bbb'),
 'address': {'buildingName': None,
             'city': 'Gaodhail',
             'country': 'GB',
             'floor': None,
             'geoSpatial': ['-5.9556', '56.5056'],
             'locality1': 'Scotland',
             'locality2': 'Argyll and Bute',
             'postalCode': 'PA72',
             'roomAptCondoFlat': None,
             'streetAddress': '1790 Short Woods Circle'},
 'name': {'first': 'Charisse',
          'last': 'Ross',
          'middle': 'X',
          'suffix': None,
          'title': 'Ms'}}
ked@ked-VirtualBox:~/learn-mongodb/chapters/12$
```

Let's now look at configuring an application to use a user with privileges restricted to a single database.

Limiting an application to a single database

In the following example, the database application is configured to use a user, bgReadWrite. This database user has limited access to the database as it is assigned the readWrite built-in role. Further, this user is only able to use the biglittle database. In the example you see in the next line, we create a MongoClient instance assigned the bgReadWrite user:

```python
import pprint
from pymongo import MongoClient
client = MongoClient(
    'localhost',
    username='bgReadWrite',
    password='password',
    authSource='biglittle',
    authMechanism='SCRAM-SHA-256');
```

All the data is first removed from test_collection in the biglittle database and a single document is inserted. This demonstrates the ability of the application to write:

```python
client.biglittle.test_collection.delete_many({})
insert_result = client.biglittle.test_collection.insert_one( {
    'first_name' : 'Fred',
    'last_name'  : 'Flintstone'
})
```

This is followed by a call to pymongo.collection.find_one(), demonstrating the ability to read, as seen here:

```python
find_result = client.biglittle.test_collection.find_one()
pprint.pprint(insert_result)
pprint.pprint(find_result)
```

The results of this successful operation are shown in the following screenshot:

```
File  Edit  View  Search  Terminal  Help
ked@ked-VirtualBox:~/learn-mongodb/chapters/12$ python3 test_read_write_biglittle.py

Insert Result:
<pymongo.results.InsertOneResult object at 0x7fb1df1749c8>

Find Result:
{'_id': ObjectId('5df46013069247257698cad0'),
 'first_name': 'Fred',
 'last_name': 'Flintstone'}
ked@ked-VirtualBox:~/learn-mongodb/chapters/12$
```

However, if in the same application we modify the code to access the
`sweetscomplete` database, with the same user, the results show a failure, as in the
following screenshot:

```
File  Edit  View  Search  Terminal  Help
pymongo.errors.OperationFailure: not authorized on sweetscomplete to execute command { delete:
 "test_collection", ordered: true, lsid: { id: UUID("93dd48b7-a61e-44dd-8095-49743fece532") },
 $db: "sweetscomplete", $readPreference: { mode: "primary" } }
ked@ked-VirtualBox:~/learn-mongodb/chapters/12$ python3 test_read_write_sweetscomplete.py ▮
```

Next, we will examine how to configure an application with expanded rights.

Creating applications with expanded rights

When discussing expanded rights, there are three aspects we must consider, expanding the
rights to include the following:

- Any database
- Database and collection administration
- Database user administration

All of these involve MongoDB role-based access control. In all of these cases, the primary
question you need to ask is *what is the application expected to do*? Let's first examine
expanding the rights to incorporate multiple databases.

Applications with rights to any database

Strictly following the principle of least privilege means not assigning a role that
encompasses more than one database. For practical reasons, however, the administrative
cost involved in creating and maintaining a large number of least-privilege database users
in every single database might be excessive. Another factor to consider is that the more
database users you need to create, the greater the chance of making a mistake is. Mistakes
could include assigning an incorrect mix of users, roles, and applications. This could not
only result in applications not working but could also cause an application to have far
greater rights than are merited. The greater security problem posed in such a situation
might outweigh the benefit of strict adherence to the principle of least privilege.

There are two primary built-in roles that allow a database user to read and/or write to all databases: `readAnyDatabase` and `readWriteAnyDatabase`, described in the previous chapter. Here is an example of a `mongo` shell command that creates a database user, `read_all`, assigned the `readAnyDatabase` role:

```
doc = {
  user: "read_all",
  pwd: "password",
  roles: [ "readAnyDatabase" ],
  mechanisms: [ "SCRAM-SHA-256" ]
}
db.createUser(doc);
```

The following code example is configured to use the `read_all` user. In this example, we retrieve the first document from the `users` collection in the `biglittle` database. Immediately after, you see a `find_one()` method call that retrieves the first document from the `customers` collection in the `sweetscomplete` database:

```
import pprint
from pymongo import MongoClient
client = MongoClient(
     'localhost',
     username='read_all',
     password='password',
     authSource='admin',
     authMechanism='SCRAM-SHA-256');
result_biglittle =
client.biglittle.users.find_one({},{"name":1,"address":1})
result_sweets    = client.sweetscomplete.customers.find_one({},
     {"firstName":1,"lastName":1,"city":1,"locality":1,"country":1})
pprint.pprint(result_biglittle)
pprint.pprint(result_sweets)
```

Here is a screenshot of the results of the test code proving that this application can read from multiple databases:

```
File  Edit  View  Search  Terminal  Help
ked@ked-VirtualBox:~/learn-mongodb/chapters/12$ python3 test_read_all.py

Find Result:
{'_id': ObjectId('5dc4fd41541e1aa5a62f7bbb'),
 'address': {'buildingName': None,
             'city': 'Gaodhail',
             'country': 'GB',
             'floor': None,
             'geoSpatial': ['-5.9556', '56.5056'],
             'locality1': 'Scotland',
             'locality2': 'Argyll and Bute',
             'postalCode': 'PA72',
             'roomAptCondoFlat': None,
             'streetAddress': '1790 Short Woods Circle'},
 'name': {'first': 'Charisse',
          'last': 'Ross',
          'middle': 'X',
          'suffix': None,
          'title': 'Ms'}}
{'_id': ObjectId('5dd36806834d9fa421215dc4'),
 'city': 'Washington',
 'country': 'US',
 'firstName': 'Patrina',
 'lastName': 'Yoder',
 'locality': 'District of Columbia'}
ked@ked-VirtualBox:~/learn-mongodb/chapters/12$
```

Please refer to the following links for more information:

- readAnyDatabase : https://docs.mongodb.com/manual/
  reference/built-in-roles/#readAnyDatabase
- readWriteAnyDatabase: https://docs.mongodb.com/manual/
  reference/built-in-roles/#readWriteAnyDatabase

Next, we will look at applications that perform database and collection administration.

Applications that perform database and collection administration

There are three primary roles, discussed in Chapter 11, *Administering MongoDB Security*, that allow the assigned database user to perform database administration:

- dbAdmin: Please refer to https://docs.mongodb.com/manual/reference/built-in-roles/#dbAdmin.

- dbOwner: Please refer to https://docs.mongodb.com/manual/reference/built-in-roles/#dbOwner.

- dbAdminAnyDatabase: Please refer to https://docs.mongodb.com/manual/reference/built-in-roles/#dbAdminAnyDatabase.

Their abilities include operations such as gathering statistics, running queries, and managing indexes, all the way up to dropping the database.

> It should be noted, however, that if the database is dropped, a database-specific user (for example, a user assigned either dbAdmin or dbOwner) would cease to exist, and accordingly, their rights would also no longer apply.

In this mongo shell script example, a user, biglittle_owner, is created and assigned the dbOwner role to the biglittle database:

```
doc = {
  user: "biglittle_owner",
  pwd: "password",
  roles: [ { role : "dbOwner", db : "biglittle" } ],
  mechanisms: [ "SCRAM-SHA-256" ]
}
db.createUser(doc);
```

We next create a demo program associated with this user, as follows:

1. The first thing to do is to define a pymongo.mongo_client.MongoClient instance:

```
import pprint
import pymongo
from pymongo import MongoClient
client = MongoClient(
    'localhost',
    username='biglittle_owner',
```

```
password='password',
authSource='biglittle',
authMechanism='SCRAM-SHA-256');
```

2. Next, in order to demonstrate the abilities of the application, we have it create two indexes, get a count of the number of loans, and gather statistics on the `loans` collections:

```
client.biglittle.loans.create_index([
    ('borrowerKey', pymongo.ASCENDING),('lenderKey',
pymongo.ASCENDING)])
count_result = client.biglittle.loans.count_documents({});
stats = client.biglittle.command('collstats', 'loans')
print("\nNumber of Documents: " + str(count_result))
pprint.pprint(stats)
```

3. Because the `biglittle_dbowner` user is assigned the `dbOwner` built-in role, the operations succeed, as shown in the following screenshot:

```
File Edit View Search Terminal Help
Number of Documents: 230
{'avgObjSize': 5843,
 'capped': False,
 'count': 230,
 'indexBuilds': [],
 'indexDetails': {'_id_': {'LSM': {'bloom filter false positives': 0,
                                   'bloom filter hits': 0,
                                   'bloom filter misses': 0,
                                   'bloom filter pages evicted from cache': 0,
                                   'bloom filter pages read into cache': 0,
                                   'bloom filters in the LSM tree': 0,
                                   'chunks in the LSM tree': 0,
                                   'highest merge generation in the LSM tree': 0,
                                   'queries that could have benefited from a Bloom filter that did not exist': 0,
                                   'sleep for LSM checkpoint throttle': 0,
                                   'sleep for LSM merge throttle': 0,
                                   'total size of bloom filters': 0},
                           'block-manager': {'allocations requiring file extension': 0,
                                             'blocks allocated': 0,
                                             'blocks freed': 0,
                                             'checkpoint size': 4096,
                                             'file allocation unit size': 4096,
                                             'file bytes available for reuse': 0,
                                             'file magic number': 120897,
                                             'file major version number': 1,
                                             'file size in bytes': 20480,
                                             'minor version number': 0},
                           'btree': {'btree checkpoint generation': 9,
                                     'column-store fixed-size leaf pages': 0,
                                     'column-store internal pages': 0,
                                     'column-store variable-size RLE encoded values': 0,
                                     'column-store variable-size deleted values': 0,
                                     'column-store variable-size leaf pages': 0,
                                     'fixed-record size': 0,
```

Finally, we will look at applications that involve the administration of database users.

Applications that perform database user management

As you can imagine, allowing an application to actually create database users and assign rights presents a potentially massive security risk. However, there may be situations in which you need to create a GUI frontend that allows customers to easily manage these aspects of their database. Accordingly, we will now turn our attention to the following built-in roles: userAdmin and userAdminAnyDatabase.

Please refer to the following links for more information:

- userAdmin: https://docs.mongodb.com/manual/reference/ built-in-roles/#userAdmin
- userAdminAnyDatabase: https://docs.mongodb.com/manual/ reference/built-in-roles/#userAdminAnyDatabase

It is much better to assign the application userAdmin rights to a specific database, rather than create an application assigned userAdminAnyDatabase. The only exception might be a situation where you need an application that can perform as a *superuser*. In such cases, it's extremely important to severely limit access to this application.

This is perhaps the most dangerous type of privilege level out of everything we have so far discussed. Please exercise extreme caution when creating applications able to create database users and to assign rights!

As in the previous sub-section, we will define a simple application to demonstrate the concept. From the mongo shell, we first define a user, biglittle_admin, assigned the dbOwner role, to the biglittle database. In this case, we do not wish to use userAdmin as the application also needs the ability to read the database as well. A user assigned the dbOwner role works well in this case because this role encompasses both read and userAdmin for a specific database:

```
doc = {
  user: "biglittle_owner",
  pwd: "password",
  roles: [ { role : "dbOwner", db : "biglittle" } ],
  mechanisms: [ "SCRAM-SHA-256" ]
}
db.createUser(doc);
```

We then create a demonstration command-line application that allows the administrator running the application to define a user assigned either the `read` or `readWrite` roles:

1. As with the other demo applications shown in this section, we first start by defining a client assigned to the new database user:

```
import sys
import pprint
from pymongo import MongoClient
client = MongoClient(
    'localhost',
    username    = 'biglittle_owner',
    password    = 'password',
    authSource = 'biglittle',
    authMechanism = 'SCRAM-SHA-256');
```

2. We then process the command-line arguments: `username`, `password`, and the letters `r` or `rw`. If the admin enters `rw`, the `readWrite` role is assigned. Otherwise, the `read` role is assigned:

```
if len(sys.argv) != 4 :
    print("Usage: test_db_admin.py <username> <password> <r|rw>\n")
    sys.exit()
if sys.argv[3] == 'rw' :
    newRole = 'readWrite'
else :
    newRole = 'read'
```

3. Finally, we build the arguments needed to create a user using the `pymongo.database.command()` method. Each command has its own arguments. These need to be identified using Python `**kwargs` (keyword arguments). In the example shown here, we are using four such arguments: `createUser`, `pwd`, `roles`, and `mechanisms`:

```
roleDoc = [ { "role" : newRole, "db" : "biglittle" } ]
mechDoc = [ "SCRAM-SHA-256" ]
result = client.biglittle.command("createUser",
createUser=sys.argv[1], pwd=sys.argv[2], roles=roleDoc,
mechanisms=mechDoc)
print ("\nResult:\n")
pprint.pprint(result);
```

Any commands issued using `pymongo.database.command()` ultimately map to the native MongoDB `db.runCommand()` method. For more information on database commands that can be executed using `db.runCommand()`, check out `https://docs.mongodb.com/manual/reference/command/#database-commands`.

4. From the command-line test run of the demo application, you can see the new user added to the database:

```
File Edit View Search Terminal Help
ked@ked-VirtualBox:~/learn-mongodb/chapters/12$ python3 test_db_admin.py "read_user_test_1" "password" "r"

Result:

{'ok': 1.0}
ked@ked-VirtualBox:~/learn-mongodb/chapters/12$
```

5. Returning to the `mongo` shell as a superuser, we can verify that the new user was added:

```
File Edit View Search Terminal Help
> use admin;
switched to db admin
> db.system.users.findOne({"user":"read_user_test_1"});
{
        "_id" : "biglittle.read_user_test_1",
        "userId" : UUID("d9f78179-a955-42e7-a5f7-fb46c5ec77f8"),
        "user" : "read_user_test_1",
        "db" : "biglittle",
        "credentials" : {
                "SCRAM-SHA-256" : {
                        "iterationCount" : 15000,
                        "salt" : "8q28TecdhBXBGjwCc7lKcPMTCqAvRzmr26Tu8Q==",
                        "storedKey" : "9dNge/GWIJQglfkE0qsWzDdltcvBkTU+FE+IhnJ77Zg=",
                        "serverKey" : "X9GSszNrjASyv/4c4OP7tu3AbMVAos/317S3ZI1UvdY="
                }
        },
        "roles" : [
                {
                        "role" : "read",
                        "db" : "biglittle"
                }
        ]
}
>
```

There is a PyMongo method, `pymongo.database.add_user()`, that would have been ideal for the task illustrated in this sub-section. In PyMongo version 3.9, however, this (and similar) methods have been deprecated in favor of the more generically useful `pymongo.database.command()` method.

In the next section, we will look at configuring applications to communicate using TLS.

Configuring applications to communicate using TLS

In order to have your applications communicate with the MongoDB database using a secured TLS connection, you first need to either generate or otherwise obtain the appropriate x.509 certificates. You then need to configure your MongoDB server to communicate using TLS, as described in `Chapter 11`, *Administering MongoDB Security*. In this section, we will examine how to configure the PyMongo client for TLS communications. First, let's look at the configuration options.

PyMongo client TLS configuration options

The TLS configuration options for `pymongo.mongo_client.MongoClient()` are in the form of keyword arguments (for example, `**kwargs`). Here is a summary of the possible key/value pairs:

Key	Value
`host`	Typically, the DNS address of the server running the `mongod` instance to which you want to connect. It's extremely important that the hostname specified here matches one of the DNS settings in the `[ alt_names ]` section of the configuration file used to generate the server certificate.
`tls`	Set this key to `True` if you wish to have the client connect via TLS. The default is `False`.
`tlsCAFile`	Specifies the certificate authority file, typically in PEM format.
`tlsCertificateKeyFile`	Specifies the file containing the client certificate and private key, signed by the **certificate authority** (**CA**) specified previously.
`tlsAllowInvalidCertificates`	If this option is set to `True`, it instructs MongoDB to allow the connection to be made regardless of certificate validity. This is useful for test and development, but not recommended for production.
`tlsAllowInvalidHostnames`	If this option is set to `True`, certificate validation is disabled. This is useful for test and development, but not recommended for production.
`tlsInsecure`	Implies `tlsAllowInvalidCertificates = True` and `tlsAllowInvalidHostnames = True`.

| tlsCertificateKeyFilePassword | If your certificate is encrypted with a password, the password can be placed here. |
| tlsCRLFile | Use this option if you using certificate revocation lists. |

Now that you have an idea of the possible TLS settings, let's look at configuring an application to communicate using TLS.

Sample app connecting with TLS

Here is an example of a client application connecting using TLS:

```
import pprint
from pymongo import MongoClient
client = MongoClient(
    username='read_all',
    password='password',
    authSource='admin',
    authMechanism='SCRAM-SHA-256',
    host='server.biglittle.local',
    tls=True,
    tlsCAFile='/etc/.certs/test-ca.pem',
    tlsCertificateKeyFile='/home/ked/test-client.pem');
result_biglittle =
client.biglittle.users.find_one({},{"name":1,"address":1})
pprint.pprint(result_biglittle)
```

Here is a screenshot showing the results of the test run:

```
File Edit View Search Terminal Help
ked@ked-VirtualBox:~/learn-mongodb/chapters/12$ python3 test_read_all_tls.py
Find Result:
{'_id': ObjectId('5dc4fd41541e1aa5a62f7bbb'),
 'address': {'buildingName': None,
             'city': 'Gaodhail',
             'country': 'GB',
             'floor': None,
             'geoSpatial': ['-5.9556', '56.5056'],
             'locality1': 'Scotland',
             'locality2': 'Argyll and Bute',
             'postalCode': 'PA72',
             'roomAptCondoFlat': None,
             'streetAddress': '1790 Short Woods Circle'},
 'name': {'first': 'Charisse',
          'last': 'Ross',
          'middle': 'X',
          'suffix': None,
          'title': 'Ms'}}
ked@ked-VirtualBox:~/learn-mongodb/chapters/12$
```

For this example, we modified `/path/to/repo/chapters/12/test_read_all.py` (described earlier in this chapter) to now incorporate TLS options. As you can see from the modified version of the file shown here, we configure the application to use the `read_all` user, assigned the `readAnyDatabase` role. TLS settings include enabling TLS, identifying the server, the location of the CA, and the client key.

> The Docker container used for this course already has a self-signed x.509 certificate installed. If you wish to experiment and create your own, an excellent article on how to create an x.509 client certificate based on a self-signed server certificate is available in the *Appendix* of the MongoDB security documentation. Be sure to go through all three appendices: A, B, and C.
>
> Use the results of *Article B, OpenSSL Server Certificates for Testing*, for the server configuration (`https://docs.mongodb.com/manual/appendix/security/appendixB-openssl-server/#appendix-server-certificate`). Use the results of *Article C, OpenSSL Client Certificates for Testing*, for the client (`https://docs.mongodb.com/manual/appendix/security/appendixC-openssl-client/`).

That concludes our discussion of configuring an application to communicate securely using x.509 certificates.

Summary

In this chapter, you learned how to configure an application to connect as a database user. As you learned, the security of your application is contingent upon which database user is selected and what roles are assigned to these users. You were shown examples with minimal rights, followed by another section demonstrating applications with expanded rights. In the last section of this chapter, you learned how to configure an application to connect over a TLS connection using x.509 certificates.

In the next chapter, you will learn how to safeguard your database by deploying a **replica set**.

13
Deploying a Replica Set

In this chapter, you will learn how to deploy a replica set. Initially, the focus of this chapter is the bigger picture: what a replica set is, how it works, and why use it. In the first section, you are given information to assist in developing a business rationale for deploying a replica set. Following that, you are shown how to use Docker to model a replica set for testing and development purposes. The remainder of the chapter focuses on configuration and deployment.

In this chapter, we will cover the following topics:

- Replication overview
- Using Docker to model a replica set
- Configuring replica set members
- Deploying a replica set

Technical requirements

The minimum recommended hardware is as follows:

- A desktop PC or laptop
- 2 GB free disk space
- 4 GB of RAM
- 500 Kbps or faster internet connection.

The software requirements are as follows:

- **OS (Linux or macOS)**: Docker, Docker Compose, and Git (optional)
- **OS (Windows)**: Docker for Windows and Git for Windows
- Python 3.x, a PyMongo driver, and Apache (already installed in the Docker container used for this book)

The installation of the required software and how to restore the code repository for this book is explained in `Chapter 2`, *Setting Up MongoDB 4.x*. To run the code examples in this chapter, open a Terminal window (Command Prompt) and enter these commands:

```
cd /path/to/repo/chapters/13
docker-compose build
docker-compose up -d
```

This brings up three Docker containers that are used as part of the demonstration replica set configuration. Further instructions on the replica set setup are given in this chapter. When you are finished working with the examples covered in this chapter, return to the Terminal window (Command Prompt) and stop Docker, as follows:

```
cd /path/to/repo/chapters/13
docker-compose down
```

The code used in this chapter can be found in the book's GitHub repository at `https://github.com/PacktPublishing/Learn-MongoDB-4.x/tree/master/chapters/13`.

MongoDB replication overview

As you will learn in this section, a **replica set** refers to three or more `mongod` instances hosting copies of the same databases. Although there are significant advantages to this arrangement, implementing a replica set incurs higher costs. Therefore, before you deploy a replica set, it's important to establish a solid business rationale. To start this discussion, let's have a look at what exactly a replica set is and what its potential benefits are.

What is a replica set?

A replica set is the concrete implementation of the larger topic of **replication**. Replication is the process of creating copies of a dataset. Each replica set consists of a **primary** and one or more **secondaries**. The recommended minimum is three `mongod` instances: one primary and two secondaries.

As you can see from the following diagram, the application client (for example, a Python application using the PyMongo client) conducts queries (reads) and saves data (writes) to the MongoDB database without being aware that a replica set is involved:

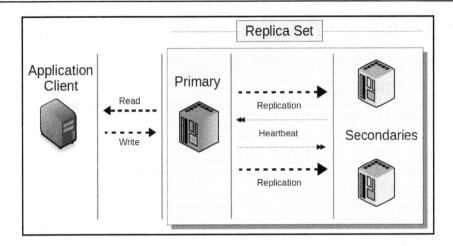

The read and write requests are by default accepted by the *primary*. Confirmation of availability between members of the replica set is maintained by the equivalent of a TCP/IP **ping**, referred to in the MongoDB documentation as a **heartbeat**. The primary maintains a log of performed operations called the **operations log (oplog)**, which is shared between replica set members. The process of **synchronization** involves secondaries performing operations based on the shared oplog.

You should recognize that the mongod instance could be running on another physical computer, or it could very well be running in a container in a cloud environment. For this reason, in this chapter, you will see references to both mongod *instances* as well as *servers*.

It's also important to note that replication is not a good solution if you need to scale up to handle massive amounts of data. For that situation, a better solution is to deploy a *sharded cluster*. This is discussed in more detail in Chapter 15, *Deploying a Sharded Cluster*.

Please refer to the following links for more information:

- **Replication**: https://docs.mongodb.com/manual/replication/#replication
- **Primary**: https://docs.mongodb.com/manual/core/replica-set-members/#primary
- **Secondaries**: https://docs.mongodb.com/manual/core/replica-set-members/#secondaries

Let's now have a look at the benefits of implementing a replica set.

What are some potential replica set benefits?

There are four primary benefits gained by deploying a replica set:

- **High availability**: In an ideal scenario, data is available 24/7 (which is 24 hours a day, 7 days a week). Realistically, however, electrical power and network connections go down occasionally (depending on where you live, perhaps more often than not!). Also, of course, servers crash and there are hardware failures. If, however, your data is replicated on other servers, possibly located in different geographic areas, the probability of the data being completely unavailable is considerably reduced.

- **Data redundancy**: Another aspect of a properly deployed replica set is that data is duplicated, thereby effectively providing a continuous online form of backup. Although it's still considered best practice to make regular offline backups, a properly designed replication strategy can help minimize the need to restore from a backup. The difference between data that is *replicated* versus data that is *backed up* is that replicated data is dynamic, always changing, and continuously available. In contrast, data that is backed up represents only a single snapshot in time and is immediately out of date the second the backup is completed.

- **Optimizing application performance**: Although by default the application is unaware of the underlying replica set, it is possible to add parameters to an application client connection, referred to as *read concerns* and *write concerns*, that influence which member of the replica is preferred to respond.

- **No downtime due to server maintenance**: An additional benefit to running a replica set is the ability to perform server maintenance with zero-downtime. If you want to upgrade MongoDB, you can do it in a rolling fashion without having to bring your application or database down.

 Replica sets are not designed as a load balancing solution! The main purpose of implementing a replica set is to ensure high availability (that is, failover handling). Trying to do load balancing using replica sets because you find that your read or write requests overwhelm a single server is not the ideal reason to implement a replica set. If you are in this situation and one of the servers in the replica set goes down, you're back to where you started!

Now, we can have a look at how a replica set responds when one of its member servers goes down.

What happens if a server goes down?

If a replica set member fails to respond to the regularly scheduled **heartbeat** within a predefined interval, the other members of the replica set assume it has failed and hold an **election** (https://docs.mongodb.com/manual/core/replica-set-elections/#replica-set-elections). A new primary is elected, which immediately starts accepting read and write requests and replicating its database. The following diagram illustrates the situation before and after an election:

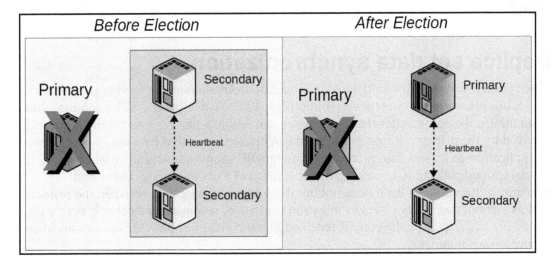

Each member server has one *vote*. On the left side of the preceding diagram, the primary has gone offline. The remaining secondaries in the replica set hold an election. After the election, as shown on the right side of the diagram, the topmost secondary is elected as the new primary.

In order to avoid a voting deadlock (for example, where an even number of remaining servers are split over who becomes the new primary), it is possible to establish **arbiters**. An arbiter is a server that is part of the replica set, has a vote, but does not retain a copy of the data. Through replica set configuration (discussed later in this chapter), it is possible to add *non-voting* replica set members. You can also assign members a *priority* to influence the election process and to make it proceed more quickly.

Please refer to the following links for more information:

- **Arbiters**: https://docs.mongodb.com/manual/core/replica-set-arbiter/index.html#replica-set-arbiter
- **Non-voting**: https://docs.mongodb.com/master/tutorial/configure-a-non-voting-replica-set-member/#configure-non-voting-replica-set-member

Now, let's have a closer look at how data is synchronized between replica set members.

Replica set data synchronization

When the replica set is first initialized, all the data from all member servers is copied to all the replica set members. The only exception is the `local` database, which contains, among other things, the oplog. After the initial synchronization, when read and write requests are received by the primary, it first performs the operation requested by a `mongo` shell instance or application code, and then records the MongoDB commands actually executed in a special capped collection, `local.oplog.rs`, referred to as the oplog. The actual synchronization process itself does not involve copying actual data between the primary and its secondaries. Rather, synchronization consists of secondaries receiving a copy of `local.oplog.rs`. Once the copy is received, the secondary applies the operations in an asynchronous manner.

Please refer to the following link for more information on oplogs: https://docs.mongodb.com/manual/core/replica-set-oplog/#replica-set-oplog.

Let's now have a look at replica set design.

Proper replica set design

As mentioned earlier, the ideal replica set would consist of a minimum of three computers. To give an example of a bad design, consider three old computers, each with an old, outdated OS, sitting on a shelf next to each other in the same room. As you can deduce from this simple example, the number of computers in a replica set is not the only consideration. What we need to look at now are positive suggestions on what would comprise a smoothly operating replica set.

The elements of a proper replica set design are listed here:

- **Identical host environments**: The *environment* could be a physical server, or it could be a virtualized host instance. The elements of the host environment would include an OS upon which a mongod instance is running and a suitable filesystem. What is important in this context is that the environment does not *have to be* identical for all replica set hosts. However, the more closely the environment of each member of the replica set resembles the other, the fewer problems you will have with replica set synchronization and performance.

- **Up-to-date OSes**: For the best security and performance, you need to ensure that the OS you choose is up to date. Please note that this does not necessarily mean that you always have the latest version of all components and drivers. In fact, in some cases, the latest version of one driver might not be compatible with the latest version of a software component you plan to use. Ultimately, you need to have a proper balance of up-to-date software where all the components and drivers are compatible with each other so that the OS performs smoothly. Also, very importantly, all security patches and bug fixes need to be applied as they become available.

- **Number of replica set members**: MongoDB recommends that any given replica set should have an *odd number* of members. The minimum recommended number of member servers is 3. The reason for this recommendation is that when the replica set first comes online, an *election* process takes place, resulting in one of the servers being chosen as the *primary*, and the others becoming secondaries. With an odd number of servers, a deadlock situation cannot occur.

- **Dispersed geographic location**: It's very possible that all members of a replica set might consist of virtualized instances all running on the same physical computer. When you stop and consider such a situation, however, it becomes obvious that if the host computer goes down, the entire replica set goes down, negating the benefits of having a replica set in the first place! Accordingly, it is recommended that you disperse replica set members to different locations. At a minimum, at least have each replica set member reside on a different physical computer! In the case of a company that has multiple offices in different cities, housing a replica set member in an office in each city might make sense. If network connections between offices suddenly drop, for example, the stranded replica set member switches to read-only mode. When connections are restored, another vote takes place and only one primary is once again elected. Its oplog is shared, and all database modifications are applied in their proper order.

If the majority of replica set members are not visible to a stranded member, this member switches to read-only mode. It could conceivably be converted to a standalone replica set, but this would have to be done manually, and could cause synchronization issues when connections are restored.

It would be impossible to *always* have an odd number of servers in the replica set, however, because you cannot predict exactly when, or how many, servers might go offline. Accordingly, other factors can be built into your replica set configuration that influence an election in order to prevent a deadlock.

One such factor that can be set manually in the configuration is **priority**. Other information that comes into play is performance metrics, to which the replica set members have access. For more information on the election process, see the *Replica Set Elections* article in the MongoDB documentation.

Please refer to the following links for more information:
Replica set dispersal: `https://docs.mongodb.com/manual/tutorial/deploy-replica-set/#requirements`.
Replica set elections: `https://docs.mongodb.com/manual/core/replica-set-elections/#replica-set-elections`.

As you can imagine, hosting your database on a replica set can become potentially quite expensive. The next sub-section shows you how to establish a solid business rationale.

Establishing a business rationale

By this point, you are most likely convinced that implementing your database using a replica set makes good sense. In order to justify the purchase of the extra resources required to support a replica set, however, you need to determine the following:

- **What is the cost of downtime?**: A related question to ask is *how much money is lost if the database goes down*? Unfortunately, you cannot give a blanket answer to this question. You need to do further research to find out how much your business makes, on average, every second. This might seem like a ridiculous figure to have to calculate, but consider this: downtime is usually (hopefully!) measured in *seconds*. So, if your server goes down for 30 seconds, you need to know, on average, how much money is lost.

- **How important is uninterrupted service?**: An often overlooked hidden cost is what is informally described as the **customer irritation factor**. When the database goes down or is unavailable for some reason, any operation a customer is performing that requires database access is interrupted. This often results in the customer having to start all over, raising the amount of irritation they experience. If such a loss of service occurs frequently, eventually, the customer irritation factor rises to a level sufficient to drive them away from your website, and right into the arms of your competitor! This means a loss of business: a serious consequence, the cost of which needs to be factored into the business rationale for implementing a replica set.

- **What is the value of your data?**: Another factor to consider is how much time (which eventually translates into money) is lost on the part of both customers and staff when data is lost. Staff need to spend time (costing the company money) restoring the data. Customers might lose money and time because any backup is not up-to-the-second, which means the customer needs to back-track and fill in the gap between when the backup took place up to the present moment.

- **How would a loss of reputation affect your business?**: Excessive downtime, loss of service, and loss of data can all add up to yet another hidden cost: loss of reputation. Eventually, word gets around the customer community that your services are unreliable, ultimately resulting in a loss of business, which in turn translates to lost revenue. Furthermore, you gain new customers more slowly, if at all, due to a loss of reputation. This is a long-term consideration as it takes quite a bit of time and effort to restore a company's reputation.

In most cases, the amount of money associated with these four factors outweighs the costs involved in implementing a replica set. For a start-up company, however, where cash and resources are tight, it might not make business sense to invest in the extra resources needed for a replica set. Also, for companies that offer a public service, or for non-profit organizations, the considerations listed previously might not matter as customers are not paying money for the service, which means that any downtime, although not appreciated, is considered acceptable (or even expected!) to website visitors.

Next, we will have a look at how a developer can model a replica set using Docker.

Modeling a replica set using Docker

A classic problem facing MongoDB developers is the lack of enough hardware to properly set up and test a replica set. Furthermore, even though some developers either have the requisite hardware or have access to a properly equipped lab, the computers available are often old and dissimilar, leading to inconsistent results during development and testing. One possible solution to this problem is to use virtualization technology to model a replica set.

 As of the time of writing, **MongoDB Atlas** offers a free sandbox tier that provides a 3-member replica set across availability zones and 500 MB of storage. For more information, refer to `https://www.mongodb.com/cloud/atlas`.

The way to start this setup is to create a **Dockerfile** for each member of the replica set, examined next. You can find these files in `/path/to/repo/chapters/13`.

Creating the Dockerfile

In order to build a Docker container, you generally need two things: an image and a file that describes a series of commands that populates the image with the software needed to perform the functions required of the new container. In this case, we're using `mongo`, the official Docker image published by MongoDB (`https://hub.docker.com/_/mongo`).

In the Dockerfile used for replica members, the `FROM` directive tells Docker what image to use as a basis for the new container. Next, we use the `RUN` directive to install the custom software packages needed for our replica set member. We first update `apt-get` and perform any upgrades needed. After that, a common set of network tools is installed, including `ping`:

```
FROM mongo
RUN \
  apt-get update && \
  apt-get -y upgrade && \
  apt-get -y install vim && \
  apt-get -y install inetutils-ping && \
  apt-get -y install net-tools
```

Here is the generic `mongod.conf` file that is linked to the `/etc` folder of the new container (MongoDB configuration is discussed in `Chapter 2`, *Setting Up MongoDB 4.x*):

```
systemLog:
   destination: file
   path: "/var/log/mongodb/mongodb.log"
   logAppend: true
storage:
   dbPath: "/data/db"
   journal:
      enabled: true
net:
   bindIp: 0.0.0.0
   port: 27017
replication:
   replSetName: learn_mongodb
```

Here is the `hosts` file, whose entries are later linked to `/etc/hosts` inside the Docker container:

```
127.0.0.1     localhost
::1     localhost ip6-localhost ip6-loopback
fe00::0     ip6-localnet
ff00::0     ip6-mcastprefix
ff02::1     ip6-allnodes
ff02::2     ip6-allrouters
172.16.0.11     member1.biglittle.local
172.16.0.22     member2.biglittle.local
172.16.0.33     member3.biglittle.local
```

 For more information on the makeup of a `hosts` file, see the entry for `/etc/hosts` on the *Ubuntu* manpage (`http://manpages.ubuntu.com/manpages/focal/man5/hosts.5.html`).

Next, we will examine how to bring up multiple containers using **Docker Compose**.

Bringing up multiple containers using Docker Compose

Docker Compose is a scripting language designed to facilitate creating, running, and maintaining Docker containers. The main configuration file is `docker-compose.yml`. It is in **YAML (YAML Ain't Markup Language)** format, meaning key directives are followed by a colon (`:`). Sub-keys are indented directly underneath parent keys.

Here is the `docker-compose.yml` file used to bring up three `mongod` instances that simulate a replica set. This file is broken down into five main blocks: three to define each of the replica set members, a fourth to describe the virtual network, and a fifth block to describe volumes. At the top of the file, we place the `version` directive. This informs Docker Compose that this file follows directives compatible with version 3:

```
version: "3"
```

We then start with the definition for `member1`. In this definition, you will see the Docker container name, the hostname, the source image, port mappings, the location of the Dockerfile, and network information.

 The command appends an `-f /etc/mongod.conf` argument to the container startup script. In the case of the `mongo` base Docker image, its startup script launches a `mongod` instance. Thus, Docker Compose causes the `mongod` instance to run using `/etc/mongod.conf` as its config file. See the documentation at `https://hub.docker.com/_/mongo`.

The `networks` directive is used to assign an IP address to the container. Under `volumes`, you can see the permanent location of the data, as well as mappings to a common `/etc/hosts` file and `/etc/mongod.conf` file:

```
services:
  learn-mongodb-replication-member-1:
    container_name: learn-mongo-member-1
    hostname: member1
    image: mongo/bionic:latest
    volumes:
    - db_data_member_1:/data/db
    - ./common_data:/data/common
    - ./common_data/hosts:/etc/hosts
    - ./common_data/mongod.conf:/etc/mongod.conf.
    ports:
    - 27111:27017
    build: ./member_1
    restart: always
    command: -f /etc/mongod.conf
    networks:
      app_net:
        ipv4_address: 172.16.0.11
```

 We have included a common volume, `/data/common`, which points to the `/path/to/repo/chapters/13/common_data` folder. This allows us to copy the sample data from the host computer, making it available to any of the member servers.

The definition blocks for `member2` and `member3` (not shown) are identical except that references to `1` are changed to `2` or `3`. Finally, the last two definition blocks describe the virtual network and volumes. The `volumes` directive represents external directories on the host computer so that when the simulated replica set is brought up or down, data is retained. The `networks` directives describe the overall network into which each member server fits, as shown in the following code snippet:

```
networks:
  app_net:
    ipam:
      driver: default
      config:
        - subnet: "172.16.0.0/24"

volumes:
  db_data_member_1: {}
  db_data_member_2: {}
  db_data_member_3: {}
```

We can then use the `docker-compose up` command to build (if necessary) and bring the three containers online, as shown in the following screenshot:

```
File  Edit  View  Search  Terminal  Help
jed@jed:~/Repos/learn-mongodb/chapters/13$ docker-compose up
Starting learn-mongo-member-3 ... done
Starting learn-mongo-member-1 ... done
Starting learn-mongo-member-2 ... done
Attaching to learn-mongo-member-3, learn-mongo-member-1, learn-mongo-member-2
```

Now that you have an idea of how to model a replica set using Docker, let's turn our attention to replica set member configuration and deployment.

Replica set configuration and deployment

In order to deploy the replica set, the `mongod.conf` file for each `mongod` instance needs to be updated with the appropriate replication settings. In this section, we will examine important settings in the `mongod.conf` files for each type of replica set member (for example, primary and secondary). After that is an overview of the steps needed to get a replica set up and running.

Replication configuration settings

Although you could use command-line options to bring up members of a replica set, aside from temporary testing and development, the preferred approach would be to add the appropriate replication options to the `mongod.conf` file for each member of the replica set.

The following table summarizes the more important options:

mongod.conf Directive	Command Line	Notes
`net.bindIp`	`--bind_ip`	The `mongod` instance running on each replica set member must be configured to listen either on a specific IP address or all IP addresses. This is required so that replica set members can both synchronize data and send each other *heartbeat* transmissions.
`replication.oplogSizeMB`	`--oplogSize`	If desired, define this setting with an integer value representing the desired oplog size in megabytes. In order for this directive to be effective, the option needs to be set before the replica set is first initialized. The `replSetResizeOplog` administrative method can be used to dynamically resize the oplog later as the replica set is running (described in the next sub-section).
`replication.replSetName`	`--replSet`	This mandatory directive indicates the name of the replica set. The value for this directive needs to be the same on all replica set members.
`replication.enableMajorityReadConcern`	`--enableMajorityReadConcern`	For most situations, avoid disabling support for the majority read concern. The default for this setting is `true`. If you wish to disable this support, configure this setting to a value of `false`.
`storage.oplogMinRetentionHours`	`--oplogMinRetentionHours`	By default, MongoDB does not set a minimum time to retain oplog entries. Introduced in MongoDB 4.4, this setting allows you to override the default behavior and force MongoDB to retain oplog entries for a minimum number of hours.

The main reason why majority read concern support might need to be disabled is in a situation where MongoDB *cache pressure* (work load) could end up immobilizing a deployment in an endless cycle of cache reads and writes. Have a look at this Stack Exchange article, *Why did MongoDB reads from disk into cache suddenly increase?*, for a good idea of what cache pressure appears in terms of performance metrics (`https://dba. stackexchange.com/questions/250939/why-did-mongodb-reads-from-disk-into-cache-suddenly-increase`).

Another consideration, not discussed in this chapter, is to configure replica set members to communicate internally using a secured (for example, TLS) connection. Only use a TLS connection between replica set members if internal communications absolutely need to be secured, as it slightly degrades database performance. Otherwise, use appropriate network and routing techniques to isolate the network segment used by replica set members to communicate with each other. Please refer to the MongoDB documentation on *Internal Membership Authentication* (`https://docs.mongodb.com/manual/core/security-internal-authentication/#internal-membership-authentication`). Also, for more information, refer to `Chapter 11`, *Administering MongoDB Security*.

Please refer to the following links for more information:

- **Replication options**: `https://docs.mongodb.com/manual/reference/configuration-options/index.html#replication-options`
- **Majority read concern**: `https://docs.mongodb.com/manual/reference/read-concern-majority/#read-concern-majority`

Now, it's time to have a look at the steps required to deploy a replica set.

Deploying the replica set

The overall procedure to get a replica set up and running is as follows:

1. Set up and test the network connectivity between the members of the replica set.
2. Configure the replication settings in the `mongod.conf` file for each member.
3. Start the `mongod` instances on each server in the replica set one by one.
4. Initiate the replica set via the `mongo` shell.
5. Confirm the replica set status.
6. Restore or upload database documents as needed.
7. Troubleshoot, reconfigure, and restart as needed.

For more information on testing connections, refer to `https://docs.mongodb.com/manual/tutorial/troubleshoot-replica-sets/#test-connections-between-all-members`.

Let's start by confirming network connectivity between replica set members.

Network connectivity between replica set members

For the purposes of this chapter, as described earlier, we will use `docker-compose up` to bring the replica set members online. As our replica set consists of Docker containers, we can create a simple shell script (`/path/to/repo/chapters/13/verify_connections.sh`) to confirm connectivity between members:

```bash
#!/bin/bash
docker exec learn-mongo-member-1 /bin/bash -c "ping -c1
member2.biglittle.local"
docker exec learn-mongo-member-2 /bin/bash -c "ping -c1
member3.biglittle.local"
docker exec learn-mongo-member-3 /bin/bash -c "ping -c1
member1.biglittle.local"
```

Here is the expected output from this sequence of `ping` commands:

```
File  Edit  View  Search  Terminal  Help
jed@jed:~/Repos/learn-mongodb/chapters/13$ ./verify_connections.sh
PING member2.biglittle.local (172.16.0.22): 56 data bytes
64 bytes from 172.16.0.22: icmp_seq=0 ttl=64 time=0.083 ms
--- member2.biglittle.local ping statistics ---
1 packets transmitted, 1 packets received, 0% packet loss
round-trip min/avg/max/stddev = 0.083/0.083/0.083/0.000 ms
PING member3.biglittle.local (172.16.0.33): 56 data bytes
64 bytes from 172.16.0.33: icmp_seq=0 ttl=64 time=0.083 ms
--- member3.biglittle.local ping statistics ---
1 packets transmitted, 1 packets received, 0% packet loss
round-trip min/avg/max/stddev = 0.083/0.083/0.083/0.000 ms
PING member1.biglittle.local (172.16.0.11): 56 data bytes
64 bytes from 172.16.0.11: icmp_seq=0 ttl=64 time=0.084 ms
--- member1.biglittle.local ping statistics ---
1 packets transmitted, 1 packets received, 0% packet loss
round-trip min/avg/max/stddev = 0.084/0.084/0.084/0.000 ms
jed@jed:~/Repos/learn-mongodb/chapters/13$
```

If it appears that the members are not able to ping each other, confirm that the network configuration in the `docker-compose.yml` file is correctly stated. Also, confirm that Docker is installed correctly on the development server, and that Docker has set up its own internal virtual network. Next, we will quickly revisit configuring replica set members.

Replica set member configuration

We have already examined replication settings in detail in the *Replication configuration settings* section of this chapter. As we are now ready to deploy the replica set, a simple Docker command shows the current `mongod.conf` file settings. As a minimum, we need to ensure that database is listening on the correct IP address (or all IP addresses, as seen in the preceding screenshot). We also need to confirm that all replica set members have the `replication.replSetName` parameter set to the same name. The next screenshot shows the result of the following Docker command:

```
docker exec -it learn-mongo-member-1 /bin/bash -c "cat /etc/mongod.conf"
```

Here is the result of the preceding command:

```
File Edit View Search Terminal Help
jed@jed:~$ docker exec -it learn-mongo-member-1 /bin/bash -c "cat /etc/mongod.conf"
systemLog:
    destination: file
    path: "/var/log/mongodb/mongodb.log"
    logAppend: true
storage:
    dbPath: "/data/db"
    journal:
        enabled: true
net:
    bindIp: 0.0.0.0
    port: 27017
replication:
    replSetName: learn_mongodb
jed@jed:~$
```

Assuming all the replica set members have the appropriate configuration, we will now have a look at starting (or restarting) the `mongod` instances involved in the replica set.

Starting mongod instances

At this point, you can start the `mongod` instance on each server according to the recommended procedure for the host OS. Thus, on a Windows server, for example, the `mongod` instance would be started as a service using the Windows Task Manager or an equivalent GUI option. In a Linux environment, you could simply type the following:

```
mongod -f /etc/mongod.conf
```

As mentioned earlier, when starting the `mongod` instance manually (or using a shell script), you can specify replication options as command-line switches. The preferred approach, however, is to place replication settings into the `mongod.conf` (or equivalent) file on each member of the replica set.

For the purposes of this chapter, as mentioned previously, we will start the three Docker containers using the following command:

```
docker-compose up
```

The result is shown here:

```
File  Edit  View  Search  Terminal  Help
jed@jed:~/Repos/learn-mongodb/chapters/13$ docker-compose up
Starting learn-mongo-member-3 ... done
Starting learn-mongo-member-1 ... done
Starting learn-mongo-member-2 ... done
Attaching to learn-mongo-member-3, learn-mongo-member-1, learn-mongo-member-2
```

We are now ready to *initiate* the replica set.

Initiating the replica set

Initiating a replica set involves connecting to one of the member servers in the replica using the `mongo` shell. If database security has been activated, be sure to log in as a user with the appropriate *cluster administration role*. The key command used to initiate the replica set is `rs.initiate()`. The generic syntax is as follows:

```
rs.initiate( <DOCUMENT> );
```

The `rs.initiate()` shell method is actually a wrapper for the `replSetInitiate` database command. If preferred, you can directly access this command using the following syntax from the shell (or from programming application code):

```
db.runComand( replSetInitiate : <DOCUMENT> )
```

The replica set configuration document supplied as an argument is in JSON format and can contain any of the following replica set configuration fields. We will first summarize the main replica set configuration fields.

Please refer to the following links for more information:

- **Cluster administration role**: `https://docs.mongodb.com/manual/reference/built-in-roles/#cluster-administration-roles`
- **Replica set configuration fields**: `https://docs.mongodb.com/manual/reference/replica-configuration/#replica-set-configuration`

Replica set configuration fields

Here is a summary of the top-line replica set configuration fields:

Directive	Data Type	Notes
`_id`	String	Replica set name; corresponds to the `replication.replSetName` parameter in the `mongod.conf` file.
`version`	Integer	Represents a value designed to increment as the replica set is revised
`protocolVersion`	Number	As of MongoDB version 4.0, the **protocol version** must be set to 1 (default). This setting is in place so that future enhancements to the replication process can be made backward-compatible.
`writeConcernMajorityJournalDefault`	Boolean	The default value of `true` instructs MongoDB to acknowledge the write operation after a majority of replica set members have written a write request to their respective **journals**. All replica set members need to have *journaling* active if this value is set to **true**. However, if any member uses the **in-memory storage engine**, the value must be set to `false`.
`configsvr`	Boolean	If this replica set is used to host the config server for a **sharded cluster**, set the value to `true`. The default for this setting is `false`.
`members`	Document	JSON sub-document describing each member server in turn.
`settings`	Document	JSON sub-document describing additional settings.

Please refer to the following links for more information:

- **Protocol version**: https://docs.mongodb.com/master/reference/replica-set-protocol-versions/#replica-set-protocol-version
- **Journaling**: https://docs.mongodb.com/master/core/journaling/index.html#journaling
- **In-memory storage engine**: https://docs.mongodb.com/master/core/inmemory/#in-memory-storage-engine

For more information on *config servers* and *sharded clusters*, please refer to `Chapter 15`, *Deploying a Sharded Cluster*.

As you will have noticed, both `members` and `settings` are themselves JSON documents with their own settings. Let's now see an example of the `members` settings.

Replica set members configuration directives

Here is a summary of the directives available in the `members` sub-document. As a minimum, you need to specify the `_id` and host for each replica set member:

Directive	Data Type	Notes
`_id`	Integer	Represents an arbitrary unique identifier for this member within the replica set. Values can range from 0 to 255 (mandatory).
`host`	String	Either the hostname or `host name:port` for this replica set member. This hostname must be *resolvable* (for example, reachable using `ping`) (mandatory).
`arbiterOnly`	Boolean	Set this value to `true` if you wish this replica set member to serve as an arbiter (does not have a copy of the database, but participates in elections in order to help prevent a deadlock). The default is `false`.
`buildIndexes`	Boolean	Set this value to `false` if you do not wish to have indexes built on this replica set member. Bear in mind that this is only in cases where this member only serves a backup/high-availability purpose and does not receive any client queries. If you choose to set this value to `false`, the MongoDB team recommends also setting `members.priority` to 0 and `members.hidden` to `true`. The default for this setting is `true`.
`hidden`	Boolean	Set this value to `true` if you do not want this replica set member to receive queries. The default is `false`.

priority	Number	Values can range from 0 to 1000. Use this setting to influence the eligibility of this replica set member for becoming a *primary* if an election occurs. The higher the value, the greater the chance this member is elected the primary. The default value is 0 for an arbiter and 1 for primary/secondaries.
tags	Document	You can define a JSON document in the form { "tag":"value" [, "tag":"value"] }, allowing you to sub-group replica set members. Tags can then be used to influence read and write operations (see the *Application program code access* section later in this chapter).
slaveDelay	Integer	This value represents the number of seconds that updates to this replica set member should lag. Use this setting to create a **delayed replica set member**. This allows you to store active data representing some time period in the past. The default value is 0 (that is, no delay).
votes	Number	Represents the number of votes this replica set member can produce in the case of an election. Set this value to 0 to establish a *non-voting member*. Note that non-voting members must also have their members.priority value set to 0. The default value for members.votes is 1.

For more information on arbiters, please refer to the following link: https://docs.mongodb.com/manual/core/replica-set-arbiter/ #replica-set-arbiter.

Finally, let's have a look at the settings directives.

Replica set settings configuration directives

Here is a summary of the directives available in the settings sub-document:

Directive	Data Type	Notes
chainingAllowed	Boolean	Set this value to false if you wish secondary replica set members to replicate *only* from the primary. The default value is true, meaning secondaries can update each other, referred to as **chained replication**.
heartbeatIntervalMillis	Integer	This value is set internally.
heartbeatTimeoutSecs	Integer	The number of seconds allowed without receiving a heartbeat before a replica set member is considered unreachable. The default value is 10.

electionTimeoutMillis	Integer	The number of milliseconds allowed to elapse without receiving a heartbeat before an election is called. This only applies to members using protocolVersion set to 1. The default value is 10000 (that is, 10 seconds). Set this value high if you have frequent communication interruptions between replica set members (and fix the communication problems!). Set this value lower if communications are reliable (increases availability).
catchUpTimeoutMillis	Integer	Represents the time in milliseconds that a newly elected primary is allowed to synchronize its data with the other replica set members. The default is −1 (infinite time). Set a value above 0 for situations where network connections are slow. Warning: the higher the value, the greater the interval during which the new primary is unable to process write requests.
catchUpTakeoverDelayMillis	Integer	This value represents the number of milliseconds a secondary must wait before it is allowed to initiate a **takeover**. A takeover occurs when the secondary detects its data is ahead of the primary, at which point the secondary initiates a new election in an effort to become the new primary. The default value is 30000 (30 seconds).
getLastErrorDefaults	Document	A JSON document specifying the write concern for the replica set. This document is in the format { w: <value>, j: <boolean>, wtimeout: <number> }. For more information, see the documentation page for *Write Concern Specification*. This document is only used when no other write concerns have been expressed for a given request.
getLastErrorModes	Document	Allows you to group write concerns by tag. The JSON document takes the form { getLastErrorModes: { <LABEL>: { "<TAG>": <int> } } }, where <int> represents the number of different tag values needed to satisfy the write concern. For more information, see *Custom Multi-Datacenter Write Concerns*.
replicaSetId	ObjectId	Automatically generated, cannot be changed via settings.

For more information, please refer to the following links:

Chained replication: `https://docs.mongodb.com/master/tutorial/manage-chained-replication/`

Write concern specification: `https://docs.mongodb.com/master/reference/write-concern/#write-concern-specification`

Custom multi-data center write concerns: `https://docs.mongodb.com/master/tutorial/configure-replica-set-tag-sets/#custom-multi-datacenter-write-concerns`

At this point, now that you have an idea of what goes into a replica set configuration document, we will turn our attention to the **BigLittle, Ltd**. replica set example.

Replica set configuration example

For the purposes of this illustration, we will use an extremely simple example configuration, only specifying the name of the replica set and the host entries for each of the members:

```
doc = {
    _id : "learn_mongodb",
    members: [
        { _id: 1, host: "member1.biglittle.local" },
        { _id: 2, host: "member2.biglittle.local" },
        { _id: 3, host: "member3.biglittle.local" },
    ]
}
rs.initiate(doc);
```

We then connect to one of the member servers, run the `mongo` shell, and execute the command, as shown here:

```
File Edit View Search Terminal Help
> rs.initiate(doc);
{
        "ok" : 1,
        "$clusterTime" : {
                "clusterTime" : Timestamp(1585109854, 1),
                "signature" : {
                        "hash" : BinData(0,"AAAAAAAAAAAAAAAAAAAAAAAAAAA="),
                        "keyId" : NumberLong(0)
                }
        },
        "operationTime" : Timestamp(1585109854, 1)
}
learn_mongodb:SECONDARY>
```

After a brief period of time, database synchronization is complete. As we are operating from an empty database directory, only the default databases appear at first. In a later section, we will restore the sample data and observe the replication process.

Only run `rs.initiate()` or the `replSetInitiate` database command once!

Now that the replica set has been initiated, let's have a look at its status.

Confirming the replica set status

First of all, you can immediately distinguish the *primary* from a *secondary* from the mongo shell prompt. Otherwise, you can use the `rs.status()` command to determine how well (or badly!) the replica set is performing:

```
File  Edit  View  Search  Terminal  Help
learn_mongodb:PRIMARY> rs.status();
{
        "set" : "learn_mongodb",
        "date" : ISODate("2020-03-25T05:29:35.815Z"),
        "myState" : 1,
        "term" : NumberLong(1),
        "syncingTo" : "",
        "syncSourceHost" : "",
        "syncSourceId" : -1,
        "heartbeatIntervalMillis" : NumberLong(2000),
        "majorityVoteCount" : 2,
        "writeMajorityCount" : 2,
        "optimes" : {
                "lastCommittedOpTime" : {
                        "ts" : Timestamp(1585114168, 1),
                        "t" : NumberLong(1)
                },
                "lastCommittedWallTime" : ISODate("2020-03-25T05:29:28.209Z"),
                "readConcernMajorityOpTime" : {
```

The `rs.status()` command includes some interesting sub-documents, including details on the election process, as captured here:

```
"electionCandidateMetrics" : {
        "lastElectionReason" : "electionTimeout",
        "lastElectionDate" : ISODate("2020-03-25T04:40:47.469Z"),
        "electionTerm" : NumberLong(1),
```

```
        . . .
        "newTermStartDate" : ISODate("2020-03-25T04:40:48.014Z"),
        "wMajorityWriteAvailabilityDate" :
ISODate("2020-03-25T04:40:48.860Z")
    },
```

You can also see detailed information on the members of the replica set. In this code block, you can see the details for the `member1` server, which also happens to be the primary:

```
"members" : [ {
        "_id" : 1,
        "name" : "member1.biglittle.local:27017",
        "health" : 1,
        "state" : 1,
        "stateStr" : "PRIMARY",
        "uptime" : 2968,
        . . .
        "electionTime" : Timestamp(1585111247, 1),
        "electionDate" : ISODate("2020-03-25T04:40:47Z"),
        "configVersion" : 1,
        "self" : true,
        "lastHeartbeatMessage" : ""
    }, . . .
```

Next, we will have a look at restoring the sample data and reviewing data replication.

Restoring the sample data

In order to verify the functionality of the replica set, we need to restore the sample data for the `biglittle` database onto the primary. To restore the sample data, proceed as follows:

1. Copy the sample data from `/path/to/repo/sample_data/biglittle*.js` to `/path/to/repo/chapters/13/common_data`.

2. Open a shell to `member1` (or whichever server was elected the primary):

   ```
   docker exec -it learn-mongo-member-1 /bin/bash
   ```

3. Use the `mongo` shell on `member1` to restore the data for `biglittle`:

   ```
   mongo /data/common/biglittle_common_insert.js
   mongo /data/common/biglittle_users_insert.js
   mongo /data/common/biglittle_loans_insert.js
   ```

4. Enter the mongo shell and confirm that the data has been restored:

```
File  Edit  View  Search  Terminal  Help
learn_mongodb:PRIMARY> show dbs;
admin       0.000GB
biglittle   0.000GB
config      0.000GB
local       0.001GB
learn_mongodb:PRIMARY> use biglittle;
switched to db biglittle
learn_mongodb:PRIMARY> show collections;
common
loans
users
learn_mongodb:PRIMARY> db.loans.find().count();
230
learn_mongodb:PRIMARY>
```

5. Next, open a mongo shell on either of the secondaries.

6. Repeat these commands to confirm that the data has been synchronized:

```
rs.slaveOk();
use biglittle;
show collections;
db.loans.find().count();
```

You should see something similar to the following screenshot:

```
File  Edit  View  Search  Terminal  Help
learn_mongodb:SECONDARY> rs.slaveOk();
learn_mongodb:SECONDARY> show dbs;
admin    0.000GB
config   0.000GB
local    0.000GB
learn_mongodb:SECONDARY> show dbs;
admin       0.000GB
biglittle   0.000GB
config      0.000GB
local       0.001GB
learn_mongodb:SECONDARY> use biglittle;
switched to db biglittle
learn_mongodb:SECONDARY> show collections;
common
loans
users
learn_mongodb:SECONDARY> db.loans.find().count();
230
learn_mongodb:SECONDARY>
```

By default, the secondary does not accept read requests directly. Use `rs.slaveOk()` from the `mongo` shell to temporarily enable this ability. When using programming language drivers, or when running a shell command, you can allow secondaries to accept read requests by setting read preferences. This is covered in more detail in the next chapter, `Chapter 14`, *Replica Set Runtime Management and Development*.

Summary

In this chapter, you learned about replica sets in general: what they are, how they work, and what their benefits are. You were also given guidelines to establish a solid business rationale pursuant to implement a replica set. After that, you were shown how a DevOp might model a replica set using Docker and Docker Compose. Next, you learned about how to configure and deploy a replica set, with detailed coverage of the various replication configuration settings for each replica set member, as well as the overall settings.

In the next chapter, you will learn how to monitor and manage replica sets, as well as addressing program code that interacts with a replica set.

14
Replica Set Runtime Management and Development

In this chapter, you'll learn how to monitor and manage a replica set, with a discussion of replica set runtime considerations, including member synchronization, backup, and restore. In addition, this chapter teaches you how to write programming code that addresses a replica set.

In this chapter, we cover the following topics:

- Managing a running replica set
- Managing replica set impact on application program code

Technical requirements

Minimum recommended hardware:

- Desktop PC or laptop
- 2 GB free disk space
- 4 GB of **random-access memory (RAM)**
- 500 Kbps or faster internet connection

Software requirements:

- OS (Linux or Mac): Docker, Docker Compose, Git (optional)
- OS (Windows): Docker for Windows and Git for Windows
- Python 3.x, PyMongo driver, and Apache (already installed in the Docker container used for the book)

Chapter 2, *Setting Up MongoDB 4.x* covers required software installation and restoring the code repository for the book.

To run the code examples in this chapter, open a Terminal window (Command Prompt) and enter these commands:

```
cd /path/to/repo/chapters/14
docker-compose build
docker-compose up -d
```

This brings up three Docker containers that are used as part of the demonstration replica set configuration. Further instructions on the replica set setup are given in the chapter. When you have finished working with the examples covered in this chapter, return to the Terminal window (Command Prompt) and stop Docker, as follows:

```
cd /path/to/repo/chapters/14
docker-compose down
```

The code used in this chapter can be found in the book's GitHub repository at `https://github.com/PacktPublishing/Learn-MongoDB-4.x/tree/master/chapters/14`.

Managing a running replica set

There are several aspects of a running replica set of which MongoDB DevOps need to be aware. Considerations include the oplog size, monitoring, synchronization, backup, and restore. First, we'll have a look at managing the size of the replica set *oplog*.

Managing the oplog size

As mentioned earlier in this book, synchronization is accomplished using the *oplog* (operations log). When the primary accepts any request causing the database to be modified, the actual operation performed is recorded in its oplog. The oplog is actually a *capped collection* named `local.oplog.rs`. This collection cannot be dropped.

The size of the oplog varies between host operating systems. For macOS systems, the default is set to 192 **megabytes** (**MB**). On Linux or Windows servers, conversely, the *WiredTiger* engine typically uses 5% of its available disk space, with a minimum setting of 990 MB and a maximum of 50 GB.

If using the *In-Memory* storage engine, on the other hand, the default for Windows and Linux servers is 5% of the available memory, with minimum and maximum boundaries of 50 MB to 50 GB.

> For more information on default oplog sizes, see the following *Replica Set Oplog Size* documentation page: `https://docs.mongodb.com/manual/core/replica-set-oplog/#replica-set-oplog-sizing`.

The default is adequate for most purposes. There are some types of operations, however, that tend to consume a greater amount of oplog space. These include the following:

- **Many update requests that affect multiple documents**: Multi-document requests are broken down into single update operations in the oplog to maintain idempotency. Interestingly, although the oplog takes a hit, database size and disk usage are not adversely affected.
- **Deletions**: Although a deletion reduces the size of the database, a delete request actually *increases* the size of the oplog as it must be recorded as a separate operation.
- **Updates**: Although an update does not affect the size of the database, an update request *increases* the size of the oplog as it must be recorded as a separate operation.

> If you wish to change the size of the oplog, perform the change on all secondaries first, and then do the same for the primary.

The size of the oplog can be adjusted, as follows:

1. Use the `mongo` shell to connect to one of the *secondary* members of the replica set. If **role-based access control** (**RBAC**) has been activated for the database, you must be a user assigned to at least the `clusterManager` role.
2. Confirm the current size of the oplog, as follows:

```
use local
db.oplog.rs.stats().maxSize;
```

3. Use the `replSetResizeOplog` database admin command to change the size of the oplog.

Please note that the `minRetentionHours` directive is only available on MongoDB 4.4 and above. Also, note that this command only affects `mongod` instances using the WiredTiger storage engine.

The generic form of this command is shown as follows:

```
db.adminCommand({
    replSetResizeOplog: <int>,
    size: <double>,
    minRetentionHours: <double>
})
```

The directives are summarized in this table:

Directive	Type	Notes
`replSetResizeOplog`	*integer*	Set this value to 1.
`size`	*double*	New size of the oplog in MB. Minimum value: 990 MB. Maximum value: 1 **petabyte (PB)**.
`minRetentionHours`	*double*	The minimum number of hours this log size is retained. If you enter a value of 0, the current size is maintained and any excess is truncated (oldest operations first). Otherwise, decimal values represent fractions of an hour. Example: 0.5 indicates 30 minutes; 1.5 represents one hour and a half, and more.

4. As an example, to double the current size of the oplog, issue the following command:

```
use local;
current = db.oplog.rs.stats().maxSize;
db.adminCommand({ replSetResizeOplog:1, size: current*2/1000000 });
```

Example results are shown in this screenshot:

```
File Edit View Search Terminal Help
learn_mongodb:SECONDARY> use local;
switched to db local
learn_mongodb:SECONDARY> db.oplog.rs.stats().maxSize;
NumberLong("4676677376")
learn_mongodb:SECONDARY> current = db.oplog.rs.stats().maxSize;
NumberLong("4676677376")
learn_mongodb:SECONDARY> db.adminCommand({replSetResizeOplog: 1, size: current*2/1000000});
{
        "ok" : 1,
        "$clusterTime" : {
                "clusterTime" : Timestamp(1585194951, 1),
                "signature" : {
                        "hash" : BinData(0,"AAAAAAAAAAAAAAAAAAAAAAAAAAA="),
                        "keyId" : NumberLong(0)
                }
        },
        "operationTime" : Timestamp(1585194951, 1)
}
learn_mongodb:SECONDARY> db.oplog.rs.stats().maxSize;
NumberLong("9807331328")
learn_mongodb:SECONDARY>
```

5. You might also consider using the `compact` database run command against the `local.oplog.rs` collection to reclaim disk space if the oplog size was reduced, as follows:

```
use local
db.runCommand({ "compact" : "oplog.rs" } )
```

6. Repeat *Steps 1-5* for each secondary. When the oplog size has been modified as desired, repeat *Steps 1-5* on the primary.

Please refer to the following links for more information:
Capped Collections—https://docs.mongodb.com/manual/core/capped-collections/#capped-collections
Oplog Size—https://docs.mongodb.com/manual/core/replica-set-oplog/#oplog-size
`replSetResizeOplog` *database admin command*—https://docs.mongodb.com/manual/reference/command/replSetResizeOplog/#replsetresizeoplog

Next, we'll have a look at the overall replication status.

Monitoring the replication status

To monitor the overall replication status, use the `rs.status()` shell method. This shell method is actually a wrapper for the `replSetGetStatus` database command. Here is a summary of the information keys to monitor:

Key	What to look for
`optimes`	There should not be a large difference between the values for `lastCommittedOpTime` and `appliedOpTime`.
`electionCandidateMetrics`	Introduced in MongoDB 4.2.1, this metric is only available when connecting to the primary. The difference between `lastElectionDate` and `newTermStartDate` should be a few milliseconds. If not, the election process is being drawn out. Consider adding an arbiter to the replica set.
`members`	Information presented under this key is further subdivided by a replica set member. For each given member, things to monitor include the following: `state` should be 1 (primary), 2 (secondary), or 7 (arbiter). A value of 0 or 5 means the replica set is starting up. If you see a value of 3 (recovering), 6 (unknown), or 8 (down), further investigation on that replica set member is warranted. See the *Replica Set Member States* link in the next information box for more information. `strState` gives you state-related messages for unhealthy servers. `health` should be 1. If not, connect to that replica set member to see what's going on. `uptime` is in seconds, and the values for all replica set members should be within a few seconds of each other. If you notice one member with a substantially smaller value, that server is a problem. `lastHeartbeatMessage`, if not empty, gives you information on potential communication problems between replica set members.
`initialSyncStatus`	This information block is only available during replica set initialization. If the replica set is not initializing properly, this information can provide hints as to the nature of the problem. If `failedInitialSyncAttempts` is high, check for connection problems. Under `initialSyncAttempts`, look for error messages in the `status` sub-key. In MongoDB 4.4, a new `operationsRetried` sub-key was added, which, if high, gives you a clue that there are communication problems with the server, indicated by the `syncSource` value.

Here is a partial screenshot, showing the unhealthy state of `member1`:

```
File  Edit  View  Search  Terminal  Help
        "members" : [
                {
                        "_id" : 1,
                        "name" : "member1.biglittle.local:27017",
                        "health" : 0,
                        "state" : 8,
                        "stateStr" : "(not reachable/healthy)",
                        "uptime" : 0,
                        "optime" : {
                                "ts" : Timestamp(0, 0),
                                "t" : NumberLong(-1)
                        },
                        "optimeDurable" : {
                                "ts" : Timestamp(0, 0),
                                "t" : NumberLong(-1)
                        },
                        "optimeDate" : ISODate("1970-01-01T00:00:00Z"),
                        "optimeDurableDate" : ISODate("1970-01-01T00:00:00Z"),
                        "lastHeartbeat" : ISODate("2020-03-28T03:55:48.165Z"),
                        "lastHeartbeatRecv" : ISODate("2020-03-28T03:55:18.883Z"),
                        "pingMs" : NumberLong(0),
                        "lastHeartbeatMessage" : "Couldn't get a connection within the time limit",
                        "syncingTo" : "",
                        "syncSourceHost" : "",
                        "syncSourceId" : -1,
                        "infoMessage" : "",
                        "configVersion" : -1
                },
```

For more information on replica set member states, refer to the following link: `https://docs.mongodb.com/master/reference/replica-states/#replica-set-member-states`.

Next, we'll have a look at the topic of replication lag.

Controlling replication lag

Replication lag refers to the amount of time a secondary is behind the primary. The greater the lag time, the greater the possibility read operations could return outdated data. An excessive lag time can be the result of several factors, including the following:

- **Network latency**: Check **local area network (LAN)** configuration, firewalls, routing, and communications media.
- **Slow disk throughput**: Check the state of the filesystem on the server or container hosting the replica set member. Consider using a faster and more up-to-date filesystem.

- **Concurrency**: Resource-intensive applications could tie up the primary, causing replication to secondaries to bottleneck. You may need to refactor such applications. One possibility is to add an appropriate *write concern* to force acknowledgments, allowing secondaries to catch up. This is covered in more detail in the next section of this chapter.

To get information on oplog data synchronization for a given server, use the `rs.printReplicationInfo()` shell method, as seen in the following screenshot:

```
File Edit View Search Terminal Help
learn_mongodb:PRIMARY> rs.printReplicationInfo();
configured oplog size:   4460.0146484375MB
log length start to end: 85575secs (23.77hrs)
oplog first event time:  Wed Mar 25 2020 04:40:36 GMT+0000 (UTC)
oplog last event time:   Thu Mar 26 2020 04:26:51 GMT+0000 (UTC)
now:                     Thu Mar 26 2020 04:26:57 GMT+0000 (UTC)
learn_mongodb:PRIMARY>
```

To get information on data synchronization between the primary and its secondaries, from the primary, use the `rs.printSlaveReplicationInfo()` shell method. Pay attention to the `syncedTo` information, which tells you the amount of time a secondary is lagging behind a primary, as seen here:

```
File Edit View Search Terminal Help
learn_mongodb:PRIMARY> rs.printSlaveReplicationInfo();
source: member1.biglittle.local:27017
        syncedTo: Thu Mar 26 2020 04:27:21 GMT+0000 (UTC)
        0 secs (0 hrs) behind the primary
source: member3.biglittle.local:27017
        syncedTo: Thu Mar 26 2020 04:27:21 GMT+0000 (UTC)
        0 secs (0 hrs) behind the primary
learn_mongodb:PRIMARY>
```

As a final consideration, MongoDB 4.2 introduced a new *flow control* mechanism. Enabled by default, as the lag time approaches the value of `flowControlTargetLagSeconds`, any write requests to the primary must first acquire a *ticket* before being allowed a lock to write. The flow control mechanism limits the number of tickets granted, thus allowing secondaries a chance to catch up on applying oplog write operations.

Although flow control is enabled by default in MongoDB 4.2 and above, set the values summarized in the following table to ensure flow control implementation:

Setting	Default	Setting to enable flow control
`replication.enableMajorityReadConcern`	`true`	`true`
`setParameter.enableFlowControl`	`true`	`true`
`setParameter.flowControlTargetLagSeconds`	`10`	Set to the desired value
`setFeatureCompatibilityVersion`	`--`	`4.2`

To make these settings effective, add the following code to the `mongod.conf` file, where `<int>` is the desired maximum target lag value (the default is 10):

```
replication:
    enableMajorityReadConcern: true
setParameter:
    enableFlowControl: true
    flowControlTargetLagSeconds: <int>
```

You would also need to set the feature compatibility value by connecting to a server in the replica set and issuing the following command as a user with sufficient rights:

```
db.adminCommand( { setFeatureCompatibilityVersion: "4.2" } )
```

> Please refer to the following links for more information:
> *MongoDB Server Parameters*—`https://docs.mongodb.com/manual/reference/parameters/#mongodb-server-parameters`
> `setFeatureCompatibilityVersion`—`https://docs.mongodb.com/manual/reference/command/setFeatureCompatibilityVersion/#setfeaturecompatibilityversion`
> *Check the Replication Lag*—`https://docs.mongodb.com/master/tutorial/troubleshoot-replica-sets/#check-the-replication-lag`
> *Write Concern*—`https://docs.mongodb.com/manual/reference/write-concern/#write-concern`

Let's now have a look at backup and restore.

Backing up and restoring a replica set

Backing up from and restoring to a replica set is not as simple as the basic backup procedure described in Chapter 3, *Essential MongoDB Administration Techniques*. One problem when using `mongodump` for backup is that all data needs to pass through the server memory, severely impacting performance.

Another consideration is that the very second the backup has completed, it's already out of date! To avoid overwriting newer data with stale data restored from a backup, you need to carefully consider your backup strategy. The choices for backing up a replica set are as follows:

- Using a filesystem snapshot
- Using `mongodump`

The tools needed to create a filesystem snapshot depend on the operating system. As such, we do not cover that aspect in this book. If you wish to pursue this approach, a good starting point is the MongoDB documentation, *Back Up and Restore Using Filesystem Snapshots*.

Please refer to the following links for more information:
Restore a Replica Set from MongoDB Backups—https://docs.mongodb.com/manual/tutorial/restore-replica-set-from-backup/#restore-a-replica-set-from-mongodb-backups
Back Up and Restore with Filesystem Snapshots: https://docs.mongodb.com/manual/tutorial/backup-with-filesystem-snapshots/#back-up-and-restore-with-filesystem-snapshots

It is also important to discuss the temptation to not do backups at all and simply rely upon the fact that the replica set itself serves as a live and dynamic backup. Although there is a certain amount of merit to this philosophy, please do your best to avoid falling into this trap. The proper approach to safeguarding your data would be to preferably use external operating system tools to capture regular filesystem snapshots. Barring that, use `mongodump` on a regular basis to back up your data. Remember: if the servers hosting the replica set are in the same building, what happens if the building is flooded and all servers are lost? Or if there is an earthquake? Or a wildfire? In the face of natural or man-made disasters, even the best replica set design can fail. Unless you've made regular backups, and—very importantly—store the backups offsite, you stand to lose all your data.

Next, we'll have a look at using `mongodump` to back up a replica set.

Backing up using mongodump

When using `mongodump` to back up a replica set, you need to use either the `--uri` or the `--host` option to specify a list of hosts in the replica set. Using our Docker replica set model as an example, the string would appear as follows:

```
mongodump --uri="mongodb://member1.biglittle.local:27017,\
    member2.biglittle.local:27017,member1.biglittle.local:27017/? \
    replicaSet=learn_mongodb"
```

Here is an alternate syntax that does the same thing:

```
mongodump --host="learn_mongodb/member1.biglittle.local:27017,\
    member2.biglittle.local:27017,member1.biglittle.local:27017"
```

Data is normally read from the primary. If you wish to allow the backup to be read from a secondary, add the `--readPreference` option. If all you want to do is to indicate that the backup can occur from a secondary, add the following option to either of the preceding examples:

```
mongodump <...> --readPreference=secondary
```

A final consideration when using `mongodump` to back up a replica set is to include the `--oplog` option. This allows MongoDB to establish the backup within a specific point in time, which is highly advantageous when performing a restore.

There are several caveats when using this option, summarized here:

- You must use the `--oplogReplay` option when restoring using `mongorestore`.
- Any `renameCollection()` operations during the actual dump process cause the backup to fail.
- Any `aggregate()` operations using the `$out` option during the actual dump process cause the backup to fail.
- The `--oplog` option only works on a *full* database dump: you cannot dump just a single database or collection.

If only one server in the replica set has gone down, there is no need to back up or restore. Just bring the server back online and allow the normal resynchronization process to take place. For more information on how this works, see the sub section entitled *Resynchronizing a replica set member*, presented later in this chapter.

Here is an example using our replica set model:

```
File  Edit  View  Search  Terminal  Help
root@member2:/# mkdir /data/common/backup
root@member2:/# mongodump --oplog --out=/data/common/backup --uri="mongodb://member1.biglittle
.local,member2.biglittle.local,member1.biglittle.local/?replicaSet=learn_mongodb"
2020-03-28T03:04:44.305+0000    writing admin.system.version to
2020-03-28T03:04:44.308+0000    done dumping admin.system.version (1 document)
2020-03-28T03:04:44.315+0000    writing biglittle.loans to
2020-03-28T03:04:44.318+0000    writing biglittle.users to
2020-03-28T03:04:44.321+0000    writing biglittle.common to
2020-03-28T03:04:44.338+0000    done dumping biglittle.users (289 documents)
2020-03-28T03:04:44.338+0000    done dumping biglittle.common (8 documents)
2020-03-28T03:04:44.354+0000    done dumping biglittle.loans (230 documents)
2020-03-28T03:04:44.356+0000    writing captured oplog to
2020-03-28T03:04:44.357+0000        dumped 1 oplog entry
root@member2:/#
```

The backup should have a directory tree resembling this:

```
/data/common/backup/
|-- admin
|    |-- system.version.bson
|    `-- system.version.metadata.json
|-- biglittle
|    |-- common.bson
|    |-- common.metadata.json
|    |-- loans.bson
|    |-- loans.metadata.json
|    |-- users.bson
|    `-- users.metadata.json
`-- oplog.bson
```

You can see that there is no backup of the `local` database. Also, you might note that the inclusion of the oplog is critical when restoring to a replica set. Next, we'll have a look at restoring data to a replica set.

```
db.collection.renameCollection()—https://docs.mongodb.com/
manual/reference/method/db.collection.renameCollection/#db.
collection.renameCollection
db.collection.aggregate()—https://docs.mongodb.com/manual/
reference/method/db.collection.aggregate/#db.collection.
aggregate
```

Restoring a replica set

Before we get into the details on how to restore a replica set, it must be made clear that if any members of a replica set survive some future disaster, you need to perform *synchronization* instead of restoration. The reason for this is that the data found on a functioning member of the replica set should be more up to date than the data in a backup. The next subsection deals with resynchronization.

You cannot restore a replica set by simply copying the backup files to the data directories of each member server and then starting the `mongod` instances. It's not that simple! The procedure outlined in this chapter is based upon the recommended procedure described in the *Restore a Replica Set from MongoDB Backups* documentation page mentioned earlier.

Assuming that no member of the replica set still remains functional, we'll now examine what is involved in restoring the replica set. Let's have a look at the process step by step, as follows:

1. Make sure that any former replica set members are not up and running `mongod`. To be on the safe side, once you've brought the member server online, remove all files and directories that were part of the previous database directory path. This path is the one indicated by the `--dbPath` command-line parameter, or, in the `mongod.conf` file, it's the value assigned to the `storage.dbPath` key. Do not wipe out the backup files!

2. Obtain the backup files from the appropriate source. As you'll recall from the discussion earlier in this section, backup files could come from a filesystem snapshot or from the output of the `mongodump` command. A *hot* backup is one generated while a `mongod` instance is running. A *cold* backup is one taken when the `mongod` instance is down. If you are restoring from a *cold* backup, for MongoDB version 4.2 and above, you have the option of starting the `mongod` instance with the `--eseDatabaseKeyRollover` command-line option to avoid the reuse of keys, maintaining database security. For MongoDB version 4.0 and earlier, backups of encrypted storage engines using *AES256-GCM* should use `mongodump` to produce the backup.

If you are running MongoDB Enterprise, you might be using an *encrypted storage engine*. If this is the case, and your encryption algorithm is *AES256-GCM*, you need to determine whether the backup was *hot* or *cold* (`https:/ /docs.mongodb.com/manual/core/security-encryption-at-rest/ #encrypted-storage-engine`).

3. Choose one of the member servers to process the restore. For the purposes of this procedure, we refer to that server as the *restore target*. On that server, copy the backup files to the database directory path.

4. If the backup files contain a backup of the `local` database, you need to drop it. To drop the local database, proceed as follows:

    1. Start `mongod` on the restore target, indicating the path to the database files. See the following code snippet for an example:

    ```
    mongod --dbpath /path/to/backup
    ```

    2. Open a `mongo` shell on the instance, drop the local database, and bring the instance down, as follows:

    ```
    > use local;
    > db.dropDatabase();
    > use admin;
    > db.shutdownServer();
    ```

5. On the restore target server, start a `mongod` instance with the previous configuration.

6. Reinitialize the replica set, using the restore target server as a *single-node primary,* as illustrated in the following code snippet:

```
doc = {
    _id : "learn_mongodb",
    members: [
        { _id: 1, host: "member1.biglittle.local" }
    ]
}
rs.initiate(doc);
```

The following screenshot shows the result of running `rs.status().members` on the restore target:

```
File Edit View Search Terminal Help
learn_mongodb:PRIMARY> rs.status().members;
[
        {
                "_id" : 1,
                "name" : "member1.biglittle.local:27017",
                "health" : 1,
                "state" : 1,
                "stateStr" : "PRIMARY",
                "uptime" : 69,
                "optime" : {
                        "ts" : Timestamp(1585371093, 1),
                        "t" : NumberLong(1)
                },
                "optimeDate" : ISODate("2020-03-28T04:51:33Z"),
                "syncingTo" : "",
                "syncSourceHost" : "",
                "syncSourceId" : -1,
                "infoMessage" : "could not find member to sync from",
                "electionTime" : Timestamp(1585371042, 2),
                "electionDate" : ISODate("2020-03-28T04:50:42Z"),
                "configVersion" : 1,
                "self" : true,
                "lastHeartbeatMessage" : ""
        }
]
learn_mongodb:PRIMARY>
```

7. Restore the data to the restore target server by either manually copying files from a filesystem snapshot or by using `mongorestore` if the source of your data is from `mongodump`. If you copy the files manually, in the database directory structure on the restore target server, make the following permission changes:

    1. Change the owner to `mongodb`.
    2. Change the group to `mongodb`.
    3. On all directories, assign the owner read, write, and execute permissions (`700` in Linux).
    4. On all files, assign the owner read and write permissions (`600` in Linux).

8. Even though it's a replica set with only a single node, you still need to confirm that the *replication* process has completed. Accordingly, it's wise to wait a few minutes before starting to reconstruct the rest of the replica set. You can also use `rs.status()` to confirm the single-node replica set is functioning properly.

Once satisfied the restore target server MongoDB installation is functioning properly, having confirmed data was restored successfully, you can now start restoring the remaining servers back into the replica set. To do this, on each remaining unrestored former member of the replica set, proceed as follows:

1. Make sure `mongod` is not running.
2. If `mongod` is running, delete all files and directories from the database directory (indicated by the `dbPath` directive).
3. If the database is large, copy files from the database directory of the restore target server into the database directory of the server you are currently restoring. Be sure to set the permissions, as indicated in the preceding *Step 7*. Otherwise, just leave the database directory blank, and allow MongoDB to recreate and resynchronize automatically.
4. Bring the `mongod` instance up, using the previous configuration from the old replica set.
5. Open a `mongo` shell on the restore target server and issue the `rs.add()` command. Here is the generic syntax:

   ```
   rs.add(host, arbiterOnly);
   ```

   `host` is a **JavaScript Object Notation (JSON)** document that contains exactly the same directives as described in the *Initiating the replica set: Replica set members configuration directives* sub section from the previous chapter, `Chapter 13`, *Deploying a Replica Set*.

   `arbiterOnly` is a Boolean value; set it to `true` if you don't want this server to have voting rights, but do want it to synchronize data.

6. As an example, we add `member2` to the restored replica set, using this command:

   ```
   member = { _id: 2,host: "member2.biglittle.local" }
   rs.add(member);
   ```

This screenshot shows the results:

```
File Edit View Search Terminal Help
learn_mongodb:PRIMARY> member2 = { _id: 2, host: "member2.biglittle.local" }
{ "_id" : 2, "host" : "member2.biglittle.local" }
learn_mongodb:PRIMARY> rs.add(member2);
{
        "ok" : 1,
        "$clusterTime" : {
                "clusterTime" : Timestamp(1585373004, 1),
                "signature" : {
                        "hash" : BinData(0,"AAAAAAAAAAAAAAAAAAAAAAAAAAA="),
                        "keyId" : NumberLong(0)
                }
        },
        "operationTime" : Timestamp(1585373004, 1)
}
learn_mongodb:PRIMARY>
```

7. You can then repeat *Steps 1-6* for each of the remaining secondaries.

Now, we'll turn our attention to resynchronization.

Resynchronizing a replica set member

At some point, be it through network disruptions, natural disasters, or any number of other factors outside your control, it's possible for the data on a replica set member to become *stale*. What this means is that the data is so radically out of date that it is impossible for it to catch up through the normal synchronization process. If you have determined that network communications are not at fault, it is time to consider performing a manual resynchronization (https://docs.mongodb.com/manual/tutorial/resync-replica-set-member/).

> If in the process of monitoring the replica set status, you notice one of the replica members has an increase in *replication lag*, you need to flag that server for closer inspection. If the lag time continues to increase, first check all network connections, operating system logs, and router logs. It's possible the problem is physical—perhaps even something as simple as a bad cable.

There are two primary techniques used to perform resynchronization. These techniques are similar to the techniques discussed earlier in the chapter, in the *Restoring a replica set* sub section, and are detailed as follows:

- Force an *initial sync* to occur
- Manually copy files from another replica set member

Let's look at the initial sync approach first.

Resynchronizing by forcing an initial sync

This technique requires a minimal number of manual steps as it relies upon the normal replica set synchronization process. The main disadvantage is that when you have a large amount of data, it inevitably generates a large amount of network traffic. This technique also impacts the primary by generating a larger-than-ordinary number of requests to the primary. Further, this technique can only work if other replica set members are up to date.

 An *initial sync* is what occurs when either a new replica set is first brought online or when a member is added to a replica set. During this process, MongoDB clones the database onto the new member and then applies any operations in the *oplog* (`https://docs.mongodb.com/manual/core/replica-set-sync/#initial-sync`).

To resynchronize by forcing an initial sync, proceed as follows:

1. Perform a graceful shutdown of the `mongod` instance on the server needing synchronization. You can either open a `mongo` shell onto this server and use the `db.shutDown()` method or, if running on Linux, issue the `mongod --shutdown` command from the command line.
2. Delete all files and directories from the directory structure indicated by the `dbPath` value.
3. Restart the `mongod` instance. You must retain all the existing replication-related directives just as they were prior to beginning this operation.

Once the server is back up and running, an *initial sync* request is issued, and data starts flowing to the restarted server. You can monitor the progress using the commands described earlier in this chapter. Now, let's look at resynchronization by manually copying files.

Resynchronizing by manual file copy

In this approach, the files from the data directory of an existing replica set member are directly copied to the replica set member in need of resynchronization. This approach is best in situations where the amount of data is sizeable, and where network communications would be adversely affected following the initial sync approach described in the previous sub section.

Unfortunately, trying to obtain a direct file copy while a `mongod` instance is running might prove difficult as database files are frequently locked, held open, and contents are changing. Accordingly, you either need to shut down the `mongod` instance in order to obtain a copy or use external operating system tools to obtain a filesystem snapshot.

Unlike the process of restoring a replica set described earlier in this chapter, you *must* include the `local` database in the copy.

 Also, it is important to note that you cannot use `mongodump` and `mongorestore` for resynchronization.

In this procedure, we refer to the replica set member with up-to-date data as the source server. We refer to the server needed to be resynchronized as the *target* server. To resynchronize using the file copy approach, proceed as follows:

1. From the *source* server, obtain a copy of all files and directories in its database directory path (indicated by the `dbPath` directive). This is accomplished either by a filesystem snapshot or by shutting down the `mongod` instance and copying the files. If you shut down the `mongod` instance on the source server, be sure to restart it after the copy is complete.
2. Shut down the `mongod` instance on the *target* server.
3. Wipe out all files and directories in the database directory on the target server.
4. Copy the database files from the source server to the database directory on the target server. Be sure to copy the complete directory structure.
5. In the database directory structure on the target server, make the following permission changes:
    1. Change the *owner* to `mongodb`.
    2. Change the *group* to `mongodb`.
    3. On all *directories*, assign the owner read, write, and execute permissions (`700` in Linux).
    4. On all *files*, assign the owner read and write permissions (`600` in Linux).
6. Restart the `mongod` instance on the target server.

Once the server is back up and running, you might notice a small amount of synchronization traffic. This is normal, as it's possible changes occurred in the midst of the copy process. You can monitor the replica set status using the commands described earlier in this chapter.

Now that you understand how to create and manage a replica set, we'll turn our attention to potential program code modifications.

Managing replica set impact on application program code

In practice, the application program code written for MongoDB does not need to change when performing reads and writes to a replica set. If your application code works well for a single `mongod` instance, it works well when accessing a replica set. In other words, just because you move your working application code over to a replica set does not mean it suddenly stops working!

There are some considerations, however, that might improve application performance by taking advantage of potential performance enhancements made possible by the replica set. There is also the question of *which server to connect*: should your application connect to the *primary*, or is it OK to connect to a *secondary*? Let's start with a basic replica set connection.

Basic replica set connection

Let's say that you've developed a fantastic application. The original code, however, was designed to work with a single server running MongoDB. Now, the company has implemented a replica set. The first question on your mind might be: *Will my existing code work on a replica set*? Let's have a look.

Here is a simple test program that connects to a single server and returns the result from a simple query:

```
import pprint
from pymongo import MongoClient
hostName = 'member1.biglittle.local';
client   = MongoClient(hostName);
result = client.biglittle.users.find_one({},
{"businessName":1,"address":1})
pprint.pprint(result)
```

The source code can be found at
/path/to/repo/chapters/14/test_replica_set_connect_using_one_server.py.

From this screenshot, you can see that even though the `member1` *mongod* instance is part of a replica set, the return results are the same, just as if `member1` had been a standalone *mongod* instance:

```
File Edit View Search Terminal Help
jed@jed:~/Repos/learn-mongodb/chapters/13$ python test_replica_set_connect_using_one_server.py
{'_id': ObjectId('5e7ae2d344defb0ed1a92185'),
 'address': {'buildingName': None,
             'city': 'Gaodhail',
             'country': 'GB',
             'floor': None,
             'geoSpatial': ['-5.9556', '56.5056'],
             'locality1': 'Scotland',
             'locality2': 'Argyll and Bute',
             'postalCode': 'PA72',
             'roomAptCondoFlat': None,
             'streetAddress': '1790 Short Woods Circle'},
 'businessName': 'Green Associates Company'}
jed@jed:~/Repos/learn-mongodb/chapters/13$
```

However, since the server is a member of a replica set, a slight modification causes your code to take advantage of high availability. All you need to do is to change the hostname to a list, and add a `**kwarg` parameter with the name of the replica set, as shown here:

```
import pprint
from pymongo import MongoClient
hosts    = ['member1.biglittle.local','member2.biglittle.local',\
            'member3.biglittle.local']
replName = 'learn_mongodb'
### Note the use of the "replicaSet" **kwarg: ###
client   = MongoClient(hosts, replicaSet=replName);
result = client.biglittle.users.find_one({}, \
         {"businessName":1,"address":1})
pprint.pprint(result)
```

As you can see from the screenshot shown here, the results are exactly the same:

```
File Edit View Search Terminal Help
jed@jed:~/Repos/learn-mongodb/chapters/13$ python test_replica_set_connect_using_list.py
{'_id': ObjectId('5e7ae2d344defb0ed1a92185'),
 'address': {'buildingName': None,
             'city': 'Gaodhail',
             'country': 'GB',
             'floor': None,
             'geoSpatial': ['-5.9556', '56.5056'],
             'locality1': 'Scotland',
             'locality2': 'Argyll and Bute',
             'postalCode': 'PA72',
             'roomAptCondoFlat': None,
             'streetAddress': '1790 Short Woods Circle'},
 'businessName': 'Green Associates Company'}
jed@jed:~/Repos/learn-mongodb/chapters/13$
```

The difference, however, is that now, your application code has a list of fallback servers, any of which can be contacted, should the first on the list fail to deliver results. If you need to be more selective about which server delivers results, we'll next look at setting *read preferences*.

Manipulating read preferences

By default, all read operations (for example, database queries) are directed to the *primary*. On a busy website, however, this might not yield optimal performance, especially where the replica set is geographically dispersed. Accordingly, it's possible to specify a *read preference* when initiating a query.

In the case of a `pymongo.mongo_client.MongoClient` instance, simple read preferences are expressed using the following values in conjunction with the `readPreference **kwarg` parameter:

Preference	Notes
primary	All read requests are directed to the primary. This is the default.
primaryPreferred	If this preference is set, requests are directed to the primary. If the primary is unavailable, however, the request falls back to one of the secondaries.
secondary	This preference causes requests to be directed to any secondary replica set member, but *not* to the primary.
secondaryPreferred	This read preference causes requests to be directed to any secondary replica set member, but if no secondaries are available, the request goes to the primary.
nearest	This preference causes MongoDB to ignore the primary or secondary state. Instead, the requests are directed to the replica set member with the least network latency.

These class constants can be used when establishing the database connection using the `readPreference **kwarg` parameter. Here is a slightly modified version of the example `test_replica_set_connect_using_one_server.py` shown earlier in this chapter, specifying a preference for the nearest server (`/path/to/repo/chapters/14/test_replica_set_connect_secondary_read_pref.py`):

```
import pprint
from pymongo import MongoClient
hosts     = ['member1.biglittle.local','member2.biglittle.local',\
             'member3.biglittle.local']
replName = 'learn_mongodb'
readPref = 'nearest'
```

```
client    = MongoClient(hosts, replicaSet=replName,
readPreference=readPref);
result    = client.biglittle.users.find_one({}, \
            {"businessName":1,"address":1})
```

You can further influence operations where the read preference is anything other than *primary,* by adding the `maxStalenessSeconds **kwarg` parameter. The default value is `-1`, indicating there is no maximum. The implication of the default is that any data is acceptable, even if extremely stale.

> You receive a `pymongo.errors.ConfigurationError`:
> `maxStalenessSeconds must be at least heartbeatFrequencyMS`
> `+ 10 seconds` message if the connection settings are not 10 seconds greater than the replica set heartbeat frequency. Use replica set monitoring techniques (described in the *Monitoring the replication status* section earlier in this chapter) to find the value for `heartbeatFrequencyMS`.
>
> As the value is in *milliseconds,* you need to divide it by 1,000 to find the equivalent in seconds. Set <code>heartbeatFrequencyMs</code> to the value thus derived as a minimum acceptable value for the `maxStaleSeconds` setting.

Here is an example of setting the `**kwargs` parameter, `maxStalenessSeconds`. In this example, pay attention to the line where a `MongoClient` instance is created. After the `hosts` parameter, the remaining parameters in the constructor are `**kwargs` parameters:

```
import pprint
from pymongo import MongoClient
hosts    = ['member1.biglittle.local','member2.biglittle.local',\
            'member3.biglittle.local']
replName = 'learn_mongodb'
readPref = 'nearest'
maxStale = 90
client   = MongoClient(hosts, replicaSet=replName, \
            readPreference=readPref, maxStalenessSeconds=maxStale);
result = client.biglittle.users.find_one({}, \
            {"businessName":1,"address":1})
```

This example can be found at `/path/to/repo/chapters/14/`
`test_replica_set_connect_read_pref_max_staleness.py`.

You can also configure groups of replica set members by implementing tag sets. For more information on this technique, see the following documentation:

Read Preference Tag Sets—https://docs.mongodb.com/manual/core/read-preference-tags/#read-preference-tag-sets

Read Preference Modes—https://docs.mongodb.com/manual/core/read-preference/#read-preference-modes

pymongo.mongo_client.MongoClient read preferences—https://api.mongodb.com/python/current/api/pymongo/read_preferences.html#module-pymongo.read_preferences

Now, let's have a look at manipulating write operations.

Managing write concerns

The nature of *write concern* management is quite different from that of establishing read preferences. The difference is that *all* replica set members are ultimately involved in a write operation. When the database is modified, the changes need to be synchronized by the entire replica set. More importantly, what is at stake in a write operation is how many (if any) *acknowledgments* must be received before the write operation is considered a success (https://docs.mongodb.com/manual/reference/write-concern/#acknowledgment-behavior).

As with read preferences, write concerns are expressed in the form of keyword arguments when defining a pymongo.mongo_client.MongoClient instance. The following list summarizes the applicable **kwargs parameters:

- w=<int>|<string>: Values can be either an integer or a string. If you set a value of 0, write acknowledgment is disabled. This increases the speed of write operations but sacrifices replica set synchronization integrity. If the value is greater than 0, the primary is always involved. A value of 1 is the default for MongoDB. It indicates that a successful acknowledgment for a write operation should be received from the primary only. Any N value greater than 1 must be equal to or less than the total number of replica set members excluding arbiters, and indicates that acknowledgments must be received from N replica set members before the write operation is considered successful.

Alternatively, you can set the value to be `majority`, in which case the write is considered a success if the majority of replica set members acknowledge the write operation. Thus, for example, if you have five members in the replica set, a `w=3` value would be the same as a value of `w='majority'`. Note that the timing behind the majority setting is controlled by the `writeConcernMajorityJournalDefault` setting when establishing the replica set. Also, note that starting in MongoDB 4.2, the current value used to determine the majority is revealed by `rs.status().writeMajorityCount`.

- `wTimeoutMs=<int>`: Set this value to an integer to indicate how many milliseconds your application is willing to wait for the write operation to complete. A value of `0` indicates the application is willing to wait forever.
- `journal=<boolean>`: Set this value to `True` if you wish maximum data integrity at the cost of performance. In this case, the write operation is not considered successful until the data has been written to the server's disk journal. If the value is `False`, data written to the server memory is considered sufficient for write acknowledgment. This option is only available if replica set members are using *journaling*. This option is mutually exclusive of `fsync`.
- `fsync=<boolean>`: If this value is set to `True`, and the replica set member does not have journaling, acknowledgment is not given until the server has synced all the MongoDB data file to disk. If replica set members use journaling, this option has the same application as `journal`. You cannot use both the `fsync` value and `journal` for a single connection.

In the following example, a connection is made to the `learn_mongodb` replica set, consisting of a list of three members. The write preference is set to `majority`, `fsync` is `True`, and the maximum allowed acknowledgment time is `100` milliseconds:

```
import pprint
from pymongo import MongoClient
hosts    = ['member1.biglittle.local','member2.biglittle.local',\
             'member3.biglittle.local']
replName = 'learn_mongodb'
writePref = 'majority'
writeMs   = 100
client    = MongoClient(hosts, replicaSet=replName, w=writePref, \
                        wTimeoutMs=writeMs, fsync=True)
```

Here is a list of sample data documents set to be inserted:

```
insertData = [
    { "fname":"Fred","lname":"Flintstone"},
    { "fname":"Wilma","lname":"Flintstone"},
    { "fname":"Barney","lname":"Rubble"},
    { "fname":"Betty","lname":"Rubble"} ]
```

We then drop the `test` collection, insert the sample data, and use `find()` to retrieve results, as illustrated in the following code snippet:

```
client.biglittle.test.drop()
client.biglittle.test.insert_many(insertData)
result = client.biglittle.test.find()
for doc in result :
    pprint.pprint(doc)
```

The full example is available here: `/path/to/repo/chapters/14/test_replica_set_connect_with_write_concerns.py`

You can also establish *custom write concerns*, where different groupings of replica set members in a large replica set can be assigned different sets of write concerns using *tag sets*. For more information, see the *Custom Multi-Datacenter Write Concerns* documentation (`https://docs.mongodb.com/manual/tutorial/configure-replica-set-tag-sets/#configure-custom-write-concern`).

Read preferences and write concerns can also be set at the collection level. For more information, see the PyMongo documentation on the `with_options()` method (`https://pymongo.readthedocs.io/en/stable/api/pymongo/collection.html#pymongo.collection.Collection.with_options`). Also, have a look at `writeConcernMajorityJournalDefault` (`https://docs.mongodb.com/manual/reference/replica-configuration/#rsconf.writeConcernMajorityJournalDefault`).

Summary

In this chapter, you were shown how to monitor replica set status and were given tips on factors to watch that might indicate a potential performance problem. Aspects covered in this section included managing the size of the *oplog* and viewing the overall replica set status.

You were also shown how to back up and restore a replica set, as well as how to resynchronize a replica set member.

You then learned that the application program code does not need to change when addressing a replica set. You were shown how to enhance your applications to take advantage of a replica set in terms of failover, directing preferences for read operations, and ensuring data integrity by manipulating write concerns.

In the next chapter, we will show you how to scale your database horizontally by implementing a *sharded cluster*.

Deploying a Sharded Cluster

15

At some point, the amount of data in a given collection might become so large that the company notices significant performance issues. In this chapter, you'll learn how to deploy a sharded cluster. The first section in this chapter focuses on the big picture: what a sharded cluster is, how it works, and what the business rationale might be for its deployment. The other two sections of this chapter show you how to configure and deploy a sharded cluster.

In this chapter, we'll cover the following topics:

- MongoDB sharding overview
- Sharded cluster configuration
- Sharded cluster deployment

Let's get started!

Technical requirements

The following is the minimum recommended hardware for this chapter:

- Desktop PC or laptop
- 2 GB free disk space
- 4 GB of RAM
- 500 Kbps or faster internet connection

The following are the software requirements for this chapter:

- OS (Linux or Mac): Docker, Docker Compose, Git (optional)
- OS (Windows): Docker for Windows and Git for Windows
- Python 3.x, PyMongo driver, and Apache (already installed in the Docker container used for this book)

The installation of the required software and how to restore the code repository for this book was explained in `Chapter 2`, *Setting Up MongoDB 4.x*. To run the code examples in this chapter, open a Terminal window (Command Prompt) and enter these commands:

```
cd /path/to/repo/chapters/14
docker-compose build
docker-compose up -d
```

This brings up three Docker containers that are used as part of the demonstration replica set configuration. Further instructions on the replica set's setup are provided in this chapter. When you have finished working with the examples covered in this chapter, return to the Terminal window (Command Prompt) and stop Docker, as follows:

```
cd /path/to/repo/chapters/14
docker-compose down
```

The code used in this chapter can be found in this book's GitHub repository at `https://github.com/PacktPublishing/Learn-MongoDB-4.x/tree/master/chapters/15`.

MongoDB sharding overview

MongoDB has a *horizontal scaling* capability known as *sharding*. In this section, we'll introduce you to the key terms, how sharding works, why you might want to deploy it, as well as how to establish a business rationale for deployment. Let's start with the basics: learning what a *sharded cluster* is.

What is a sharded cluster?

In order to properly understand a MongoDB *sharded cluster*, you need to know something about its key components: *shards, mongos,* and the *config server*. Here is a brief summary of the key components:

- **Shards**: A *shard* is literally a fragment of a database collection. Shards are used when the amount of data in a single collection is extremely large. Accordingly, the best practice is for each shard to reside on a replica set (see `Chapter 13`, *Deploying a Replica Set*). A *sharded cluster* is a term used to represent the group of servers or replica sets that, as a whole, represent the fragmented collection.

- **mongos**: A mongos instance routes database queries to the appropriate shard. It is started much like a mongod (the MongoDB server daemon) instance, with many of the same startup options. Also, much like mongod, you can open a shell on a mongos instance in order to monitor and manage a sharded cluster. Program application queries must be directed to a mongos instance rather than individual mongod instances in order to access all sharded data. It is a common practice to start the mongos instance on the same server where the application code is running. It's possible to have more than one mongos instance pointing to the same sharded cluster.

- **Config server**: In order to maintain order, the shards and routers are controlled by a replica set of *config servers*. As of MongoDB version 3.4, sharded cluster config servers must be implemented as a replica set. The config server replica set stores metadata associated with the location of blocks of data, referred to as *chunks*, within the shards, and the *ranges* used to control the size of each chunk. In addition, the config server replica set can hold authentication information associated with internal server communications, as well as role-based access control information.

- **Sharded cluster balancer**: The MongoDB *sharded cluster balancer* runs as a background process on the primary server of the config server replica set. Its responsibility is to monitor the distribution of data between shards. A value known as the *migration threshold* is automatically set (or can be set manually, as described in `Chapter 16`, *Sharded Cluster Management and Development*). When this value is exceeded, the sharded cluster balancer automatically kicks in and moves chunks of data between shards until an equal number of chunks per shard has been achieved.

Please refer to the following links for more information:
Shard: `https://docs.mongodb.com/manual/core/sharded-cluster-shards/#shards`.
Mongos: `https://docs.mongodb.com/manual/core/sharded-cluster-query-router/#mongos`.
*Config
Servers:* `https://docs.mongodb.com/manual/core/sharded-cluster-config-servers/#config-servers`
Sharded Cluster Balancer: `https://docs.mongodb.com/manual/core/sharding-balancer-administration/#sharded-cluster-balancer`.

Here is a diagram to give you a better idea of how a sharded cluster might appear conceptually. The yellow circles in the diagram shown here represent replica sets:

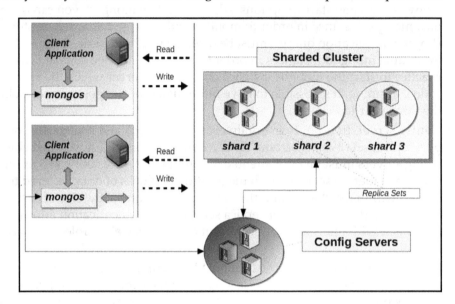

The client application connects to a *mongos* instance. The *mongos* instance conveys read and write requests to the sharded cluster. The *config servers* (deployed as a replica set) convey metadata information to the *mongos* instances and also manage information on the state of the sharded cluster. Now, let's have a look at why you might want to deploy a sharded cluster.

 It's also worth noting that replica sets hosting shards can host other collections as well, even if they're not sharded.

Why deploy a sharded cluster?

In order to understand why you might want to deploy a sharded cluster, we need to further expand upon two concepts you might have heard of before in other contexts: *vertical scaling* and *horizontal scaling*.

Vertical scaling

Vertical scaling is when you increase the power and capacity of a single server. Upgrades could include a more powerful CPU, adding memory, and increasing storage capacity. For example, you might have a server using a second-generation Intel Xeon Scalable processor. Vertical scaling could involve updating the CPU from 16 cores and 32 threads to 24 cores and 48 threads, as an example. Vertical scaling could also involve more obvious upgrades such as increasing the amount of server RAM, adding more and larger hard drives or SSDs to a RAID array, and so on.

Vertical scaling has an advantage in that you only need to worry about a single server: no additional network cards, IP addressing or routing problems, no further strain on server room power requirements, and no additional worries about heat buildup. Also, in some cases, costs might actually be lower for vertical scaling as you are not adding entire additional computers to your network, just internal components.

Vertical scaling has a serious disadvantage, however, in that there is a limit as to how far you can expand. Even the computational power and tremendous storage capacity of today's computers might *still* not be enough to handle massive amounts of streaming data. As just one example of massive data, imagine you are working with health officials to design an application that accepts mobile phone tracking information in order to help prevent the spread of a pandemic virus. If the health officials planned to implement this in a country as populous as India, how long would it be before a single server, even one that has been vertically scaled, became overwhelmed?

The answer to this would be to implement *horizontal scaling*, which we'll discuss next.

Horizontal scaling

Horizontal scaling involves setting up additional servers that are collectively referred to as a *cluster*. The cluster is then managed by a control mechanism that is responsible for assigning tasks to cluster member servers. One example of horizontal scaling that many of you will be familiar with is the concept of a *web server farm*: where you have a group of web servers that service requests on the backend but appear to the internet as a single server. The control mechanism would be a scheduling daemon that determines which server in the web server farm has the most available bandwidth, and directs the next request to that server.

In the context of MongoDB, horizontal scaling is achieved by splitting data in a large collection into fragments referred to as *shards*. All the shards together are referred to as a *sharded cluster*.

The advantage of horizontal scaling is that you can simply continue to add servers as needed, giving you virtually unlimited storage. Another benefit is that read and write operations can be performed in parallel, producing a faster response time. Even if a server or replica set hosting a shard becomes unavailable, the remaining shards can still respond to read and write requests.

In many cases, the cost of adding additional off-the-shelf computers to create a sharded cluster is less expensive than vertically scaling an existing computer. Horizontal scaling can also lead to potentially better performance in that a certain amount of parallel processing is possible. However, please bear in mind that breaking up a collection into shards is risky if the server hosting one of the shards goes down. This is why MongoDB recommends that you implement each shard as a replica set. Accordingly, even though the cost of a *single* server might be less than gaining the equivalent power through vertical scaling efforts, in reality, you probably need to add at least *three* servers per shard, which might negate the cost differential.

With these considerations in mind, let's turn our attention to developing a business rationale for deploying a sharded cluster.

Developing a business rationale for sharded cluster deployment

As we mentioned earlier in this chapter, the main driving factor is the amount of data your website needs to handle. Any time you need to handle large amounts of streaming data where the data needs to be stored for later review or analysis, the size of the database could quickly increase as the number of users increases. A decision point arrives when one of two things occurs:

- The amount of data exceeds the existing server's hard drive capacity
- Sluggish and unacceptable performance is experienced as read and write requests are bottlenecked

Once either or both of these obstacles are encountered, you need to decide between *vertical* and *horizontal* scaling solutions. If you strongly feel that the amount of data does not increase exponentially, or if you feel that, by adding a bit more RAM, the current server could handle the load, a vertical solution might be in order. If, on the other hand, there is no end in sight for data expansion, and the server is already at its maximum RAM, a horizontal scaling solution is called for.

Implementing a sharded cluster can be accomplished *incrementally*. You can add a single replica set and split collections within the large database into two shards as a first phase. Later, if the database continues to grow in size, as a second phase, you can add another replica set, and further split the collections across three shards, and so on and so forth.

As we mentioned earlier, building a sharded cluster should be implemented in the form of one replica set per shard. Accordingly, right away, you need to calculate the cost of three additional servers (or Docker containers in a cloud environment), as well as the additional time needed for normal server maintenance. Each server has an operating system that needs to be updated periodically. Each server has to be added to the overall network infrastructure, configured for routing, their log files monitored, and so forth.

Next, you need to calculate the costs associated with the following:

- **What is the minimum unit of data that has business value?**: You need to determine what minimum units of data are of value to your business. For example, if your business consists of streaming videos, what is the average size of the videos that are posted? If your business is developing metrics to track the spread of a virus, what is the minimum time unit that is considered of value (minutes, hours, days, weeks, months, and so on)? Based on that time unit, how much data are you collecting that is associated with that time unit?
- **What is the cost of losing a minimum unit of data?**: Once you have determined the minimum unit of data, you need to determine the cost associated with losing that unit of data. The reason why you might lose data is because your server is overloaded and unable to handle the additional amount. For example, your company offers the *latest videos* from a popular provider. However, if that video is suddenly no longer available, will customers become annoyed and move to a competitor?

  Using the second example mentioned here, let's say that your local government depends on virus information that is updated hourly. If the last hour of data is not available, the local government would be caught unaware if a new outbreak developed, costing dozens of lives!

- **What is the cost of losing a customer?**: If your company starts to lose business because it can no longer handle the amount of data produced, how much money does each lost customer represent to your company? However, you also need to balance these costs against the nature of your business. If your business is made available for free, there might not be any monetary cost associated with the loss of a customer because the customers are not paying anything.

However, you may be receiving advertising revenue by attracting people to the site, in which case fewer people visiting means loss of advertising revenue.

The process of establishing a business rationale for deploying a sharded cluster thus becomes a matter of balancing the costs associated with adding three new servers against the potential costs associated with the considerations listed here.

Now, let's have a look at sharded cluster configuration.

Sharded cluster configuration

As with establishing replica sets, setting up a sharded cluster involves making modifications to the `mongod.conf` files of the servers going to be involved in the deployment. Before we get into the configuration details, however, it's extremely important that you understand the importance of the *shard key*. The choice of shard key can make or break the success of a sharded cluster implementation.

Understanding the shard key

The *shard key* is a field (or fields) in a collection document. At this point, you're probably wondering: *okay... so, what's the big deal?* To answer this rhetorical question: the shard key is used by the *sharded cluster balancer* (https://docs.mongodb.com/manual/core/sharding-balancer-administration/#sharded-cluster-balancer) to determine how documents are distributed into chunks, and how chunks are distributed between shards. This can be quite a weighty decision as a poor choice of shard key could result in an uneven distribution of documents, leading to poor performance and an overwhelmed server. What's even more important is that the choice of shard key *cannot be changed* once the sharded cluster is initiated. Accordingly, you need to put careful thought into your choice for a shard key!

As of MongoDB 4.2, although you are still not allowed to change the choice of shard key, you can now change the value of the data in the shard key. This feature is not available in earlier versions of MongoDB. As of MongoDB 4.4, a new database command called `refineCollectionShardKey` is available, allowing you to add to an existing shard key.

In MongoDB 4.2 and below, the shard key size cannot exceed 512 bytes. In MongoDB 4.4 this limit has been removed.

In order to discuss shard keys, let's have a look at document distribution within shards.

Document distribution within shards

For greater efficiency, the MongoDB sharded cluster balancer (the sharding mechanism) doesn't operate at the document level. Rather, documents are grouped into *chunks*. Which documents are placed into a given chunk is determined by the shard key range. *Mongo* shell helper methods exist that let you view the size of individual shards and manually move one or more chunks from one shard to another. This process is known as *chunk migration* (covered in the next chapter, Chapter 16, *Sharded Cluster Management and Development*).

Here is a diagram that illustrates the shard key's distributional role:

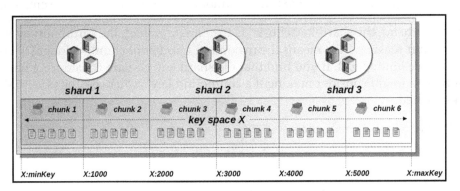

The overall diagram represents a sharded cluster. In this diagram, we're assuming the shard key is a field named **X**. Within each shard is a distribution of two chunks. Within each chunk is a group of documents. The *key space* for **X** ranges from its smallest value, labeled **X:minKey**, to its largest, labelled **X:maxKey**. The range of documents placed into **chunk 1** are those where the value of the document field, **X**, is less than 1,000. In **chunk 2**, the documents have a value for **X** greater than or equal to 1,000, but less than 2,000, and so forth.

Next, we'll have a look at the guidelines for choosing a shard key.

Choosing a shard key

As mentioned earlier, the value of the field or fields chosen as the shard key influences how the balancer decides to distribute documents into chunks. In order to choose a shard key that produces the evenest distribution of documents within chunks, careful consideration needs to be made of the following: cardinality, frequency, and change. Let's have a look at cardinality first.

Cardinality

According to the Merriam-Webster Dictionary, cardinality is defined as follows:

> *Definition of cardinality: The number of elements in a given mathematical set*
> *"Cardinality." Merriam-Webster.com Dictionary, Merriam-Webster,* https://www.
> merriam-webster.com/dictionary/cardinality.

So, for example, if you have a Python list of x = [1,2,3], its *cardinality* would be three. The cardinality of the set represented by the shard key represents the maximum number of shards possible. As an example, if your shard key is a country field containing ISO 3 country codes, the cardinality of this set would be the number of ISO 3 country codes (https://unstats.un.org/unsd/tradekb/Knowledgebase/Country-Code). This number at the time of writing, and thus the maximum number of chunks that could be created, is 246.

If, on the other hand, the shard key chosen is the seasons field, the cardinality would be 4 as there are four seasons. This means the maximum number of chunks MongoDB could create would be limited to 4. If you find that the shard key you are considering has low cardinality, you might consider making it a compound key to increase the potential number of chunks created.

Now, let's have a look at frequency.

Frequency

Shard key frequency refers to how many documents have the same shard key value. As an example, if you choose the ISO 3 country code as a shard key, let's say that 67% of your customers are in India and the other 33% are evenly spread across the collection. When the sharding process completes, you might end up with a skewed distribution that, visually, might appear like this:

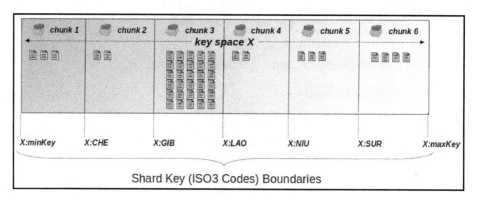

Shard Key (ISO3 Codes) Boundaries

In this example, the value of **minKey** would be *ABW* (Aruba), while the value of **maxKey** would be *ZWE* (Zimbabwe). To avoid the effect of uneven frequency, combining the ISO 3 country code field with another field such as telephone area dialing code might produce a more even frequency. Finally, we'll examine how the rate of change affects the sharding process.

Change

If the field chosen as the shard key is incremented or decremented, or changes *monotonically* in some form (for example *AA, AB, AC, AD ... AZ, BA, BB*, and so on) as new documents are added to the collection, you are at risk of having all the new documents being inserted into the last chunk if the shard key value increases, or the first chunk where the value of the shard key decreases. The reason this happens is that when the collection is initially sharded, shard key ranges are established and documents are inserted into chunks. As new data is added to the collection, the values of *minKey* and *maxKey* are continually upgraded, but the chunk shard key ranges remain the same.

To illustrate this point, in the following diagram, the initial sharding process produces shard key boundaries that range from 101 to 130. As the client application continues to write data, the value of X increments, so the value of *maxKey* is continuously updated. However, as the boundaries have already been set, additional documents all end up in chunk 6:

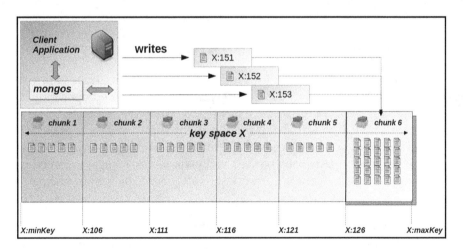

For more information, see the documentation article on *Shard Keys* (https://docs.mongodb.com/manual/core/sharding-shard-key/#shard-keys). Pay close attention to the sub-section on choosing a shard key.

Now that you have an idea of how to choose a shard key, let's have a look at shard key strategies.

Shard key strategies

Before you start configuring your database for sharding, another choice needs to be made: *hashed sharding* or *ranged sharding*. In either case, you still need to wisely choose a shard key. However, the role of the key differs, depending on the shard key strategy chosen. First, let's have a look at hashed sharding.

Hashed sharding

Hashed sharding causes MongoDB to create a *hashed index* on the value of the shard key, rather than using the key as is. Hashed sharding tends to produce a more even distribution of data within the sharded cluster, and is an ideal solution if the shard key has high cardinality or is a field that automatically increments. Because the hashed value is used to perform distribution rather than the actual value of the shard key field itself, a shard key that automatically increments does not cause MongoDB to eventually dump all data into the last shard (as discussed earlier).

The performance we get when using hashed sharding is very slightly lower. When a new document is added to the sharded collection, its shard key value needs to be hashed before it's added to the hashed index. However, the overall gain in performance resulting in even distribution could potentially compensate for this loss.

The main disadvantage of using hashed sharding is a decreased number of *targeted operations* and an increased number of *broadcast operations*. A targeted operation is where the *mongos* instance has already identified the exact shard where the requested data resides. This operation is immediately available if the shard key is not hashed as *mongos* is able to determine the exact shard due to its knowledge of shard key ranges assigned to shards. Because of the anonymous nature of a hashed value, however, *mongos* is forced to perform a broadcast operation instead, sending a request to all the shards first in order to locate the requested data.

Please refer to the following links for more information:
Hashed Sharding: `https://docs.mongodb.com/manual/core/hashed-sharding/`
Targeted Operations: `https://docs.mongodb.com/manual/core/sharded-cluster-query-router/#targeted-operations`
Broadcast Operations: `https://docs.mongodb.com/manual/core/sharded-cluster-query-router/#broadcast-operations`

Now, let's have a quick look at range-based sharding.

Range-based sharding

In *range-based sharding* (`https://docs.mongodb.com/manual/core/ranged-sharding/`
`#ranged-sharding`), MongoDB simply uses the direct value of the chosen shard key to determine the shard boundaries. This has a distinct advantage in that application program code can perform read and write operations directly specifying the value of the shard key. This, in turn, allows *mongos* to perform a targeted operation, resulting in a performance boost. The disadvantage of this strategy is that the onus is on the MongoDB DevOp to wisely choose a shard key, paying close attention to cardinality, frequency, and the rate of change. A bad choice of shard key quickly results in poor performance and the uneven distribution of data among the shards.

In a range-based sharding implementation, because the shard key is known, the application code can include the shard key in its queries. If the shard key is compound (for example, ISO 3 country code plus telephone area dialing code), the query could alternatively include just the prefix (for example, just the ISO 3 country code). This information is enough to allow the *mongos* instance to conduct a targeted operation. If the shard key information is unknown, missing, or hashed, the *mongos* instance is forced to perform a broadcast in order to locate the appropriate shard to handle the request.

Now, we'll turn our attention to configuration.

Sharded cluster member configuration

The sharding options in the `mongod.conf` file include three `sharding.*` directives, as well as two related directives (`replication.replSetname` and `net.bindIp`), as summarized here:

- `sharding.clusterRole`: The value for this directive is in the form of a string and can be either `configsvr` or `shardsvr`. If the value is set to `configsvr`, the *mongod* instance starts listening on port `27019` by default. The instance then starts acting as a *config server*, handling shard-related metadata. If, on the other hand, the value is set to `shardsvr`, the *mongod* instance listens on port `27018` by default and is able to handle MongoDB data, including the ability to handle *shards*.

- `sharding.archiveMovedChunks`: This directive is in the form of a Boolean value. The default is `false`, meaning that documents residing in chunks that have *migrated* out of the shard, either through a manual operation (discussed in the next chapter) or by means of the balancer, are not saved on this server.
- `sharding.configDB`: This setting is a string representing the name of the replica set that hosts the config server, and a list of `servers:ports` in that replica set. An example would be as follows, where `<REPLICA_SET>` is the name of the replica set:

```
sharding:
    configDB: <REPLICA_SET>/server1.example.com:27019,\
        server2.example.com:27019
```

- `replication.replSetname`: Although only required for a config server, as discussed earlier in this chapter, it is a best practice to configure shard *mongod* instances as a replica set in light of the massive amount of data to be handled. Accordingly, you would also indicate the name of the replica set that this server belongs to.
- `net.bindIp`: It's important to ensure that members of the sharded cluster, as well as replica set members, are able to communicate with each other over the network. Accordingly, you need to set this directive to the appropriate *IP address* or have it listen to *all* IP addresses (for example, use a setting of `0.0.0.0`).

Now that the `mongod.conf` file has been updated with the appropriate configuration settings, it's time to have a look at sharded cluster deployment.

 Please refer to the following link for more information on *sharding options*: `https://docs.mongodb.com/manual/reference/configuration-options/index.html#sharding-options`.

Sharded cluster deployment

The following is a general overview of the steps needed to deploy a sharded cluster. This overview is loosely based on the MongoDB documentation article *Deploy a Sharded Cluster*. The steps are summarized here:

1. Confirm network communications
2. Deploy the *config server* replica set
3. Deploy the *shard* replica sets

4. Start a *mongos* instance to represent the sharded cluster
5. Restore data
6. Add shards to the cluster
7. Enable sharding for the target database
8. Choose a shard key
9. Shard the target collection

Before we start down this list, however, let's take a quick look at how to use Docker to model a sharded cluster for testing and development purposes.

Deploy a Sharded Cluster: `https://docs.mongodb.com/manual/tutorial/deploy-shard-cluster/#deploy-a-sharded-cluster`.

Using Docker to model a sharded cluster

As we covered in `Chapter 13`, *Deploying a Replica Set*, Docker technology can be used to model a sharded cluster as well. In this case, however, in order to minimize the impact it has on a single development computer, all replica sets are modeled as *single-node replica sets*. This greatly reduces system requirements and allows DevOps to model the replica set on computers that are potentially less powerful than the ones used in actual production.

A *single node* replica set is simply a single *mongod* instance assigned a unique replica set name. When initiating the replica set, the replica set configuration document, which is supplied as an argument to `rs.initiate()`, only contains a single member name. When the *mongod* instance is started, seeing as there is only one member, it automatically elects itself as the primary.

In this sub-section, we'll examine the Docker Compose configuration, individual *Dockerfiles*, a common *hosts* file, and the `mongod.conf` files used to create the sharded cluster model. Let's start with the `docker-compose.yml` file.

Understanding the docker-compose.yml file

As in `Chapter 13`, *Deploying a Replica Set*, we start by defining a `docker-compose.yml` file that defines the five containers we use in this illustration to model a sharded cluster: one container to host the config server, one to host the *mongos* instance, and three to represent shards. At the top of the file, we place the `version` directive.

This informs Docker Compose that this file follows directives compatible with version 3:

```
version: "3"
```

We then define `services`, starting with the first member of the sharded cluster. In this definition, you can see that we assign a name of *learn-mongo-shard-1* to the Docker container and a hostname of *shard1*. The `image` directive defines a name for the image (ultimately derived from the official MongoDB Docker image) created by Docker Compose, as shown here:

```
services:
  learn-mongodb-shard-1:
    container_name: learn-mongo-shard-1
    hostname: shard1
    image: learn-mongodb/shard-1
```

Under `volumes`, we define the location of the database, a common volume, and mappings to the `hosts` and `mongod.conf` files:

```
volumes:
  - db_data_shard_1:/data/db
  - ./common_data:/data/common
  - ./common_data/hosts:/etc/hosts
  - ./common_data/mongod_shard_1.conf:/etc/mongod.conf
```

> Note that the port is mapped to `27018` and not `27017` because this is a shard, not a normal *mongod* instance.

Additional directives define the location of the *Dockerfile*, how to restart it, and what options to provide to the startup entry point shell script. Finally, you'll see a `networks` directive that defines the IP address of the *learn-mongodb-shard-1* container:

```
ports:
  - 27111:27018
build: ./shard_1
restart: always
command: -f /etc/mongod.conf
networks:
  app_net:
    ipv4_address: 172.16.1.11
```

The definitions for the next two containers (not shown) are identical, except the identifying numbers are 2 and 3 instead of 1.

Also, the outside port mapping needs to be different, as well as the IP address. You can view the entire file at /path/to/repo/chapters/14/docker-compose.yml.

Next, still under the services key, we have the *config server* definition. Note that it's almost identical to the definition for a shard, except that the port maps to 27019. Also, of course, the hostname, volume name, and IP address values are different:

```
learn-mongodb-config-svr:
  container_name: learn-mongo-config
  hostname: config1
  image: learn-mongodb/config-1
  volumes:
   - db_data_config:/data/db
   - ./common_data:/data/common
   - ./common_data/hosts:/etc/hosts
   - ./common_data/mongod_config.conf:/etc/mongod.conf
  ports:
   - 27444:27019
  build: ./config_svr
  restart: always
  command: -f /etc/mongod.conf
  networks:
    app_net:
      ipv4_address: 172.16.1.44
```

The last entry under services defines the *mongos* instance. Although this daemon could be run just about anywhere, including the application server, for the purposes of this chapter, we'll define it in its own container. Aside from the unique hostname, port, and IP address, this definition lacks a /data/db volume mapping. That is because a *mongos* instance does not handle MongoDB data: rather, it serves as a router between the application's MongoDB driver and the sharded cluster. It also communicates with the config server to gather sharded cluster metadata:

```
learn-mongodb-mongos:
  container_name: learn-mongo-mongos
  hostname: mongos1
  image: learn-mongodb/mongos-1
  volumes:
   - ./common_data:/data/common
   - ./common_data/hosts:/etc/hosts
   - ./common_data/mongos.conf:/etc/mongos.conf
  ports:
   - 27555:27020
  build: ./mongos
  restart: always
  command: mongos -f /etc/mongos.conf
```

```
            networks:
              app_net:
                ipv4_address: 172.16.1.55
```

The last two sections in the `docker-compose.yml` file define the overall virtual network over which the containers communicate, as well as external volumes mapped to the `/data/db` folders:

```
    networks:
      app_net:
        ipam:
          driver: default
          config:
            - subnet: "172.16.1.0/24"

    volumes:
      db_data_shard_1: {}
      db_data_shard_2: {}
      db_data_shard_3: {}
      db_data_config: {}
```

Next, we'll look at the additional configuration files needed to complete the model.

Additional files needed by Docker and MongoDB

Each container needs to have a *Dockerfile* in order for *docker-compose* to work. Under `/path/to/repo/chapters/14`, you will see the following directory structure, where a directory exists for each type of container to be built:

```
├── docker-compose.yml
├── config_svr
│   └── Dockerfile
├── mongos
│   └── Dockerfile
├── shard_1
│   └── Dockerfile
├── shard_2
│   └── Dockerfile
└── shard_3
    └── Dockerfile
```

Each *Dockerfile* is identical and contains instructions for the source Docker image, as well as information regarding what additional packages to install.

An example of this is as follows:

```
FROM mongo
RUN \
    apt-get update && \
    apt-get -y upgrade && \
    apt-get -y install inetutils-ping && \
    apt-get -y install net-tools
```

In the `/path/to/repo/chapters/14/common_data` directory, there are JavaScript files used to restore the *BigLittle, Ltd.* sample data. In addition, there are sample JavaScript commands to initialize the shards and replica sets. In this directory, you will also find a common `/etc/hosts` file and `mongod.conf` files for each respective container. This system of five containers (three shards, one config server, and one *mongos* instance) can be brought online using the `docker-compose up` command, as shown in the following screenshot:

```
File Edit View Search Terminal Help
jed@jed:~/Repos/learn-mongodb/chapters/14$ docker-compose up
Starting learn-mongo-mongos  ... done
Starting learn-mongo-shard-3 ... done
Starting learn-mongo-shard-1 ... done
Starting learn-mongo-config  ... done
Starting learn-mongo-shard-2 ... done
Attaching to learn-mongo-shard-1, learn-mongo-shard-3, learn-mongo-mongos, learn-mongo-shard-2, learn-mongo-config
learn-mongo-mongos         | 2020-04-07T08:30:16.697+0000 W  SHARDING [main] Running a sharded cluster with fewer
 than 3 config servers should only be done for testing purposes and is not recommended for production.
```

A warning is displayed because running a config server with less than three members in its replica set is not considered a best practice, and should not be used for production. Also, bear in mind that none of the single-node replica sets have been initialized yet.

Now, let's take a look at the list of steps mentioned in the overview for this section, starting with confirming network communications.

Confirming sharded cluster network communications

As you might have noticed, we installed the `inetutils-ping` Linux package, which allows us to run the `ping` command. For that purpose, you'll find a script called `/path/to/repo/chapters/14/verify_connections.sh` that tests network communications between servers, as shown here:

```
#!/bin/bash
docker exec learn-mongo-shard-1 /bin/bash -c \
```

```
    "ping -c1 shard2.biglittle.local"
docker exec learn-mongo-shard-2 /bin/bash -c \
    "ping -c1 shard3.biglittle.local"
docker exec learn-mongo-shard-3 /bin/bash -c \
    "ping -c1 config1.biglittle.local"
docker exec learn-mongo-config /bin/bash -c \
    "ping -c1 mongos1.biglittle.local"
docker exec learn-mongo-mongos /bin/bash -c \
    "ping -c1 shard1.biglittle.local"
```

After running `docker-compose up`, from another Terminal window, here is the output from the verification script:

```
File Edit View Search Terminal Help
jed@jed:~/Repos/learn-mongodb/chapters/14$ ./verify_connections.sh
PING shard2.biglittle.local (172.16.1.22): 56 data bytes
64 bytes from 172.16.1.22: icmp_seq=0 ttl=64 time=0.120 ms
--- shard2.biglittle.local ping statistics ---
1 packets transmitted, 1 packets received, 0% packet loss
round-trip min/avg/max/stddev = 0.120/0.120/0.120/0.000 ms
PING shard3.biglittle.local (172.16.1.33): 56 data bytes
64 bytes from 172.16.1.33: icmp_seq=0 ttl=64 time=0.113 ms
--- shard3.biglittle.local ping statistics ---
1 packets transmitted, 1 packets received, 0% packet loss
round-trip min/avg/max/stddev = 0.113/0.113/0.113/0.000 ms
PING config1.biglittle.local (172.16.1.44): 56 data bytes
64 bytes from 172.16.1.44: icmp_seq=0 ttl=64 time=0.111 ms
--- config1.biglittle.local ping statistics ---
1 packets transmitted, 1 packets received, 0% packet loss
round-trip min/avg/max/stddev = 0.111/0.111/0.111/0.000 ms
PING mongos1.biglittle.local (172.16.1.55): 56 data bytes
64 bytes from 172.16.1.55: icmp_seq=0 ttl=64 time=0.082 ms
--- mongos1.biglittle.local ping statistics ---
1 packets transmitted, 1 packets received, 0% packet loss
round-trip min/avg/max/stddev = 0.082/0.082/0.082/0.000 ms
PING shard1.biglittle.local (172.16.1.11): 56 data bytes
64 bytes from 172.16.1.11: icmp_seq=0 ttl=64 time=0.166 ms
--- shard1.biglittle.local ping statistics ---
1 packets transmitted, 1 packets received, 0% packet loss
round-trip min/avg/max/stddev = 0.166/0.166/0.166/0.000 ms
jed@jed:~/Repos/learn-mongodb/chapters/14$
```

Now, we'll have a look at deploying the replica set to represent the *config server*.

Deploying the config server replica set

The next step is to deploy the config server. As mentioned earlier in this chapter, the config server stores sharding metadata and must be deployed as a replica set.

Accordingly, the `mongod.conf` file appears as follows:

```
systemLog:
    destination: file
    path: "/var/log/mongodb/mongodb.log"
    logAppend: true
storage:
    dbPath: "/data/db"
    journal:
        enabled: true
net:
    bindIp: 0.0.0.0
    port: 27019
replication:
    replSetName: repl_config
sharding:
    clusterRole: configsvr
```

To initialize the replica set, we use the `rs.initiate()` shell method. The following syntax is sufficient for the purposes of this illustration:

```
doc = {
    _id : "repl_config",
    members: [
        { _id: 1, host: "config1.biglittle.local:27019" }
    ]
}
rs.initiate(doc);
```

We open a *mongo* shell on the config server and issue the command shown in the preceding code block. As discussed earlier in this section, we initiate a single-node replica set for testing and development purposes only.

Note that as we configured the *mongod* instance to listen on port `27019`, we need to add the `--port 27019` option to the command to run the *mongo* shell.

Initially, the single node runs an election and votes itself as the primary.

The result is shown in the following screenshot:

```
File Edit View Search Terminal Help
root@config1:/# mongo --port 27019 --quiet
> doc = {
...      _id : "repl_config",
...      members: [
...          { _id: 1, host: "config1.biglittle.local:27019" }
...      ]
... }
{
        "_id" : "repl_config",
        "members" : [
                {
                        "_id" : 1,
                        "host" : "config1.biglittle.local:27019"
                }
        ]
}
> rs.initiate(doc);
{
        "ok" : 1,
        "$gleStats" : {
                "lastOpTime" : Timestamp(1586315206, 1),
                "electionId" : ObjectId("000000000000000000000000")
        },
        "lastCommittedOpTime" : Timestamp(0, 0),
        "$clusterTime" : {
                "clusterTime" : Timestamp(1586315206, 1),
                "signature" : {
                        "hash" : BinData(0,"AAAAAAAAAAAAAAAAAAAAAAAAAAA="),
                        "keyId" : NumberLong(0)
                }
        },
        "operationTime" : Timestamp(1586315206, 1)
}
repl_config:SECONDARY>
repl_config:PRIMARY>
```

Next, we'll examine the process of deploying *shard* replica sets.

Deploying shard replica sets

Now, we'll repeat a very similar process for each member of the sharded cluster. In our illustration, we'll initiate three single-node replica sets, each representing a shard. The `mongod.conf` file for each shard is similar to the following:

```
systemLog:
   destination: file
   path: "/var/log/mongodb/mongodb.log"
   logAppend: true
storage:
   dbPath: "/data/db"
   journal:
      enabled: true
net:
```

```
      bindIp: 0.0.0.0
      port: 27018
replication:
      replSetName: repl_shard_1
sharding:
      clusterRole: shardsvr
```

To initialize the shard replica set, as described in the previous sub-section, we open a *mongo* shell to each of the member servers in turn. It's important to add the `--port 27018` option when opening the shell. We can then initialize the replica set using the following shell commands:

```
doc = {
    _id : "repl_shard_1",
    members: [
            { _id: 1, host: "shard1.biglittle.local:27018" }
    ]
}
rs.initiate(doc);
```

The resulting output is similar to the replica set initialization shown for the config server. Once each of the shard replica sets has been initialized, we turn our attention to starting a *mongos* instance to represent the sharded cluster.

Starting a mongos instance

The `mongod.conf` file for a *mongos* instance has one major difference from the shard configuration files: there is no *storage* option. The purpose of the *mongos* instance is not to handle data, but rather to serve as an intermediary between a MongoDB application programming language driver, the config server, and the sharded cluster. The following is the `mongos.conf` file that's used to start the *mongos* instance. It gives us access to our model's sharded cluster:

```
systemLog:
    destination: file
    path: "/var/log/mongodb/mongodb.log"
    logAppend: true
net:
    bindIp: 0.0.0.0
    port: 27020
sharding:
    configDB: repl_config/config1.biglittle.local:27019
```

To start the *mongos* instance, we use the following command:

```
mongos -f /etc/mongos.conf
```

The following screenshot shows the results:

```
File Edit View Search Terminal Help
root@mongos1:/# mongos -f /etc/mongos.conf
2020-04-08T03:32:31.468+0000 W  SHARDING [main] Running a sharded cluster with f
ewer than 3 config servers should only be done for testing purposes and is not r
ecommended for production.
root@mongos1:/# ps -ax
  PID TTY       STAT   TIME COMMAND
    1 ?         Ssl    0:04 mongos -f /etc/mongos.conf
   40 pts/0     Ss     0:00 /bin/bash
   55 pts/0     R+     0:00 ps -ax
root@mongos1:/#
```

When opening a shell, we can simply indicate the port that the *mongos* instance is listening on. Here is the result:

```
File Edit View Search Terminal Help
root@mongos1:/# mongo --port 27020 --quiet
Welcome to the MongoDB shell.
For interactive help, type "help".
For more comprehensive documentation, see
        http://docs.mongodb.org/
Questions? Try the support group
        http://groups.google.com/group/mongodb-user
mongos>
```

Now, we can restore the sample data before proceeding.

Restoring the sample data

For the purposes of this example, we'll restore four collections of the *biglittle* database. The data is restored on the *shard1* server only. Later, we'll shard the last collection, *world_cities* (covered in later subsections).

For the purposes of this illustration, the sample data is mapped to the /data/common directory in the Docker container representing *shard1*. You can perform the restore by running the shell scripts from a Terminal window inside the container, as shown here:

```
docker exec -it learn-mongo-shard-1 /bin/bash
mongo --port 27018 /data/common/biglittle_common_insert.js
mongo --port 27018 /data/common/biglittle_loans_insert.js
```

```
mongo --port 27018 /data/common/biglittle_users_insert.js
mongo --port 27018 /data/common/biglittle_world_cities_insert.js
```

Alternatively, from outside the container, use the
`/path/to/repo/chapters/14/restore_sample_data.sh` script, as shown here:

```
#!/bin/bash
docker exec learn-mongo-shard-1 /bin/bash -c \
    "mongo --port 27018 /data/common/biglittle_common_insert.js"
docker exec learn-mongo-shard-1 /bin/bash -c \
    "mongo --port 27018 /data/common/biglittle_loans_insert.js"
docker exec learn-mongo-shard-1 /bin/bash -c \
    "mongo --port 27018 /data/common/biglittle_users_insert.js"
docker exec learn-mongo-shard-1 /bin/bash -c \
    "mongo --port 27018 /data/common/biglittle_world_cities_insert.js"
```

 Note that we use the backslash (\\) convention to indicate the command should be on a single line.

To confirm the restore, open a *mongo* shell on the *shard1* server and confirm that the collections exist. At this time, the data only resides on the *shard1* server. If you open a shell on either *shard2* or *shard3*, you will see that there is no *biglittle* database. Now, it's time to add shards to the cluster.

Adding shards to the cluster

To create a sharded cluster, you need to add servers configured as shard servers under the following conditions:

- You need to open a *mongo* shell on a *mongos* instance that's been configured for the sharded cluster you are building.
- You need to have been granted the appropriate cluster administration role (`https://docs.mongodb.com/manual/reference/built-in-roles/#cluster-administration-roles`).

The shell command to add a shard is `sh.addShard()`. Here is the generic syntax:

```
sh.addShard("REPLICA_SET/MEMBER:PORT[,MEMBER:PORT]");
```

`REPLICA_SET` represents the name of the replica set hosting the shard. `MEMBER` is the DNS or resolvable hostname of the replica set member. `PORT` defaults to `27018` (not the same as a standalone *mongod* instance)!

If sharding has been enabled for the database, when you execute `sh.addShard()`, the sharded *cluster balancer* immediately starts moving chunks to the new shard. Be prepared for a high initial volume of database traffic.

In our model sharded cluster, first, we open a shell on the *mongos* instance. We then issue the `sh.addShard()` shell command to add the first shard, as shown in the following screenshot:

```
File Edit View Search Terminal Help
jed@jed:~$ docker exec -it learn-mongo-mongos /bin/bash
root@mongos1:/# mongo --port 27020 --quiet
Welcome to the MongoDB shell.
For interactive help, type "help".
For more comprehensive documentation, see
        http://docs.mongodb.org/
Questions? Try the support group
        http://groups.google.com/group/mongodb-user
mongos> sh.addShard("repl_shard_1/shard1.biglittle.local:27018");
{
        "shardAdded" : "repl_shard_1",
        "ok" : 1,
        "operationTime" : Timestamp(1586407231, 1),
        "$clusterTime" : {
                "clusterTime" : Timestamp(1586407231, 1),
                "signature" : {
                        "hash" : BinData(0,"AAAAAAAAAAAAAAAAAAAAAAAAAAA="),
                        "keyId" : NumberLong(0)
                }
        }
}
mongos>
```

We then continue to add the *shard2* and *shard3* servers to complete the sharded cluster. Before actually sharding a collection, we need to enable sharding for the target database. We'll cover this next.

Enabling sharding on the database

As an additional safeguard, just setting up the sharded cluster does not automatically start the sharding process. First, you need to enable sharding at the database level. Otherwise, just imagine the chaos that might occur! The primary command to enable sharding is `sh.enableSharding()`. The generic syntax is shown here. Obviously, you need to substitute the actual name of the database:

```
sh.enableSharding("DATABASE")
```

Continuing with our sharded cluster model, still connected to a *mongos* instance, we issue the appropriate command for the *biglittle* database, as shown here:

```
File  Edit  View  Search  Terminal  Help
mongos> sh.enableSharding("biglittle");
{
        "ok" : 1,
        "operationTime" : Timestamp(1586407897, 3),
        "$clusterTime" : {
                "clusterTime" : Timestamp(1586407897, 3),
                "signature" : {
                        "hash" : BinData(0,"AAAAAAAAAAAAAAAAAAAAAAAAAAA="),
                        "keyId" : NumberLong(0)
                }
        }
}
mongos>
```

MongoDB elects a shard to serve as a primary, much as it does within members of a replica set. We use `sh.status()` to confirm that all three servers have been added to the sharded cluster, as shown here:

```
File  Edit  View  Search  Terminal  Help
mongos> sh.status();
--- Sharding Status ---
  sharding version: {
        "_id" : 1,
        "minCompatibleVersion" : 5,
        "currentVersion" : 6,
        "clusterId" : ObjectId("5e8d3fc8e5189742de6075e2")
  }
  shards:
        {  "_id" : "repl_shard_1",  "host" : "repl_shard_1/shard1.biglittle.local:27018",  "state" : 1 }
        {  "_id" : "repl_shard_2",  "host" : "repl_shard_2/shard2.biglittle.local:27018",  "state" : 1 }
        {  "_id" : "repl_shard_3",  "host" : "repl_shard_3/shard3.biglittle.local:27018",  "state" : 1 }
  active mongoses:
        "4.2.5" : 1
  autosplit:
        Currently enabled: yes
  balancer:
        Currently enabled:  yes
        Currently running:  no
        Failed balancer rounds in last 5 attempts:  0
        Migration Results for the last 24 hours:
                No recent migrations
  databases:
        {  "_id" : "biglittle",  "primary" : "repl_shard_1",  "partitioned" : true,  "version" : {  "uuid" :
UUID("a58f2a69-dcca-4d1d-9f06-0f833e82e64f"),  "lastMod" : 1 } }
        {  "_id" : "config",  "primary" : "config",  "partitioned" : true }
                config.system.sessions
                        shard key: { "_id" : 1 }
                        unique: false
                        balancing: true
                        chunks:
                                repl_shard_1      1
                        { "_id" : { "$minKey" : 1 } } -->> { "_id" : { "$maxKey" : 1 } } on : repl_shard_1 Ti
mestamp(1, 0)

mongos>
```

We are now ready to choose a shard key from the document fields of the target collection to be sharded.

Choosing a shard key

For the purposes of this example, we'll shard the *world_cities* collection. The sample data for this collection can be found at `/path/to/repo/sample_data/biglittle_world_cities_insert.js`. Here is a sample document from that collection:

```
{
    "_id" : ObjectId("5e8d4851049ec9ee6b5cfd37"),
    "geonameid" : "3040051",
    "name" : "les Escaldes",
    "asciiname" : "les Escaldes",
    "latitude" : "42.50729",
    "longitude" : "1.53414",
    "feature class" : "P",
    "feature code" : "PPLA",
    "country code" : "AD",
    "cc2" : "",
    "admin1_code" : "08",
    "admin2_code" : "",
    "admin3_code" : "",
    "admin4_code" : "",
    "population" : "15853",
    "elevation" : "",
    "dem" : "1033",
    "timezone" : "Europe/Andorra",
    "modification_date" : "2008-10-15"
}
```

To facilitate future application queries, we choose the *range-based* sharding strategy. This allows us to issue read and write requests with advance knowledge of the contents of the shard key. Some of the fields you can see here are unique (for example, `_id` and `geonameid`), giving us a maximum amount of *cardinality*. Some fields might be missing information (for example `admin3_code` and `admin4_code`), and are thus not good candidates for a shard key. The field labeled `population` is random and changes frequently, and is thus also unsuitable as a shard key.

Two potential candidates might be `timezone` and `country code`. `country code` might turn out to be too restrictive, however, as the distribution of data is skewed. On the other hand, `timezone` could potentially be extremely useful when generating requests. It is also possible to use both fields as the shard key.

If data already exists in the collection, you must create an index on the shard key field(s) before sharding the collection.

Before we perform the actual collection sharding operation, however, as data already exists, we need to create an index on `timezone`, which is the chosen shard key field. Here is the command we issue:

```
db.world_cities.createIndex( { "timezone": 1} );
```

Now, it's time to shard the collection.

Sharding the collection

To shard the collection, you must open a shell to a *mongos* instance. The command to shard a collection is `sh.shardCollection()`. The generic syntax is shown here:

```
sh.shardCollection(DB.COLLECTION, KEY, UNIQUE, OPTIONS)
```

The parameters are summarized here:

- `DB.COLLECTION` (string): This string represents the name of the database and the collection name, separated by a period.
- `KEY` (JSON document): This parameter is in the form of a JSON key-value pair, where the *key* is the name of the shard key field and the value is either `1` or `hashed`. Use *hashed* if you are using the *hashed* sharding strategy. If the index is a compound index, the key is the *prefix* index.
- `UNIQUE` (Boolean): The default value for this parameter is *false*. You need to specify *true* if the index is not hashed and enforces uniqueness. Also, if you plan to add *options* (see the following bullet), you must set this value to *true*.
- `OPTIONS` (JSON document): This parameter is also in the form of a JSON document. The subkeys include `numInitialChunks`, an integer representing how many chunks you wish to allocate immediately. The other possible key is `collation`, which is used to specify internationalized aspects of collection data. For more information, have a look at the documentation on *collation* (https://docs.mongodb.com/manual/reference/collation/).

The following is the command that's used to shard the *world_cities* collection.

In our example, this looks as follows:

```
collection = "biglittle.world_cities";
shardKey   = { "timezone" : 1 };
sh.shardCollection(collection, shardKey);
```

Here is a screenshot of the result:

```
File  Edit  View  Search  Terminal  Help
mongos> collection = "biglittle.world_cities";
biglittle.world_cities
mongos> shardKey     = { "timezone" : 1 };
{ "timezone" : 1 }
mongos> sh.shardCollection(collection, shardKey);
{
        "collectionsharded" : "biglittle.world_cities",
        "collectionUUID" : UUID("280bf009-a166-46e5-9d34-64cf65d14835"),
        "ok" : 1,
        "operationTime" : Timestamp(1586502478, 9),
        "$clusterTime" : {
                "clusterTime" : Timestamp(1586502478, 9),
                "signature" : {
                        "hash" : BinData(0,"AAAAAAAAAAAAAAAAAAAAAAAAAAA="),
                        "keyId" : NumberLong(0)
                }
        }
}
mongos>
```

And that concludes this chapter!

Summary

In this chapter, you learned about MongoDB sharded clusters. We discussed establishing a business rationale for deployment that included a review of the concept of *vertical scaling* in contrast to *horizontal scaling*. You then learned about sharded cluster configuration and the key factors involved in choosing an appropriate shard key: cardinality, frequency, and rate of change. You also learned about the difference between *hashed* and *range-based* sharding strategies.

You then learned how to deploy a sharded cluster by using Docker technology to model a sharded cluster for testing and development purposes. You also learned about how to deploy the key components of a sharded cluster: the config server, shards, and the *mongos* instance. After that, you learned that the config server must be deployed as a replica set and that each shard in the sharded cluster should also be a replica set in order to maintain data integrity and best performance.

In the next chapter, you'll learn how to manage a sharded cluster, as well as learning about programming considerations when writing code that interacts with a sharded cluster.

16
Sharded Cluster Management and Development

In this chapter, you will learn how to manage a sharded cluster. Management aspects include monitoring the shard status and exerting control over the distribution of data between shards. You will also learn about assigning and managing *chunks* and *zones*. After that, you will learn about the impact of sharding on the code of an application program.

In this chapter, we will cover the following topics:

- Managing a sharded cluster
- Sharded cluster impacts on program code

Technical requirements

Minimum recommended hardware:

- Desktop PC or laptop
- 2 **gigabytes** (**GB**) free disk space
- 4 GB of **random-access memory** (**RAM**)
- 500 **kilobits per second** (**Kbps**) or faster internet connection

Software requirements:

- OS (Linux or Mac): Docker, Docker Compose, Git (optional)
- OS (Windows): Docker for Windows and Git for Windows
- Python 3.x, PyMongo driver, and Apache (already installed in the Docker container used for the book)

Installation of the required software and how to restore the code repository for the book is explained in Chapter 2, *Setting Up MongoDB 4.x*.

To run the code examples in this chapter, open a Terminal window (Command Prompt) and enter these commands:

```
cd /path/to/repo/chapters/16
docker-compose build
docker-compose up -d
```

This brings up three Docker containers that are used as part of the demonstration replica set configuration. Further instructions on the replica set setup are given in the chapter. When you have finished working with the examples covered in this chapter, return to the Terminal window (Command Prompt) and stop Docker, as follows:

```
cd /path/to/repo/chapters/16
docker-compose down
```

The code used in this chapter can be found in the book's GitHub repository at `https://github.com/PacktPublishing/Learn-MongoDB-4.x/tree/master/chapters/16`.

Managing a sharded cluster

In order to properly manage a sharded cluster, you need to be able to monitor its status. You might also need to divide the data into *zones*, which can then be assigned to shards. You also have the ability to manually split chunks, adjust the chunk size, and handle backup and restore operations. Another topic of interest is managing the *sharded cluster balancer*, a process that ensures an even distribution of documents into chunks, and chunks into shards. The first topic we'll now examine is monitoring the sharded cluster status.

Monitoring the sharded cluster status

The most important general-purpose monitoring command is `sh.status()`. You need to run this command from a shell connected to a `mongos` instance, as a database user with sufficient cluster administration rights. A screenshot of the output for our sharded cluster model is shown in the *Enabling sharding on the database* sub-section of `Chapter 15`, *Deploying a Sharded Cluster*.

Look carefully for the following information, summarized here:

Information key	What to look for
shards	Should show a list of shards, with a list of tags exactly as you specified.
active mongoses	How many `mongos` instances are connected and their version numbers.
replica set	Whether or not `autosplit` (chunks are automatically split) is enabled on this sharded cluster.
balancer	`Failed balancer rounds in last 5 attempts` should be 0. `Migration Results for the last 24 hours:` You should see a number followed by `Success`. If you see any number greater than 0 in the `Failed balancer rounds in the last 5 seconds` value, look at the log files of each of the shard servers to determine the problem. Possible problems could include communications errors, permissions errors, incorrect shard configuration, and missing or invalid indexes.
databases	Under the key for the database you are monitoring, look for *chunks*. You should see an even distribution of chunks between shards, as per your zones. Next, you see a list of ranges. Make sure they are associated with the correct shard. You should then see a list of `tags`. These should correspond to the zones you created.

As an example of an error, note the following entry after running `sh.status()`:

```
File Edit View Search Terminal Help
  balancer:
          Currently enabled:  yes
          Currently running:  no
          Failed balancer rounds in last 5 attempts:  0
          Migration Results for the last 24 hours:
                  6 : Success
                  83 : Failed with error 'aborted', from repl_shard_1 to repl_shard_2
  databases:
```

If we then examine the log of the primary *mongod* instance in the `repl_shard_1` replica set, we spot the cause of the problem, as illustrated in the following screenshot:

```
File Edit View Search Terminal Help
2020-04-11T05:24:44.970+0000 W  SHARDING [conn16] Chunk move failed :: caused by :: IndexNotFound
: can't find index with prefix { timezone: 1.0 } in storeCurrentLocs for biglittle.world_cities
2020-04-11T05:24:44.970+0000 I  SHARDING [conn16] about to log metadata event into changelog: { _
id: "shard1:27018-2020-04-11T05:24:44.970+0000-5e91549c5167d4f87582aeeb", server: "shard1:27018",
 shard: "repl_shard_1", clientAddr: "172.16.1.44:43752", time: new Date(1586582684970), what: "mo
veChunk.from", ns: "biglittle.world_cities", details: { min: { timezone: MinKey }, max: { timezon
e: "America/Anchorage" }, step 1 of 6: 0, step 2 of 6: 1, to: "repl_shard_2", from: "repl_shard_1
", note: "aborted" } }
```

In this example, the shard migration failed due to a missing or invalid index. Other useful methods operate on the collection directly.

The `db.collection.getShardDistribution()` collection-level command gives you information on how many chunks exist and on which shards they reside. The output from our example system is shown here, using `world_cities` as the collection:

```
File Edit View Search Terminal Help
mongos> db.world_cities.getShardDistribution();

Shard repl_shard_1 at repl_shard_1/shard1.biglittle.local:27018
 data : 9.35MiB docs : 24217 chunks : 1
 estimated data per chunk : 9.35MiB
 estimated docs per chunk : 24217

Totals
 data : 9.35MiB docs : 24217 chunks : 1
 Shard repl_shard_1 contains 100% data, 100% docs in cluster, avg obj size on shard : 405B

mongos>
```

To get even more detail, look at the `shards` property from the `db.collection.stats()` output. You see a breakdown by shard, with information on indexes, caching, compression, query performance, and much more (https://docs.mongodb.com/manual/reference/command/collStats/#output).

We now have a look at sharded cluster *zones*.

Working with sharded cluster zones

An advantage of range-based sharding is that you can create *zones* (https://docs.mongodb.com/manual/core/zone-sharding/#zones) within the sharded data. A *zone* is an arbitrary tag or label you can assign to a range of documents based on the shard key value. The zones can then be later associated with specific shards to allow for the geographic distribution of data. Ranges cannot overlap. Further, the field (or fields) representing the shard key must be indexed so that you are assured the range values are consecutive.

To work with zones within a sharded cluster, you need to do two things: define the zones and then assign ranges. Let's start with the zone definition.

Defining zones

When you define a zone, you also associate a shard to the zone. This can be accomplished with the `sh.addShardToZone()` helper method. Here is the generic syntax:

```
sh.addShardToZone(SHARD, ZONE);
```

Both arguments are strings. SHARD represents the name of the shard, and ZONE is an arbitrary tag or label. In our sharded cluster model, let's say we wish to create zones based on continents and oceans. We might issue the following set of commands to create zones and distribute them between shards:

```
sh.addShardToZone("repl_shard_1", "americas");
sh.addShardToZone("repl_shard_1", "atlantic");
sh.addShardToZone("repl_shard_2", "africa");
sh.addShardToZone("repl_shard_2", "europe");
sh.addShardToZone("repl_shard_3", "asia");
sh.addShardToZone("repl_shard_3", "australia");
sh.addShardToZone("repl_shard_3", "indian");
sh.addShardToZone("repl_shard_3", "pacific");
```

Here is a screenshot of the output from the preceding command:

```
File Edit View Search Terminal Help
mongos> sh.addShardToZone("repl_shard_3", "pacific");
{
        "ok" : 1,
        "operationTime" : Timestamp(1586582099, 2),
        "$clusterTime" : {
                "clusterTime" : Timestamp(1586582099, 2),
                "signature" : {
                        "hash" : BinData(0,"AAAAAAAAAAAAAAAAAAAAAAAAAAA="),
                        "keyId" : NumberLong(0)
                }
        }
}
mongos>
```

> To get a list of shard names, execute sh.status() from a mongo shell connected to a mongos instance, and examine the shards property.

Next, we have a look at assigning ranges to the newly added zones.

Assigning ranges to zones

To assign ranges to zones, from a mongo shell connected to a mongos instance, use the sh.updateZoneKeyRange() helper method. The generic syntax for the shell helper method is as follows:

```
sh.updateZoneKeyRange(DB.COLLECTION, MIN, MAX, ZONE);
```

The arguments are summarized here:

- DB.COLLECTION (string): This string represents the name of the database, and collection name, separated by a period.
- MIN (**JavaScript Object Notation (JSON)**): You can specify a JSON document that describes the shard key field/value key pair considered the lower bound of the range. This value is *inclusive*—that is to say: *greater than or equal to*. If the shard key contains multiple fields, you can include some or all of the fields in the MIN document, as long as you always include the *prefix*.
- MAX (JSON): You can specify a JSON document that describes the shard key field/value key pair considered the upper bound of the range. This value is *exclusive*—that is to say: *less than*. If the shard key contains multiple fields, you can include some or all of the fields in the MAX document, as long as you always include the *prefix*.
- ZONE (string): This value represents the name (that is, tag or label) associated with the given range.

In this example, we divide the world_cities collection into ranges based on the timezone field. Here is how this set of commands might appear:

```
target = "biglittle.world_cities";
sh.updateZoneKeyRange(target, { "timezone":MinKey},
    { "timezone":"America/Anchorage"}, "africa");
sh.updateZoneKeyRange(target, { "timezone":"America/Anchorage"},
    { "timezone":"Arctic/Longyearbyen"}, "americas");
sh.updateZoneKeyRange(target, { "timezone":"Arctic/Longyearbyen"},
    { "timezone":"Asia/Aden"}, "atlantic");
sh.updateZoneKeyRange(target, { "timezone":"Asia/Aden"},
    { "timezone":"Australia/Adelaide"}, "asia");
sh.updateZoneKeyRange(target, { "timezone":"Australia/Adelaide"},
    { "timezone":"Europe/Amsterdam"}, "australia");
sh.updateZoneKeyRange(target, { "timezone":"Europe/Amsterdam"},
    { "timezone":"Indian/Antananarivo"}, "europe");
sh.updateZoneKeyRange(target, { "timezone":"Indian/Antananarivo"},
    { "timezone":"Pacific/Apia"}, "indian");
sh.updateZoneKeyRange(target, { "timezone":"Pacific/Apia"},
    { "timezone":MaxKey}, "pacific");
```

Here is the output from the last command listed:

```
File Edit View Search Terminal Help
mongos> sh.updateZoneKeyRange(target, {"timezone":"Pacific/Apia"}, {"timezone":MaxKey}, "paci
fic");
{
        "ok" : 1,
        "operationTime" : Timestamp(1586582678, 1),
        "$clusterTime" : {
                "clusterTime" : Timestamp(1586582678, 1),
                "signature" : {
                        "hash" : BinData(0,"AAAAAAAAAAAAAAAAAAAAAAAAAAA="),
                        "keyId" : NumberLong(0)
                }
        }
}
mongos>
```

 MongoDB version 4.0.2 and above allows you to define zones on non-sharded collections, or collections that have not yet been populated.

 To undo a range-to-zone assignment, use the `sh.removeRangeFromZone()` shell helper method.

We now need to have a look at working with the sharded cluster balancer.

Working with the sharded cluster balancer

As mentioned in the previous chapter, the *sharded cluster balancer* continuously manages the distribution of documents into chunks and chunks into shards. The goal of the balancer is to ensure an even distribution of data between servers in the sharded cluster. It's extremely important to note that, for the most part, it's best to let the balancer do its job. However, situations may arise where you wish to manually intervene in this distribution process by managing the size of chunks, splitting chunks, and manually *migrating* chunks between shards. In such cases, you need to know how to manage the balancer (https://docs. mongodb.com/manual/tutorial/manage-sharded-cluster-balancer/#manage-sharded-cluster-balancer).

The first consideration is to determine the *state* of the balancer.

Determining the balancer state

To determine the balancer state, here are two `mongo` shell helper methods that might prove to be useful. Bear in mind that these can only be issued when connected to an appropriate `mongos` instance:

- `sh.getBalancerState()`: This command returns a Boolean value. A value of `true` indicates that the balancer is stable and available. A value of `false` means it's offline and/or there is a problem.
- `sh.isBalancerRunning()`: This shell method also returns a Boolean value. A value of `false` does not indicate a problem: it only tells you that the balancer is not running at this exact moment.

Now, let's look at disabling the balancer.

Disabling the balancer

In order to disable the balancer, especially before an operation that might impact sharding, you not only need to stop the balancer but you must also wait until it has stopped running before proceeding with the operation. Examples of where this is required would be prior to performing a backup, splitting a chunk, or manually migrating a chunk.

To disable the balancer, proceed as follows:

1. Open a `mongo` shell onto a `mongos` instance with sufficient cluster administration rights.
2. Issue the `sh.stopBalancer()` command to stop the balancer. You can supply an optional argument string representing the database and collection if you wish to disable balancing only on a specific collection. The command can be seen in the following code snippet:

```
sh.stopBalancer("DB.COLLECTION");
```

3. Confirm the balancer state by running the following command. A value of `false` indicates the balancer is disabled:

```
sh.getBalancerState();
```

4. If you see a value of `true`, wait a few seconds. Do not proceed further until the preceding command returns a value of `false`.

Here is the output from the sharded cluster model used as an example in this chapter:

```
File Edit View Search Terminal Help
mongos> sh.getBalancerState();
true
mongos> sh.stopBalancer();
{
        "ok" : 1,
        "operationTime" : Timestamp(1586839271, 2),
        "$clusterTime" : {
                "clusterTime" : Timestamp(1586839271, 2),
                "signature" : {
                        "hash" : BinData(0,"AAAAAAAAAAAAAAAAAAAAAAAAAAA="),
                        "keyId" : NumberLong(0)
                }
        }
}
mongos> sh.getBalancerState();
false
mongos>
```

Of course, you also need to know how to *enable* the balancer!

Enabling the balancer

To enable the balancer, proceed as follows:

1. Open a `mongo` shell onto a `mongos` instance with sufficient cluster administration rights.
2. Issue the `sh.startBalancer()` command to enable the balancer. You can supply an optional argument string representing the database and collection if you wish to enable balancing only on a specific collection. The command can be seen in the following code snippet:

   ```
   sh.startBalancer("DB.COLLECTION");
   ```

3. Confirm the balancer state by running the following command. A value of `true` indicates the balancer is enabled:

   ```
   sh.getBalancerState();
   ```

Here is the output of this sequence from the sharded cluster model:

```
File Edit View Search Terminal Help
mongos> sh.startBalancer();
{
        "ok" : 1,
        "operationTime" : Timestamp(1586839855, 3),
        "$clusterTime" : {
                "clusterTime" : Timestamp(1586839855, 3),
                "signature" : {
                        "hash" : BinData(0,"AAAAAAAAAAAAAAAAAAAAAAAAAAA="),
                        "keyId" : NumberLong(0)
                }
        }
}
mongos> sh.getBalancerState();
true
mongos>
```

If you need to enable or disable the balancer from an application, it's best to use a direct database admin command.

Here is the command to disable the balancer: `db.adminCommand( { balancerStop: 1 } )`.

Here is the command to start the balancer: `db.adminCommand( { balancerStart: 1 } )`.

For more information, see `https://docs.mongodb.com/manual/tutorial/manage-sharded-cluster-balancer/`.

Let's now examine how to manage chunk size.

Managing chunk size

As mentioned earlier in the previous chapter, the *chunk* is the lowest block of data the sharded cluster balancer can manage. Normally, the sharded cluster balancer automatically creates a new chunk when it has grown past the default (or assigned) chunk size. The default chunk size is 64 **megabytes** (**MB**). Documents are then *migrated* between chunks until a balance occurs. Further, with zone-to-shard assignments in mind, the balancer also migrates chunks between shards to maintain a balanced load.

There are several approaches to managing chunk size: adjusting the chunk size, which triggers the automatic balancing process; or, manually *splitting* chunks. Let's now have a look at adjusting the chunk size.

Adjusting chunk size

The default chunk size of 64 MB works well for 90% of the deployments you make. However, due to the server or network constraints, you might notice that automatic migrations are producing **input/output (I/O)** bottlenecks. If you *increase* the default chunk size, you experience a lower number of automatic migrations (as there is more room within each chunk, giving the balancer more leeway in its allocations); however, each migration takes more time. On the other hand, if you *decrease* the default chunk size, you might notice an uptick in the number of automatic migrations; however, each migration occurs more rapidly.

To change the default chunk size, proceed as follows:

1. Connect to a `mongos` instance from a `mongo` shell as a user with sufficient cluster administration rights.
2. Use the `config` database by running the following command:

   ```
   use config;
   ```

3. Use the `save()` method on the `config.settings` collection. The value for `SIZE` should be an integer representing the new default chunk size in MB. The code can be seen in the following snippet:

   ```
   db.settings.save( { _id:"chunksize", value: SIZE } )
   ```

    For more information, see the following *Modify Chunk Size in a Sharded Cluster* documentation article: `https://docs.mongodb.com/manual/tutorial/modify-chunk-size-in-sharded-cluster/#modify-chunk-size-in-a-sharded-cluster`. For a general discussion on chunk size, see `https://docs.mongodb.com/manual/core/sharding-data-partitioning/#chunk-size`.

We now turn our attention to manually splitting a chunk.

Splitting a chunk

The `sh.status()` and `db.collection.getShardDistribution()` commands, described earlier, allow you to determine the actual distribution of documents between chunks. Once you have determined that the distribution is uneven, you have the option to manually *split* a chunk. The shell helper method you can use is `sh.splitAt()`. The generic syntax is shown here:

```
sh.splitAt(DB.COLLECTION, QUERY);
```

The first argument, DB.COLLECTION, is a string representing the database and collection names. The second argument, QUERY, is a JSON document that describes the lower boundary shard key at which the split should occur.

Using our sharded cluster model, we first issue the db.collection.getShardDistribution() shell command to get an idea of the current distribution, using world_cities as the collection, as illustrated in the following screenshot:

```
File Edit View Search Terminal Help
mongos> db.world_cities.getShardDistribution();

Shard repl_shard_3 at repl_shard_3/shard3.biglittle.local:27018
 data : 3.06MiB docs : 8054 chunks : 4
 estimated data per chunk : 785KiB
 estimated docs per chunk : 2013

Shard repl_shard_2 at repl_shard_2/shard2.biglittle.local:27018
 data : 3.62MiB docs : 9400 chunks : 2
 estimated data per chunk : 1.81MiB
 estimated docs per chunk : 4700

Shard repl_shard_1 at repl_shard_1/shard1.biglittle.local:27018
 data : 2.65MiB docs : 6763 chunks : 2
 estimated data per chunk : 1.32MiB
 estimated docs per chunk : 3381

Totals
 data : 9.35MiB docs : 24217 chunks : 8
 Shard repl_shard_3 contains 32.8% data, 33.25% docs in cluster, avg obj size on shard : 399B
 Shard repl_shard_2 contains 38.79% data, 38.81% docs in cluster, avg obj size on shard : 404B
 Shard repl_shard_1 contains 28.4% data, 27.92% docs in cluster, avg obj size on shard : 411B

mongos>
```

We can then use sh.status() to review shard-to-zone distribution (not shown). Assuming the distribution is uneven, we proceed to use sh.splitAt() to break up the offending chunks. Based on status and shard distribution information, we decide to break up the chunks residing on repl_shard_2. In order to effectuate the split, we need to carefully consult our notes to determine the shard key ranges associated with the second shard: africa and europe.

As it's most likely the case that the number of documents representing Europe is large, we perform a number of database queries to determine what is the best split between European time zones. However, because there are many timezones that start with Europe, we need to aggregate totals from a substring of the timezone field that includes the letters Europe followed by a slash and only the first letter. Accordingly, here is a query using the aggregation framework to accomplish that purpose:

```
doc = [
    { $match : { "timezone" : /^Europe/ }},
```

```
    { $addFields : { "count" : 1, "prefix" : \
        {"$substrBytes" : ["$timezone", 0, 8]}}},
    { $group : { _id : "$prefix", "total" : { "$sum" : "$count" }}},
    { $sort   : { "_id" : 1 }}
]
db.world_cities.aggregate(doc);
```

And here are the results of that query:

```
File  Edit  View  Search  Terminal  Help
mongos> db.world_cities.aggregate(doc);
{ "_id" : "Europe/A", "total" : 342 }
{ "_id" : "Europe/B", "total" : 1619 }
{ "_id" : "Europe/C", "total" : 70 }
{ "_id" : "Europe/D", "total" : 32 }
{ "_id" : "Europe/G", "total" : 2 }
{ "_id" : "Europe/H", "total" : 74 }
{ "_id" : "Europe/I", "total" : 393 }
{ "_id" : "Europe/J", "total" : 1 }
{ "_id" : "Europe/K", "total" : 219 }
{ "_id" : "Europe/L", "total" : 910 }
{ "_id" : "Europe/M", "total" : 1316 }
{ "_id" : "Europe/O", "total" : 33 }
{ "_id" : "Europe/P", "total" : 746 }
{ "_id" : "Europe/R", "total" : 592 }
{ "_id" : "Europe/S", "total" : 273 }
{ "_id" : "Europe/T", "total" : 31 }
{ "_id" : "Europe/U", "total" : 11 }
{ "_id" : "Europe/V", "total" : 103 }
{ "_id" : "Europe/W", "total" : 333 }
{ "_id" : "Europe/Z", "total" : 142 }
mongos>
```

As you can see from the query result, the number of documents for time zones starting with *Europe* appears to be evenly spread between the `Europe/A*` and `Europe/L*` ranges and the range from `Europe/M*` to `Europe/Z*`. In order to accomplish the split, let's first create and test a query document using `db.collection.find()`. Please note that in order for the split to be successful, the `split_query` document must identify an *exact* shard key value, not a pattern or substring. Here is an example of such a query:

```
split_query = {"timezone" : "Europe/Madrid"}
db.world_cities.find(split_query, {"timezone":1}).count();
```

If the results are within acceptable parameters, we can proceed with the split by first turning off the sharded cluster balancer. We then execute the split and turn the balancer back on. Here are the commands to issue:

```
sh.stopBalancer();
sh.getBalancerState();
sh.splitAt("biglittle.world_cities", split_query);
sh.startBalancer();
```

Here is the resulting output:

```
File  Edit  View  Search  Terminal  Help
mongos> sh.splitAt("biglittle.world_cities", split_query);
{
        "ok" : 1,
        "operationTime" : Timestamp(1586763567, 1),
        "$clusterTime" : {
                "clusterTime" : Timestamp(1586763576, 4),
                "signature" : {
                        "hash" : BinData(0,"AAAAAAAAAAAAAAAAAAAAAAAAAAA="),
                        "keyId" : NumberLong(0)
                }
        }
}
mongos>
```

If you now rerun `db.world_cities.getShardDistribution()`, you can see that there are now nine chunks in our sharded cluster model. If we ran `sh.status()` right now (not shown), we would also see three shards now residing in shard 2.

You can also manually perform a migration, covered briefly next.

Manually migrating a chunk

To manually migrate a chunk, use the following command:

```
db.adminCommand( { moveChunk : DB.COLLECTION, find : QUERY, to : TARGET } )
```

The parameters are summarized as follows:

- `DB.COLLECTION` (string): This string represents the name of the database, and collection name, separated by a period.
- `QUERY` (JSON): You can specify a JSON document that describes the shard key field/value key pair considered the lower bound of the chunk to be migrated.
- `TARGET` (string): This is a string representing the shard server to which the chunk is moved.

The process of determining the best query document is identical to that described in the previous sub-section when determining a split point.

Use this command only if system performance is severely compromised. Be sure to stop the sharded cluster balancer before performing this action. It is considered a best practice to leave chunk migration to the sharded cluster balancer.

The last topic to cover is backup and restore, featured next.

Backing up and restoring a sharded cluster

There are several considerations when performing a backup on a sharded cluster. First of all, and most importantly, backing up a sharded cluster using `mongodump` and restoring using `mongorestore` cannot be used. The reason for this is because it is *impossible* for backups created using `mongodump` to guarantee database integrity if sharded transactions are in progress. Accordingly, the only realistic backup solution for a sharded collection is to use filesystem snapshots.

Another extremely important consideration is to *disable the balancer* before the filesystem snapshot backup is taken. Once the backup is complete, you can then re-enable the balancer. For more information on disabling and re-enabling the balancer, see the *Disabling the balancer* and *Enabling the balancer* sub-sections earlier in this chapter.

It is possible to *schedule* the balancer to run only at certain times. In this manner, you are able to open a window of time for backups to occur. Here is an example that sets up the balancer to run only between 02:00 in the morning until 23:59 at night. This leaves a 2-hour window during which backups can take place:

```
use config;
db.settings.update(
    { _id : "balancer" },
    { $set : { activeWindow : { start : "02:00", stop : "23:59" } } }, true
);
```

Otherwise, bear in mind that it is also highly recommended that you implement each shard as a replica set. You are thus assured of a higher availability factor and are protected against a total crash of your database. We now turn out attention to application program development in sharded clusters.

Sharded cluster impacts on program code

The primary difference when writing application code that addresses a sharded collection is to make sure you connect to a *mongos* instance. Otherwise, as shards are normally implemented as replica sets, the same considerations with regard to access to a replica set, addressed in `Chapter 13`, *Deploying a Replica Set*, would apply. The first consideration we'll examine is how to perform a basic connection to a sharded cluster.

Connecting to a sharded cluster

There are three important differences for an application accessing data on a sharded cluster, listed as follows:

- The port is most likely not going to be the default of `27017`.
- Connecting directly to one of the shard members does not return a full dataset.
- To get the best results, connect to a *mongos* instance configured for the sharded cluster.

To modify the port number, all you need to do is to modify the connection string argument supplied to the `pymongo.mongo_client.MongoClient` instance, as shown here, in a `/path/to/repo/chapters/16/test_shard_one_server.py` file:

```
hostName = 'server1.biglittle.local:27018';
client   = MongoClient(hostName);
```

Another consideration, however, is that shards are most likely going to be configured as replica sets, requiring a further alteration of the connection string, as illustrated in the following code snippet:

```
hostName = 'repl_shard_1/server1.biglittle.local:27018, \
            server2.biglittle.local:27018';
client   = MongoClient(hostName);
```

Unless you connect to the *mongos* instance, however, you only receive results from data stored on that shard. As an example, have a look at this block of code. First, we connect to a single shard:

```
import pprint
from pymongo import MongoClient
hostName = 'shard1.biglittle.local:27018';
client   = MongoClient(hostName);
```

We then execute three queries from the `world_cities` collection, knowing the data resides on three different shards. The code for this can be seen in the following snippet:

```
count = {}
tz = 'America/Anchorage'
docs = client.biglittle.world_cities.count_documents({'timezone':tz})
count.update({tz : docs})
tz = 'Europe/Amsterdam'
docs = client.biglittle.world_cities.count_documents({'timezone':tz})
count.update({tz : docs})
tz = 'Asia/Aden'
docs = client.biglittle.world_cities.count_documents({'timezone':tz})
count.update({tz : docs})
```

And finally, here is a simple output routine:

```
print('{:>20} | {:>6s}'.format('Timezone', 'Count'))
print('{:>20} | {:>6s}'.format('--------', '-----'))
for key,val in count.items() :
    output = '{:>20} | {:6d}'.format(key, val)
    print(output)
```

When we run the example, note here that the output only draws data from the first shard:

To correct this situation, we modify the connection string to the *mongos* instance instead, as follows (full code example can be found at:
`/path/to/repo/chapters/16/test_shard_mongos.py`):

```
hostName = 'mongos1.biglittle.local:27020';
client   = MongoClient(hostName);
```

And here are the revised results:

Next, we look at manipulating *read preferences* when working with a sharded cluster.

Reading from a sharded cluster

As mentioned earlier, you need to configure your application to connect to a *mongos* instance in order to retrieve full results from all shards.

However, it is extremely important to note that although you can perform queries without using the shard key, it forces the *mongos* instance to perform a *broadcast* operation. As mentioned earlier, this operation first sends out a query to all shards before retrieving the data. In addition, if you do not include the shard key in the query, its index is not used, further degrading performance.

As an example, from the sample data, let's assume that the management wants a count of all `world_cities` documents in England. The main difference in your program code is that you need to connect to a *mongos* instance rather than a *mongod* instance. Here is a sample program that achieves this result:

```python
from pymongo import MongoClient
hostName = 'mongos1.biglittle.local:27020';
client   = MongoClient(hostName);
code     = 'ENG'
query    = {'admin1_code' : code}
proj     = {'name' : 1}
result = client.biglittle.world_cities.find(query, proj)
plan     = result.explain()
count    = 0
for doc in result :
    count += 1
print('There are {} documents for the admin code {}'.format(count, code))
pprint.pprint(plan)
```

 We purposely printed the output from the `pymongo.cursor.Cursor.explain()` method in order to gather query statistics. This command should not be used in production!

The resulting partial output appears as follows:

```
File Edit View Search Terminal Help
[engjad:~/Repos/learn-mongodb/chapters/14$ python test_shard_query_without_shard_key.py
There are 643 documents for the admin code ENG
{'$clusterTime': {'clusterTime': Timestamp(1586933954, 1),
                  'signature': {'hash': b'\x00\x00\x00\x00\x00\x00\x00\x00'
                                        b'\x00\x00\x00\x00\x00\x00\x00\x00'
                                        b'\x00\x00\x00\x00\x00',
                                'keyId': 0}},
 'executionStats': {'allPlansExecution': [{'allPlans': [],
                                            'shardName': 'repl_shard_1'},
                                           {'allPlans': [],
                                            'shardName': 'repl_shard_3'},
                                           {'allPlans': [],
                                            'shardName': 'repl_shard_2'}],
                    'executionStages': {'executionTimeMillis': 12,
```

The output produced by `pymongo.cursor.Cursor.explain()` includes
an `executeStats.executionStages.executionTimeMillis` parameter. As you can
see from the screenshot, the performance was an abysmal 12 milliseconds. To remedy the
situation, we hint to the shard by including the shard key in the query. Here is the modified
query document:

```
code    = 'ENG'
tz      = 'Europe/London'
query   = {'$and' : [{'timezone' : tz},{'admin1_code' : code}]}
```

The remaining code is exactly the same. Notice here the difference in performance:

```
File  Edit  View  Search  Terminal  Help
jed@jed:~/Repos/learn-mongodb/chapters/14$ python test_shard_query_using_shard_key.py
There are 643 documents for the admin code ENG
{'$clusterTime': {'clusterTime': Timestamp(1586934314, 1),
                  'signature': {'hash': b'\x00\x00\x00\x00\x00\x00\x00\x00'
                                        b'\x00\x00\x00\x00\x00\x00\x00\x00'
                                        b'\x00\x00\x00\x00',
                                'keyId': 0}},
 'executionStats': {'allPlansExecution': [{'allPlans': [],
                                           'shardName': 'repl_shard_2'}],
                    'executionStages': {'executionTimeMillis': 4,
                                        'nReturned': 643,
                                        'shards': [{'executionStages': {'advanced': 643,
```

The final value for the execution time dropped to 4 milliseconds, representing a 300%
improvement in response time. The complete code for the two contrasting examples just
shown can be found here:

- `/path/to/repo/chapters/16/test_shard_query_without_shard_key.py`
- `/path/to/repo/chapters/16/test_shard_query_using_shard_key.py`

You can add a `**kwarg` parameter,
`allow_partial_results=True`, when using
the `pymongo.collection.Collection.find()` method. If set to `True`,
the *mongos* instance returns partial results even if a shard is down. If set
to `False`—the default—an error is returned if one or more shards are
down.

Now, we look at how to write to a sharded cluster.

Writing to a sharded cluster

Assuming that your sharded cluster has been implemented as a series of replica sets, one per shard, writing data to a sharded cluster is no different than writing data to a replica set. Here are some considerations for `insert`, `update`, and `delete` operations:

- **Insert**: When performing an insert operation, the sharded cluster balancer decides into which chunk to place the document, based on the shard key value of the document to be added.
- **Update**: In the case of an update, bear in mind that you'll need to supply a query document in addition to the update document. Just as with a read operation, if you include the shard key in the query document, updates show improved performance. If you set the `upsert` argument to `True`, you must include a shard key in the update-document in order for the sharded cluster balancer to correctly place the inserted document.
- **Delete**: Again, as with an update, you need to supply a query document. Include the shard key in the query document to improve delete performance.

And that concludes the discussion for this chapter.

Summary

This chapter showed you how to monitor and manage the sharded cluster. You learned about creating *zones* based on the shard key, and how to manage the sharded cluster balancer, adjust chunk sizes, and split a chunk. This was followed by a discussion of the impact of sharding on your backup strategy.

The last section covered the impact of a sharded cluster on application code. You learned, first of all, that the application must connect to a *mongos* instance in order to reach the entire sharded collection. You then learned that database read operations experience improved performance if the shard key is included in the query. Likewise, you learned that it's best to include the shard key in the query document associated with `update` and `delete` operations.

Other Books You May Enjoy

If you enjoyed this book, you may be interested in these other books by Packt:

MongoDB 4 Quick Start Guide
Doug Bierer

ISBN: 978-1-78934-353-3

- Get a standard MongoDB database up and running quickly
- Perform simple CRUD operations on the database using the MongoDB command shell
- Set up a simple aggregation pipeline to return subsets of data grouped, sorted, and filtered
- Safeguard your data via replication and handle massive amounts of data via sharding
- Publish data from a web form to the database using a program language driver
- Explore the basic CRUD operations performed using the PHP MongoDB driver

Mastering MongoDB 4.x - Second Edition
Alex Giamas

ISBN: 978-1-78961-787-0

- Perform advanced querying techniques such as indexing and expressions
- Configure, monitor, and maintain a highly scalable MongoDB environment
- Master replication and data sharding to optimize read/write performance
- Administer MongoDB-based applications on premises or on the cloud
- Integrate MongoDB with big data sources to process huge amounts of data
- Deploy MongoDB on Kubernetes containers
- Use MongoDB in IoT, mobile, and serverless environments

Leave a review - let other readers know what you think

Please share your thoughts on this book with others by leaving a review on the site that you bought it from. If you purchased the book from Amazon, please leave us an honest review on this book's Amazon page. This is vital so that other potential readers can see and use your unbiased opinion to make purchasing decisions, we can understand what our customers think about our products, and our authors can see your feedback on the title that they have worked with Packt to create. It will only take a few minutes of your time, but is valuable to other potential customers, our authors, and Packt. Thank you!

Index

www.ingramcontent.com/pod-product-compliance
Lightning Source LLC
Chambersburg PA
CBHW060635060326
40690CB00020B/4405